Selected Titles in This Series

Volume

13 **Lars Gårding**
   Mathematics and mathematicians: Mathematics in Sweden before 1950
   1998

12 **Walter Rudin**
   The way I remember it
   1997

11 **June Barrow-Green**
   Poincaré and the three body problem
   1997

10 **John Stillwell**
   Sources of hyperbolic geometry
   1996

9 **Bruce C. Berndt and Robert A. Rankin**
   Ramanujan: Letters and commentary
   1995

8 **Karen Hunger Parshall and David E. Rowe**
   The emergence of the American mathematical research community, 1876–1900: J. J. Sylvester, Felix Klein, and E. H. Moore
   1994

7 **Henk J. M. Bos**
   Lectures in the history of mathematics
   1993

6 **Smilka Zdravkovska and Peter L. Duren, Editors**
   Golden years of Moscow mathematics
   1993

5 **George W. Mackey**
   The scope and history of commutative and noncommutative harmonic analysis
   1992

4 **Charles W. McArthur**
   Operations analysis in the U.S. Army Eighth Air Force in World War II
   1990

3 **Peter L. Duren et al., Editor**
   A century of mathematics in America, part III
   1989

2 **Peter L. Duren et al., Editor**
   A century of mathematics in America, part II
   1989

1 **Peter L. Duren et al., Editor**
   A century of mathematics in America, part I
   1988

MATHEMATICS *and* Mathematicians

Mathematics in Sweden before 1950

Lars Gårding

618006

History of Mathematics
Volume 13

American Mathematical Society
London Mathematical Society

Editorial Board

American Mathematical Society
George E. Andrews
Bruce Chandler, Chairman
Karen Parshall
George B. Seligman

London Mathematical Society
David Fowler
Jeremy J. Gray, Chairman
S. J. Patterson

MATEMATIK OCH MATEMATIKER: Matematiken i Sverige före 1950
(Mathematics and Mathematicians: Mathematics in Sweden before 1950)
by Lars Gårding

Copyright © 1994 by Lars Gårding
Originally published in Swedish by Lund University Press, Lund, Sweden, 1994
Translated from the Swedish by Lars Gårding

Photographs on the front cover are (left to right) Sonja Kovalevski and Gösta Mittag-Leffler; the photographs on the back cover are (left to right) Arne Beurling, Torsten Carleman, and Ivar Fredholm.

A list of photograph credits is included at the beginning of this volume.

1991 *Mathematics Subject Classification.* Primary 01–XX.

Library of Congress Cataloging-in-Publication Data
Gårding, Lars, 1919–
  Mathematics and mathematicians : mathematics in Sweden before 1950 / Lars Gårding.
    p. cm. — (History of mathematics, ISSN 0899-2428 ; v. 13)
  Includes bibliographical references (p.   –   ) and index.
  ISBN 0-8218-0612-2 (hardcover : alk. paper)
  1. Mathematics—Sweden—History.   I. Title.   II. Series.
QA27.S85G37 1997
510′.9485′09034—dc21   97-27619
                                                                                           CIP

Copying and reprinting. Individual readers of this publication, and nonprofit libraries acting for them, are permitted to make fair use of the material, such as to copy a chapter for use in teaching or research. Permission is granted to quote brief passages from this publication in reviews, provided the customary acknowledgment of the source is given.

Republication, systematic copying, or multiple reproduction of any material in this publication (including abstracts) is permitted only under license from the American Mathematical Society. Requests for such permission should be addressed to the Assistant to the Publisher, American Mathematical Society, P. O. Box 6248, Providence, Rhode Island 02940-6248. Requests can also be made by e-mail to `reprint-permission@ams.org`.

© 1998 by the American Mathematical Society. All rights reserved.
Printed in the United States of America.
The American Mathematical Society retains all rights
except those granted to the United States Government.
∞ The paper used in this book is acid-free and falls within the guidelines
established to ensure permanence and durability.
The London Mathematical Society is incorporated under Royal Charter
and is registered with the Charity Commissioners.
Visit the AMS home page at URL: `http://www.ams.org/`

10 9 8 7 6 5 4 3 2 1   03 02 01 00 99 98

Contents

Photograph Credits ix

Preface to the Swedish edition xi

Preface to the American edition xiii

Chapter 1 The eighteenth century 1
Klingenstierna 1, Bring 5, Bibliography 7.

Chapter 2 The time 1800 - 1850 9
Hill 10, The Uppsala school 12, Svanberg 13, E.G. Björling 14, Malmsten 18, Hj. Holmgren 26, Appendix: Solvable equations 27, Bibliography 30.

Chapter 3 A new time in Uppsala and Lund 1860 - 1890 33
Tidskrift för matematik och fysik 33, The 1871-73 contest for a professorship in Lund 35, Björling as a mathematician and teacher 37, Bibliography 39.

Chapter 4 Algebraic geometry in Lund before 1900 41
Projective and algebraic geometry in the nineteenth century 41, Algebraic curves 42, Algebraic surfaces 44, Algebraic geometry in Lund 1870-1900 44, Bäcklund 44, Björling 45, Björling's textbooks 46, Ruled surfaces. Bergstedt and Wiman 47, Bibliography 49.

Chapter 5 Bäcklund 51
Differential equations at the end of the nineteenth century 51, Bäcklund on partial differential equations 54, Bäcklund transformations 55, Bäcklund transformations after Bäcklund 60, Oseen's obituary of Bäcklund 61, Bibliography 62.

Chapter 6 Uppsala 1860 - 1900 65
Daug 65, Dillner 66, Falk 67, Berger 68, The 1889 contest for a professorship in Uppsala 70, Bibliography 71.

Chapter 7 Gösta Mittag-Leffler – a biography 73
Family and growing up 73, Studies in Uppsala, Paris, Berlin 74, The time in Helsingfors 76, The eventful decade 1880-1890 77, *Acta Mathematica* 77, Sonya Kovalevski 79, The royal prize 81, The time after the age of fifty 82, The will and the Gösta and Signe Mittag-Leffler foundation 83, Mittag-Leffler's students 84, Bibliography 84.

Chapter 8 Mittag-Leffler's and Sonya Kovalevski's mathematical papers 85
Gösta Mittag-Leffler 85, Mittag-Leffler's theorem about meromorphic functions 86, Mittag-Leffler's star 87, Summation of series 88, Borel summation of power series 88, Mittag-Leffler's notes 89, Sonya Kovalevski 93, Bibliography 95.

Chapter 9 Astronomy and optics ... 97

Celestial mechanics 97, Newton 97, Lagrange and Hamilton 98, Generating function 100, Astronomy in the nineteenth century, Poincaré 101, Poincaré and the three body problem 102, Lindstedt, Bohlin, Gyldén 102, Gullstrand 106, Appendix 107, Bibliography 108.

Chapter 10 Stockholm University 1880-1920 I ... 109

Bendixson 109, Phragmén 112, von Koch 115, Infinite determinants 115, Differential equations at singularities 116, von Koch's papers 117, Fredholm 119, Fredholm's construction of fundamental solutions 120, Integral equations 121, After Fredholm 124, Fredholm's obituary 126, Appendix: Fundamental solutions by distribution theory 126, Homogeneous equations, the Gelfand-Shilov formula 128, Variants and reductions 129, Hyperbolic operators 129, Bibliography 130.

Chapter 11 Stockholm University 1880-1920 II ... 133

Analysis 133, Kobb 133, Cassel 134, Petrini 134, Grönwall 135, Malmquist 136, Hille 138, Analytic number theory 138, Stridsberg 140, Wigert 140, Cramér 141, Fourier analysis 142, Riesz 142, Zeilon 146, Zeilon on fundamental solutions 147, Light in doubly refracting crystals, Lamé, Kovalevski 148, Zeilon's solution 150, Appendix: On Huygens's principle and lacunas 151, Bibliography 151.

Chapter 12 Uppsala 1900-1930 ... 155

Wiman 156, Group theory 156, Wiman on solvable equations 158, Entire functions 158, Differential equations 162, Algebraic geometry 162, Obituary 163, Wiman's students 163, Holmgren 164, Holmgren's uniqueness theorem 165, Other papers 166, Obituary 168, Holmgren's students 168, Bibliography 168.

Chapter 13 Lund 1900-1925 ... 171

Brodén 171, Differential equations of Fuchsian type and Riemann's problem 172, Discrete subgroups of the Möbius group 173, Nörlund 174, Difference equations and interpolation 175, Block 177, Theses 178, Bibliography 181.

Chapter 14 Stockholm 1925-1950 ... 185

Carleman 185, Integral equations 185, Carleman kernels 186, Carleman and the abstract theory 189, Generalized kernels 189, Applications 190, Carleman's other papers on integral equations 191, Jensen's formula in a half-plane 191, The Jensen-Carleman formula 192, Approximation by powers z^λ 193, Quasianalytic classes 193, Criteria for quasianalyticity 194, Carleman's short papers 196, Harmonic majorants and harmonic measure 196, Asymptotic paths 197, Approximation by entire functions, Lindelöf's function 200, Uniqueness 200, Mathematical physics 201, Ergodic theorem 201, The Schrödinger operator 202, Asymptotics of eigenvalues 202, The kinetic theory of gases 204, Late papers 205, Summary 206, Carleman's students 206, Pleijel 207, Carlson 208, The thesis 208, Theorems on power series 208, Theorems on Dirichlet series 211, Geometry 212, Summary 212, Appendix: The spectral theorem for self-adjoint operators 213, Hilbert 213, von Neumann 214, A proof of the Denjoy-Carleman theorem 215, Bibliography 216.

Chapter 15 Lund 1925-1950 ... 219

Riesz 219, Interpolation between inequalities, the Riesz-Thorin theorem 219, Applications 221, Conjugate functions 222, Short papers, Medd. Lunds Univ. Mat. Sem. 223, Fractional potentials 224, The wave operator 226, Spinors 229, Obituary 229, Zeilon 230, Oseen's wake theory 230, Zeilon's papers 232, Mathematicians, theses, and papers 234, Hössjer 234, V. Bergström 235, Frostman 236, Berg, Gårding, Fremberg, Malmheden, Lannér 242, Bibliography 243.

Chapter 16 Uppsala 1930 - 1950 247

Nagell 247, Generalities about rational solutions 247, Algebraic number theory 249, Nagell's work 249, Summary 250, Beurling 250, Beurling's thesis 251, Extremal distance and extremal length 251, Beurling's lemma 252, Milloux's problem 254, Complex analysis 254, Exceptional sets, outer capacity 254, Outer and inner functions 255, Beurling's primes 257, Spectral analysis 258, Spectral synthesis 263, Later papers and summary 264, Beurling's students 264, Esseen 264, Borg 266, Broman, Kjellberg 267, Nyman, Hall, Carleson 267, Appendix 268, Bibliography 268.

Mathematicians in Sweden 1700–1950 271

Postscript 283

Index 285

Photograph Credits

The AMS gratefully acknowledges the kindness of the Institut Mittag-Leffler in granting the following permissions.

Photograph of Viktor Bäcklund; from *Acta Mathematica* 1882–1912, *Table Générale*; Courtesy of the Institut Mittag-Leffler.

Photograph of Ivar Bendixson; from *Acta Mathematica* 1882–1912, *Table Générale*; Courtesy of the Institut Mittag-Leffler.

Photograph of Alexander Berger; from *Acta Mathematica* 1882–1912, *Table Générale*; Courtesy of the Institut Mittag-Leffler.

Photograph of Arne Beurling on the cover and in the photograph section; from *Acta Mathematica*, vol. 161, 1988; Courtesy of the Institut Mittag-Leffler.

Photograph of Karl Bohlin; from *Acta Mathematica* 1882–1912, *Table Générale*; Courtesy of the Institut Mittag-Leffler.

Photograph of Torsten Brodén; from *Acta Mathematica* 1882–1912, *Table Générale*; Courtesy of the Institut Mittag-Leffler.

Photograph of Torsten Carleman on the cover and in the photograph section; from *Acta Mathematica*, vol. 82, 1950; Courtesy of the Institut Mittag-Leffler.

Photograph of Matths Falk; from *Acta Mathematica* 1882–1912, *Table Générale*; Courtesy of the Institut Mittag-Leffler.

Photograph of Ivar Fredholm on the cover and in the photograph section; from *Acta Mathematica* 1882–1912, *Table Générale*; Courtesy of the Institut Mittag-Leffler.

Photograph of Allvar Gullstrand; from *Acta Mathematica* 1882–1912, *Table Générale*; Courtesy of the Institut Mittag-Leffler.

Photograph of Hugo Gyldén; from *Acta Mathematica* 1882–1912, *Table Générale*; Courtesy of the Institut Mittag-Leffler.

Photograph of Gustaf Kobb; from *Acta Mathematica* 1882–1912, *Table Générale*; Courtesy of the Institut Mittag-Leffler.

Photograph of Helge von Koch; from *Acta Mathematica* 1882–1912, *Table Générale*; Courtesy of the Institut Mittag-Leffler.

Photograph of Sonja Kovalevski on the cover and in the photograph section; from *Acta Mathematica* 1882–1912, *Table Générale*; Courtesy of the Institut Mittag-Leffler.

Photograph of Anders Lindstedt; from *Acta Mathematica* 1882–1912, *Table Générale*; Courtesy of the Institut Mittag-Leffler.

Photograph of Johannes Malmquist; from *Acta Mathematica* 1882–1912, *Table Générale*; Courtesy of the Institut Mittag-Leffler.

Photograph of Carl Johan Malmsten; from *Acta Mathematica* 1882–1912, *Table Générale*; Courtesy of the Institut Mittag-Leffler.

Photograph of Gösta Mittag-Leffler on the cover and in the photograph section; from *Acta Mathematica* 1882–1912, *Table Générale*; Courtesy of the Institut Mittag-Leffler.

Photograph of Niels Erik Nörlund; from *Acta Mathematica* 1882–1912, *Table Générale*; Courtesy of the Institut Mittag-Leffler.

Photograph of Wilhelm Oseen; from *Acta Mathematica* 1882–1912, *Table Générale*; Courtesy of the Institut Mittag-Leffler.

Photograph of Henrik Petrini; from *Acta Mathematica* 1882–1912, *Table Générale*; Courtesy of the Institut Mittag-Leffler.

Photograph of Edvard Phragmén; from *Acta Mathematica* 1882–1912, *Table Générale*; Courtesy of the Institut Mittag-Leffler.

Photograph of Marcel Riesz; from *Acta Mathematica* 1882–1912, *Table Générale*; Courtesy of the Institut Mittag-Leffler.

Photograph of Erik Stridsberg; from *Acta Mathematica* 1882–1912, *Table Générale*; Courtesy of the Institut Mittag-Leffler.

Photograph of Anders Wiman; from *Acta Mathematica* 1882–1912, *Table Générale*; Courtesy of the Institut Mittag-Leffler.

Preface to the Swedish edition

The purpose of this book is to document mathematical research in Sweden before 1950: what was done, who did it and, partly, why it was done. So far only the work of the Swedish eighteenth century mathematician Samuel Klingenstierna has been documented in this way by the physicist Carl W. Oseen and the librarian Hildebrand Hildebrandsson (1919). With the exception of Mittag-Leffler I have been restrictive with biographical detail. My purpose has been to write about the work of mathematicians for readers interested in mathematics.

The year 1950 is a natural limit. The ensuing forty-five years have given a historical perspective and the end of the war was also the end of a university organization in Sweden which had lasted for a hundred years. During that time a university professor was often the only one responsible for the teaching and examination in his subject. In addition he was expected to do research and publish new results. Sometimes he had the help of a docent on a six-year scholarship. In practice, these two categories of people were the only ones who had sufficient motivation and time for research. Because the system promoted the most successful docents to professors it is inevitable that the greater part of my book is about the work of professors of mathematics. But almost all theses in the subject from 1890 to 1950 are mentioned and the most important ones treated in more detail.

In some cases which I have thought interesting, I have gone outside the circle of mathematicians, but applied mathematics in general is outside my scope. In particular, most mathematical physicists are excluded. Probability is mentioned, but only in connection with the mathematicians who have written in this field.

All mathematical papers are of course not mentioned. Those chosen for commentary have been interesting for at least one of the following reasons. One is quality, another is a wish to give a fuller mathematical profile of the most interesting mathematicians. Sometimes a paper has been selected because its background or mistakes have been more interesting than the paper itself.

In order to break a monotonous sequence of remarks and comments I have sometimes sketched the social setting and, when possible, challenged the reader by reproducing or at least sketching some proofs.

The decisive event for the position of mathematics in Sweden happened in the 1880's when Gösta Mittag-Leffler, the first professor to be nominated at Stockholm University, gathered a circle of young mathematicians and founded an international mathematical journal, the *Acta Mathematica*. Since then Sweden has not been lacking in good mathematicians.

Many have read and criticized my first drafts and thereby saved me from many faults and slips. I thank Gunnar Blom, Yngve Domar, Bent Fuglede, Eva Gårding, Lars Hörmander, Dan Laksov, Jana Madjarova, Jaak Peetre, Ulf Persson and Anders Reiz. I owe other useful help to Sven Spanne, Michael Benedicks, Gunnar Larsson-Leander and to Harry Kumlien who designed the front page and the cover. I also thank those who lent me photos of mathematicians.

The manuscript has been written with a selection from AMSTEX chosen by Nils Dencker. Contributions to the printing have come from the Royal Physiographical Society in Lund and from the Swedish Science Foundation.

I dedicate my book to Swedish mathematicians, the circle in which I spent my active professional life.

Preface to the American edition

In this edition some printing errors and slips have been corrected. I thank those who saw them, in particular Karl Gustav Andersson. I thank also Jana Madjarova for letting me benefit from her expert knowledge of TEX and for her patient proofreading.

Finally I thank my English reader, Tom Archibald, for his very thorough work and educational remarks.

L.G.

CHAPTER 1

The Eighteenth Century

Klingenstierna Bring

Before the beginning of the nineteenth century, the main task of the two Swedish universities in Lund and Uppsala was to teach Latin to everybody, Lutheran theology, Greek and Hebrew to future servants of the church, and law to future civil servants. But there were also other disciplines, for instance astronomy, the supreme exact science, and mathematics, a subject of undisputed usefulness and with a lot of prestige borrowed from Euclidean geometry. The teaching of mathematics dealt mainly with simple arithmetic, geometry, and algebra and could be handled by interested theologians who in many cases left this occupation for more lucrative positions in the service of the church. Under these circumstances it was more by chance that two mathematicians appeared in Sweden during this time who left marks in the history of mathematics.

Samuel Klingenstierna (1698-1765) belonged to a small group of students in Uppsala who studied the exact sciences. These included the astronomer Anders Celsius, the originator of the thermometer scale bearing his name. The mathematician Klingenstierna is unknown outside Sweden, but he was very well known in his own country. Although he published very little, his manuscripts show that he was well versed in the mathematics of his time and that he could contribute to it. Klingenstierna's life and work was documented in the 1920s (see the bibliography).

The only Swedish mathematical paper from the eighteenth century that has some reputation was written by Erland Samuel Bring (1736-1798), a man of many positions at the university in Lund. He proved that the general equation of degree five can be reduced to an equation of the same degree with only three terms.

Klingenstierna

Klingenstierna started studying law in Uppsala with a future as a civil servant in mind. But he found the lectures soporific and after a scrap with a lecturer he decided to study on his own. He then ran into the works of a famous German political scientist Samuel Pufendorf, who had studied mathematics in his youth. This author had been much impressed by the strict, logical method of Euclid's Elements and tried to apply the same method of exposition in his own work. The contrast between the strict form and the necessarily somewhat nebulous content puzzled Klingenstierna who at that time had no experience of mathematics. A friend of his then advised him to study Euclid directly. This became a turning point in Klingenstierna's life. He was happy to find a science which only contained 'unmistakable truths' and added mathematics to his law studies.

Successful interventions in the public disputations in Uppsala on humanistic and mathematical subjects earned Klingenstierna a reputation for wit and learning, something which at that time counted for more than formal degrees. After some time as a clerk in the central administration in Stockholm, Klingenstierna received a grant in 1723 for further studies in Uppsala. Here he also started to give mathematical lectures. One of his subjects was the new mathematics discovered by Newton and Leibniz, which makes Klingenstierna the initiator of infinitesimal analysis in Sweden. His clear lectures attracted many students, some of whom were later to play an important part in the foundation of the Swedish Academy of Sciences.

In 1727 Klingenstierna got a travel grant that was to carry him to the European centers of learning. The first stop was Marburg in Germany where he met the professor Christian Wolff, a pupil of Leibniz who was also famous as a philosopher. Here Klingenstierna wrote a manuscript, of which a copy is preserved, in which he completed the proofs in Newton's treatise on the forms of the real plane curves of degree three. The same work had been done earlier by Stirling, but Klingenstierna, who found pleasure in the work itself, never cared much for informing himself about the latest results available. The manuscript was sent to Uppsala and served its author well when both he and Celsius applied for the vacant chair of mathematics. A very strong letter of recommendation from Wolff, who had some connections with Sweden, decided the case in favor of Klingenstierna. During his absence he had to pay his competitor for taking on the duties of the chair. Later, Celsius was appointed professor of astronomy.

From Marburg Klingenstierna left for Basel in Switzerland to see Johann Bernoulli and his sons Nicolaus and Daniel. The visitor made a favorable impression on his distinguished hosts. The problems discussed were Newton's fluxions contra Leibniz's infinitesimals and the problem of the brachistochrone, i.e., a curve with fixed endpoints along which a heavy particle glides from one end to the other in the least possible time. The study of the brachistrone was the beginning of the calculus of variations.

Klingenstierna's next stop was Paris where he met the leading mathematicians and physicists and made friends with Fontenelle, the secretary of the French Academy of Sciences. Whether the earth is flatter at the poles than at the equator was the dominating subject of discussion, later settled by the measurements in Lappland by Celsius and Maupertuis. Klingenstierna also took part in the discussion of another burning question, the nature of infinitely small quantities. As a friend of clarity he doubted their existence.

From Paris Klingenstierna went to London where he published a short paper in the Philosophical Transactions dealing with a problem that leads to the quadrature of the hyperbola. No proofs are given.

In 1731 Klingenstierna returned to Uppsala to hold the chair of Geometry. Despite the name of the post, the teaching covered all parts of mathematics and also experimental physics. In addition, Klingenstierna was also considered as an authority in philosophy where Wolff's system had gained some popularity, though he soon tired of this position. One reason was that he did not want to antagonize the theologians, another that he himself did not feel at home in theoretical philosophy where 'the thoughts are divorced from our senses'. Mathematics appealed more to him and he carried his teaching to a very high level for those who could follow him. For the others (at this time mathematics was a compulsory subject) he did his duty but at the end of his service he tired and remarked that 'he longed to hear the stroke of the clock which freed him of his own torment and from the sight of the torment of others, who with difficulty and without interest or talent listened to a matter which he had to explain with disgust to those who longed for the day when they could forget all about it'.

After Celsius's death in 1744, Klingenstierna applied for the professorship in astronomy and got a recommendation, but ceded to Mårten Strömer who had applied for but not received the chair of astronomy in Lund.

Uppsala had had a chair of physics since 1593. It covered theoretical and natural philosophy according to known authorities and was mostly run by theologians. In the middle of the eighteenth century time was ripe for physics based on mathematics and experiments. The chairs of semitic languages and poetics were turned over to other subjects and replaced by two chairs, one in physics and one in chemistry. The chair in physics went to Klingenstierna who had already begun experimental work. Mathematics was taken care of by an officer of the Engineering Corps.

Klingenstierna did not stay long as a professor of physics. The commander of the artillery wanted to have him as an advisor and Klingenstierna was not unwilling to be free of teaching and have time for mathematical research. A post was arranged in 1752, but shortly afterwards Klingenstierna was promoted to be teacher to the young Crown Prince Gustaf, later King Gustaf III, which left him some time for scientific activity. In the beginning, the new post was not without its problems. but in time Klingenstierna gained the queen's confidence and the affection of his pupil. Sickness forced Klingenstierna to leave his post in 1764 and he died one year later. The queen erected a mausoleum to Klingenstierna and a former teacher of the crown prince, the author and journalist Olof von Dalin.

In the history of mathematics Klingenstierna appears as a witness that already in 1728 Johann Bernoulli knew the differential equations of geodesics, published in 1742. This is also clear from Klingenstierna's own manuscripts from the time. His first published paper is a geometrical exercise where the problem is to find the radius of the circle which circumscribes the largest quadrangle with given sides a, b, c, d. For the radius Klingenstierna found the following formula, also found by Legendre in 1794,

$$\sqrt{\frac{(ac+bd)(ad+bc)(ab+cd)}{(a+b+c-d)\ldots}}$$

where the points mean three expressions which are permutations of the one written out.

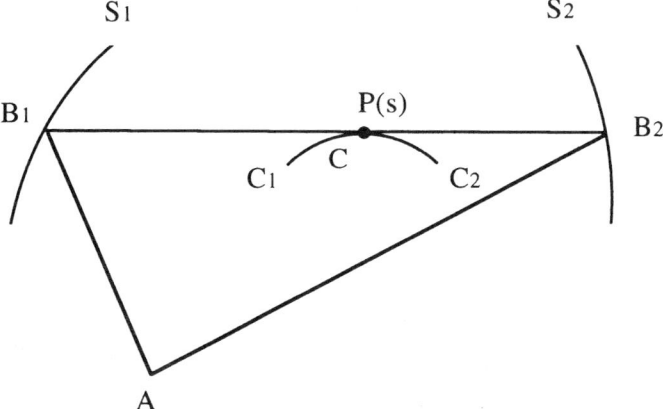

Figure 1

An interesting paper from 1749 carries the title 'Mathematical question concerning a curved line which carries a ray of light back to its origin after two reflections'. The ellipse has this property since every ray from a focus is reflected first to the other focus and then back to the first one. But as an example the ellipse is too special. The problem was first posed by Euler in 1745 and solved analytically by him and other mathematicians.

Klingenstierna's solution is a geometric construction which also gives an analytic expression. In Figure 1, the point A is fixed and the circumference of the triangle AB_1B_2 is fixed. When the line B_1, B_2 rolls on the fixed cirve C, its endpoints trace two pieces S_1 and S_2 of a mirror. The position $P = P(s)$ of the point of tangency is a function of the arclength s on C, counted from left to right. If $f_k(s)$ is the length of the curve $AB_kP(s)$, it follows that $f_1(s') = f_1(s) + s' - s$ and $f_2(s') = f_2(s) + s - s'$.

When C is a point, S_1 and S_2 are parts of an ellipse. In the general case, the usual thread construction of an ellipse can be used to construct for instance S_1 as follows. A thread from A passing the point of a pencil at B_1 runs along C to C_2 where it is fixed. When the point of the pencil moves, the thread moves freely around it and hence it will move orthogonally to the line bisecting the angle $AB_1P(s)$ so that the law of reflection is fulfilled at the mirrors.

Klingenstierna did not publish mathematical articles after 1750. Instead he turned to a controversial question in optics, the refraction of light of different colours. Newton had observed that refractive indices depend on the colour of light. From one of his experiments he had drawn the conclusion that it was impossible to eliminate this effect by a combination of refracting lenses: rays which are parallel when entering the system must also leave it as parallels if color dispersion is to be avoided. Euler's remark that the human eye refracts light without colour dispersion did not shake Newton's authority.

In a paper, printed in Swedish in the proceedings of the Swedish Academy of Sciences in 1754, Klingenstierna proved that Newton's statement is not consistent with the law of refraction. In his proof he considers a triangular prism bounded by two different refracting media and shows that the ingoing and outgoing rays with no colour dispersion do not remain parallel when the indices of refraction vary. But he remarks that the refracting angle of the prism must be large for the effect to be noticeable. This article broke a taboo. An attempt to construct systems of lenses with little or no colour dispersion was no longer hopeless.

A year later Klingenstierna wrote to a friend in London about his discovery. This correspondent showed the letter to a well-known instrument maker John Dollond who in a short time managed to put together a concave and convex lense with very little colour dispersion and immediately took credit for this success. Klingenstierna's protests had no effect. Six years later he wrote a more elaborate paper which also gave a theory of spherical aberration.

His papers give Klingenstierna priority in two important parts of optics which are important also in practice. But recognition was hard to get. The papers were written in Swedish, and they were published in a periodical that at the time was meant for a Swedish audience.

Klingenstierna left a large collection of manuscripts which shows that he dealt with and mastered the mathematical problems of his time. A large part is purely geometrical and contains theorems: for instance, that the tangents to a conic section at the three corners of an inscribed triangle meet in three points on a line (a special case of Pascal's theorem). Another theorem says that if a triangle is inscribed in a circle and circumscribes another circle, then there are infinitely many triangles with this property. The proof ends with the remark that projection makes the theorem valid for conic sections. Here Klingenstierna has priority. The theorem is usually ascribed to L'Huilier (1810) and was extended by Poncelet to n-gons which are inscribed in one conic section and circumscribe another one.

Klingenstierna worked with many of the favorite themes of his day, for instance the construction of curves whose tangents are given in some way or other. A typical case is the following (Oseen p. 56-57). A curve C and a straight line L are given. A tangent T at a point P on the curve meets the line L in a point Q. Determine the curve when the distance v from P to Q is a given function of the distance z from Q to a fixed point R on L.

If we introduce the angle α between the two line segments (see Figure 2) we get

$$vd\alpha = \sin(\pi - \alpha)dz = \sin\alpha dz$$

whence

$$dz/v = d\log\tan(\alpha/2).$$

With a first-rate intuition Klingenstierna introduces from the beginning a variable $t = \tan(\alpha/2)$ and gets $dz/v = dt/t$ by a simple argument. Two special cases are treated, one where v is proportional to z (a problem by Leibniz), and one where $v = z\sqrt{a^2 + z^2}$.

Klingenstierna's mathematical biographer, C. W. Oseen, did not finish his work, and a complete review of the manuscripts remains to be done.

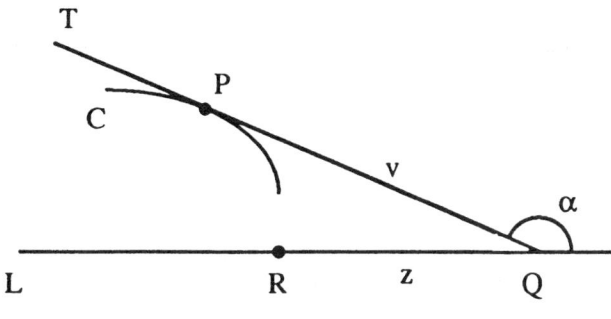

Figure 2

Bring

Compared to Klingenstierna, Bring was an amateur who did mathematics on the side while occupying various academic posts. He entered Lund University at the age of sixteen and took his law degree seven years later. After a period at the central courts of law he came back to Lund as the notary of the university and finally succeeded his father's brother as professor of history.

Bring's reputation as a mathematician rests on a single paper from 1786, a Latin dissertation with the English title: 'Mathematical studies on the transformation of algebraic equations.' The nominal author, who received a doctor's degree for defending it, was the son of a friend and colleague of Bring.

Around 1500 Cardano and Ferrari had solved the general equations of degree three and four by rational operations and root extractions. After this the solution in the same way of equations higher degree became a leading problem in mathematics.

Another important problem was the deduction of Euclid's axiom of parallels from the others. It took a long time to see that both were problems without solutions. The proofs by Bólyai and Lobachevsky and by Abel appeared almost simultaneously at the beginning of the nineteenth century.

In Bring's time it was still possible to try to solve the equations of degree five in general form. That an equation

$$(1) \qquad f(x) = 0, \quad f(x) = x^n + a_{n-1}x^{n-1} + \ldots + a_0,$$

is solvable meant then that it has at least one root which can be obtained by rational operations and root extractions starting from the coefficients $a_{n-1}, \ldots, a_0$.

Bring's main result is a reduction of the general equation of degree five to the form $y^5 + py + q = 0$ by rational operations and root extractions. The basis of his work is the following theorem: if x is a root of an algebraic equation (1), then every y of the form

$$(2) \qquad y = g(x), \quad g(x) = x^m + b_{m-1}x^{m-1} + \ldots + b_0,$$

is a root of an equation of degree at most n. Proof: all powers $x^n, x^{n+1}, \ldots$ are linear combinations of the powers $1, x, \ldots, x^{n-1}$. But then the same is true of all powers of y and hence the powers $1, \ldots, y^n$ must be linearly dependent. In order to find a polynomial of degree n of which y is a zero, it suffices to compute the greatest common divisor $R(y, a, b)$ of the polynomials $f(x)$ and $g(x) - y$,

$$f_1(x)f(x) + g_1(x)(g(x) - y) = R(y, a, b),$$

where y is a parameter and the coefficients are polynomials in y and the coefficients $a = (a_{n-1}, \ldots, a_0)$ and $b = (b_{m-1}, \ldots, b_0)$ of the polynomials $f(x)$ and $g(x)$. It is clear that

$$(3) \qquad R(y, a, b) = 0$$

when $f(x) = 0$ and $g(x) = y$, so that $R(y, a, b)$ must be a polynomial of at least degree n in y. Conversely, if $R(y, a, b) = 0$ and $g(x) = y$, then $f(x) = 0$ so that $R(y, a, b)$ has at most degree n. All this follows also from the formula for $R(y, a, b)$ given by the theory of determinants.

In 1683 Tschirnhausen used these facts to contruct a method for solving equations: if (2) and (3) can be solved by root extractions, the same is true of (1). More precisely: if (3) is solvable and the same is true of (2) with x as an unknown, then (2) has a root in common with (1) which itself is the result of rational operations and root extractions starting with the coefficients of $g(x)$ and $f(x)$. But a root of (1) can only depend on the coefficients of $f(x)$, i.e., (1) is solvable. When the method succeeds, it will in the end give n different expressions for the roots of (1) with general coefficients.

If $m \leq 4$ in (2), one of the conditions above is fulfilled by the work of Cardano and the problem is to choose the coefficients of (2) so that sufficiently many coefficients of (3) disappear to make the equations solvable. This situation opens a possibility to solve the general equation of degree five. Long and involved computations are the main obstacle.

Without mentioning Tschirnhausen Bring carries through the program above for $m = 1, 2, 3$ and $n = 2, 3, 4, 5$. The goal, a solvable equation for y, was attained when $n \leq 4$. When $n = 5$ Bring got an equation of the form $y^5 + py + q = 0$. His paper is so short and well written that it can bear a summary with unchanged notations. His first case is of course the equation of degree two $x^2 + mx + n = 0$ with an auxiliary equation $x + a + y = 0$. With $a = m/2$ we get $y^2 = \frac{m^2}{4} - n$. For an equation of any degree the same auxiliary equation can make the next to highest coefficient disappear.

For an equation of degree three, $x^3 + mx^2 + nx + p = 0$, Bring uses an auxiliary equation of degree 2, $x^2 + bx + a + y = 0$, and writes down without a proof an explicit equation for y. The coefficient of y^2 is rather simple,

$$m^2 - mb - 2n + 3a,$$

but the next coefficient has degree three and nine terms. Bring finds that a linear equation between a and b and a quadratic equation for b gives an equation $y^3 = $ const. For an equation of degree four without a second term,

(4) $$x^4 + nx^2 + px + q = 0,$$

Bring also uses a quadratic auxiliary equation $x^2 + bx + a + y = 0$. If $a = n/2$ and b satisfies a certain cubic equation, it turns out the equation for y does not have terms with y and y^3. This gives a quadratic equation for y^2 which solves the problem. Next, Bring is puzzled that the same method does not permit him to get rid of the second and third terms in (4) because then a has to satisfy an equation of degree six. Instead he tries a cubic auxiliary equation to achieve the same aim. But this requires that one coefficient again solves an equation of degree six so that also this method fails.

The partial success of the method of an auxiliary equation to simplify equations motivated Bring to apply the same method to the general equation of degree five. Here he employs an auxiliary equation of degree four and writes down the coefficients for y^4, y^3, y^2 in the equation for y. The last coefficient is a polynomial of degree four with around forty terms. It is shown, finally, that it suffices to solve equations of degree at most four in order to choose the coefficients of the auxiliary equation so that the general equation of degree five takes the form $y^5 + py + q = 0$. Bring's paper ends with a self-confident appeal to the reader to compare the solution given with Cardano's and then decide which is the most interesting one.

Bring's elegant formulas do not show how he carried out his eliminations and it is clear that he thinks that his method makes an explicit form of the zeros superfluous.

Bring's reduction of the general equation of degree five was redone in 1834 by an Englishman Jerrard and in this connection Bring's work also surfaced and made him internationally known.

Bibliography

KLINGENSTIERNA S.
Samuel Klingenstiernas levnad och verk, Biografisk skildring utgiven av K. Svenska Vetenskapsakademin I. Levnadsteckning (1919, Hildebrand Hildebrandsson), II. Vetenskapliga arbeten (1925, C.W. Oseen).

BRING E.S.
1. *Biografi*, Svenskt Biografiskt Lexikon.
2. *Meletemata quaedam mathematica circa transformationem aequationum algebraicarum*, Diss. Lund 1786.

CHAPTER 2

The Time 1800 - 1850

Hill   The Uppsala school   Malmsten   Holmgren

The beginning of the nineteenth century was the first time that mathematics in Lund and Uppsala was represented by professors who were not primarily astronomers, physicists, or theologians. This was perhaps a natural step in a progressive specialisation, and it is also not unthinkable that rumours of great progress in mathematics through Euler, d'Alembert, Lagrange, Legendre, and Laplace had spread to the peripheral European universities. But conditions at the Swedish universities were not favorable for change. All students studied all subjects of their own faculties and the theses required for the highest degree, that of *magister*, were short papers where the most spectacular feature was richly decorated compliments to some relative or benefactor, often from the nobility. A public defense of the thesis, a *disputation*, was mandatory, but it was not necessary to present new research and sometimes it sufficed for the student to defend something his professor had written.

In the beginning of the period, almost all scientific papers were published in some Swedish periodical. The most distinguished one was *Förhandlingar*[1] of the Swedish Academy of Sciences along with the more informal *Öfversigt af Förhandlingar*[2] The local scientific societies had periodicals of their own, in Lund: *Lunds Universitets Årsbok*[3] and *Kungl. Fysiografiska Sällskapets i Lund förhandlingar*,[4] and in Uppsala, *Acta* from the Royal Scientific Society, here for the time being referred to as *Nova Acta*.

The contents of these periodicals varied from semitic languages to mathematics. They were sent to other academies and universities, but must be considered as rather local phenomena where priority and competition were not always prominent features in spite of the fact that prominent members of the academies guaranteed the quality of the papers they presented.

New factors in the development were the periodicals for mathematics and physics that started publication around 1830 and were named after their first editors, (*Crelle's Journal*[5], *Grunert's Archiv*, and *Journal de Liouville*). These periodicals, which still exist under other names, were excellent places for the presentation of new research and initiated a spirit of competition in mathematics. With the growth of specialised journals during the nineteenth century, the more general publications from the academies came to play a secondary part for mathematicians, physicists, and chemists with international ambitions.

The new age turned out differently at the two universities. In Lund, time stood

---

[1] Proceedings
[2] Review of proceedings, quoted here as Öfversigt.
[3] The Yearbook of Lund University
[4] Proceedings of the Royal Physiographical Society of Lund
[5] quoted here as Crelle.

Hill

The first Swedish professor of mathematics who devoted himelf entirely to his subject was Carl Johan Danielsson Hill (1793-1875). He started his studies in Lund but considering the limited teaching of mathematics at the university in the early nineteenth century, we must look upon him as largely self-taught.

After his disputation in Lund in 1817, Hill spent periods of his life as a military surgeon, teacher of astronomy in Lund, and as professor of physics at the Technical Institute in Stockholm, a forerunner of the Royal Institute of Technology. But his favorite subject was mathematics, and after he was appointed professor of mathematics in Lund in 1830, he devoted himself exclusively to mathematics and stayed in his post till 1870. His son was the famous painter Carl Fredrik Hill.

Hill wrote many papers and had many disputations. In the first volumes of *Crelle*, which began in 1826 and in which Abel published his great work on elliptic functions, there are also papers by Hill, mostly written in Latin. The first one from 1827 deals with the so-called irreducible case for equations of degree three in which the equation has only real roots. This case seemed curious since Cardano's formula for the general case then involves third roots of complex numbers. Hill's method to avoid this difficulty was to write the equation as

$$x^3 - 3ax^2 - 3x + a = 0,$$

which could be achieved by real operations. When $a \neq 0$ is real, this equation has indeed three real, separate roots. The point of Hill's paper, which, in spite of its title has very little do do with the irreducible case, is that the roots $x = x(a)$ of this equation have the property that

(1) $$x\left(\frac{a+b}{1-ab}\right) = \frac{x(a) + x(b)}{1 - x(a)x(b)}.$$

He denotes the roots by $\sqrt[3]{a}$ with the root sign turned upside down and considers his formula simpler than that of Cardano.

The curious formula above depends on the fact that the equation for x can be written in the form

$$a = x(3 - x^2)/(1 - 3x^2) = \tan(3 \arctan x),$$

whence

$$x(a) = \tan(3^{-1} \arctan a),$$

so that (1) follows, apart from the ambiguity of the right side. Hill's discovery is just a variant of the trigonometric method for the irreducible case, known since the seventeenth century. In a later paper (1831) in *Crelle*, which also hinges on the irreducible case, Hill found functions with the same law of addition as the hyperbolic tangent function. As it turned out later, composition laws expressed in 'simple and telling' notations became a theme which never ceased to fascinate Hill.

Hill's second article in *Crelle* (1828) is written in German. In a footnote the editor says that he improved the language without entirely understanding the contents. The subject is the classical indeterminate integrals of rational functions combined with logarithms. The article, which has a long and fuzzy introduction, contains nothing new. Instead, we see Hill's intent to catalogize and systematize and his love of new notation. The functions tan and arctan appear as T and a T written upside down. This method gives difficulties with C for cos which is solved by denoting arccos by a C reversed. Misdirected systematics, wayward notation, verbosity, and lack of clarity are the characteristics of Hill's work. Most of his articles are pointless pontifications, e.g., the last three in *Crelle* in Latin with the following titles, here translated into English, 'Fragments about linear differential equations', 'Rational solutions of Riccati's equation','When does a differential equation have a complete algebraic solution, when a chiefly transcendental one, when a complete integral form?' It is apparent from the last paper that Hill has not understood the connection between a first order differential equation and the collection of its orbits. To be just, however, it should be said that the first issues of Crelle's journal contain many papers that are equally unclear in purpose and contents as those of Hill.

Sometimes Hill returned to physics and wrote about electricity and other subjects. In connection with a reform of Swedish agriculture where the old villages were broken up and each farmer settled on a piece of land entirely his own, he wrote a paper with the title *The average distance from a field to home* (Öfversigt (1849)). As an example of a piece of mathematical research, this paper made Hill and his subject targets of perhaps undeserved ridicule.

Already in his thirties Hill entered into a conflict with the Academy caused by Hill's sharp criticism of a mathematical paper printed in the *Proceedings*. At about the same time he sent one of his own papers to be judged. The Academy wrote that 'The author has proved himself versed in the higher problems of science' but that 'the safest way for the author to gain the literary fame which his work rightly deserves is attained by printing the paper at his own expense.' This insult made a deep impression on Hill, but with time the relations improved. Hill was elected member of the Academy in 1848 and, to begin with, he got a grant for the printing of his big work *Deo favente Matheseos Nova Analytica* in three volumes.[6]

In the introduction to the second volume Hill motivates his work. After a critical review of some matematical textbooks, among them one by Professor Ohm in Berlin (a brother of the physicist) he gives his own view of arithmetic and mathematics.[7]

> For these things [formulas like $a + b$ etc.] which are defined by the formula $(a|b)|c = (c|b)|a$ with the additional condition that $Ia = a$ arbitrary, or by a similar $(a|b)|c = a|(b|c)$ and, in addition, the symmetrical formula $(a|b) = (b|a)$, which, in the introduction p. 36 has proved itself the most basic and general. Certainly, the incomparable Abel has analysed a corresponding $f(x, f(y,z)) = f(f(x,y),z)$ (symmetric in x,y,z) and hereby found that $\psi(f(x,y)) = \psi(x) + \psi(y)$. However, from our synthesis we have found something similar (p. 60) if $f(x,y)$ is assumed to be $x|y$ and, in addition found the significance of his ψx, namely Λx, and that there is a truly general method of computation $x \odot \bar{a}$. Hence, instead of the usual (plus) $+$ we can use the more general $|$ ('more') and, from every aggregate we can deduce a more general product $r \odot a = a|a|a...$ and it is possible to use the habitual operational symbols $+, \times, ...$ etc. in a wider sense as $|, \odot ...$ etc. In our introduction of new notions we have not been able to avoid new symbols of which the simple ones are few but the composed ones $|_b^a, \odot_b^a, ...$ (p. 143-47) not only do replace the old ones but also almost uncountably many

---

[6] With God's help a new Analytical Mathematics.
[7] Hill's Latin is not easy. I thank Birger Bergh for help with the translation. Hill's page number refer to his book *Almän Storhetslära* (General Theory of Entities) from 1844.

new ones. Some new ones will soon make their appearance [among them some from the Zodiac].

Since Hill's Λ means logarithm, we can deduce that Hill has not understood Abel's paper where it is shown that a one-parameter Lie group can be parametrized so that the composition appears as addition of the parameter. The novelty in Hill's three volumes is the use of his new notation together with the habitual ones in a mathematics course with an eighteenth century basis.

Hill's interest in laws of composition is apparent already in his first paper in *Crelle* about the equation of degree three. It is displayed, fully fledged, in his *Prolegomena till hvarje blifvande almän storhetslära* (1844).[8]

Already from the beginning Hill was a controversial person. During his forty years as a university professor he developed into a friendly, eccentric and, with time, a more and more incomprehensible teacher. By the rules at the time, his papers were defended in public disputations by candidates for the degree of magister. If it were not for the great ability of youth to absorb and find a way to understand even rather abstruse constructions, one is tempted to pity those who defended parts of Hill's *Almän Storhetslära*.

Hill overrated himself and his abilities also in private life. He ran a business as a moneylender and tried without success to manage a sugar refinery for which he designed the machinery. Hill is the source of the anecdote about the learned professor who built himself a house without a staircase to the second floor and a fireplace without a flue. But it was the memory of Hill's curious notations that survived longest in the mythology of Lund professors.

The Uppsala school

As in Lund, mathematics in Uppsala was for a long time connected with astronomy. The last professor with this background was Jöns Svanberg (1771-1851). He received the degree of magister in 1796 with a thesis on Newton's curves of the third degree. Later he was employed by the Academy in order to check Maupertuis's 1736 degree measurements in Lapland. His result, which received some attention, was that Maupertuis had overrated the flattening of the earth. After this, Svanberg was elected secretary of the academy and nominated professor at the recently founded Surveyor Corps, but later he preferred the chair of mathematics in Uppsala. Svanberg was nominated Doctor of Theology in 1830 and left his chair in 1841 to be parson in a country parish.

Svanberg's papers deal mostly with astronomy, surveying and experiments in physics. Among his sparse mathematical papers is one about the theory of equations which, in spite of its name, *Nouvelles considérations sur la résolution des équations algébriques*, does not contain anything new, but it shows that the author has read the work by Lagrange about the solvability of equations by root extractions. Here Lagrange's resolvents are introduced and also his results about the behavior of rational functions of the zeros under permutations which laid the foundations for the work by Abel and Galois on solvable equations (see the Appendix at the end of the chapter).

Svanberg had a number of students who were curious about new mathematics from the European continent, in the first place Legendre's three volumes *Exercices de calcul intégral* (1811-1819), which, in spite of the name do not contain exercises.

---

[8]Preparations for every future general theory of entities.

It is rather an encyclopedic account of properties of integrals and more or less explicit formulas for them. For those readers who thought elliptic integrals and the Euler-Maclaurin summation formula too heavy, there were long, stimulating lists of nice explicit formulas for integrals. Later, papers by Cauchy and his didactic works were to play an important role.

The influence of Legendre is especially noticeable in the work of Adolf Fredrik Svanberg (1806- 1857), son of Jöns Svanberg, and Carl Johan Malmsten (1814-1886), while Cauchy was the great teacher for Emanuel Gabriel Björling (1808-1872).

Svanberg

In 1828 Adolf Fredrik Svanberg was made docent of mathematics at Uppsala. After an interlude at the School of Artillery in Marieberg he became professor of mathematics and mechanics in 1841. He was nominated docent on the strength of a paper on the curvature of surfaces and then wrote many papers on integrals, e.g. (1832).

Svanberg starts with the fomulas

$$\int_0^\infty 2\cos mx \, dx/(h^2+x^2) = (\pi/h)e^{-mh}, \quad m,h > 0,$$

$$\int_0^\infty 2x \sin mx \, dx/(h^2+x^2) = \pi e^{-mh},$$

proved by Lagrange before the invention of residue calculus. The fact that

$$\int_{-\infty}^{+\infty} \frac{F(x)}{x^2+h^2} dx = \pi F(ih)/h$$

for suitable analytic functions is hidden in many of Svanberg's formulas. He arrives at such formulas by several methods, for instance by summing over m in Lagrange's two integrals above and by introducing a parameter and differentiating with respect to it. At his time, the theory of analytic functions and residue calculus was not ripe for routine applications.

Svanberg's computations are mostly formal, but it happens that he has to reject some of his results, for instance the formula

$$\int_0^\infty \lambda x^2 dx/(h^2+x^2) = \pi \log(1+e^{-\lambda h}).$$

Svanberg's work illustrates the situation in analysis before strict proofs made life less adventurous. Also in other papers Swanberg returns to series expansions where convergence is hardly ever mentioned.

One of Svanberg's papers shows that the work of Abel had made an impression in Sweden. In the second volume of *Crelle*, Abel had computed relations between the integrals

$$y = \int_0^1 (x+a)^{\gamma+1} x^{\alpha-1}(1-x)^{\beta-1} dx$$

for different values of the parameters α, β, γ where α, β lie between 0 and 1. Abel's method was to deduce a second order differential equation, $y'' + py' + qy = 0$, for $y = f(a,\alpha,\beta,\gamma)$ as a function of a. It then follows that

$$z = f(a, 1-\beta, 1-\alpha, \alpha+\beta+\gamma-1)$$

satisfies the same equation. The fact that $w = y'z - z'y$ satisfies the equation $w' + pw = 0$ then produces explicit relations between y, y', z, z', all of which, except for multiplication by constants, have the same form as y. Svanberg found new, rather complicated relations of the same nature in a situation where two integrals, y, z each satisfies a second order equation.

During his time as a professor, Svanberg wrote a didactic paper (1847) about the effect of a transformation of variables in a double integral with fixed limits, a subject where formal mathematics may lead one astray. His proof uses geometrical arguments. The results are, of course, essentially right but Svanberg's formal past reveals itself in one or two mistakes which would have been unthinkable if he had visualized the integral as extended over a plane region. Such things do not appear when Svanberg is on familiar ground as when he writes about second order equations (1854).

Compared to the eccentric Hill, Svanberg was a completely normal mathematician, but nevertheless his papers show that it was easy to get lost in the many formulas of analysis before the breakthrough of strict analysis at the end of the nineteenth century.

E. G. Björling

Emanuel Gabriel Björling (1808-1872) took his magister's degree in Uppsala in 1830 under Jöns Svanberg. The subject of the thesis was the movements of rigid bodies. Later he was nominated docent and lektor[9] at the gymnasium in Västerås. Björling was a prominent pedagogue and the author of important textbooks for the teaching of elementary mathematics. He was chosen member of the Academy in 1850 and was several times rewarded for mathematical papers. He also wrote in other fields: a theological disputation in 1828 and a book on Latin etymology.

Björling's most ambitious paper (1844 a) deals with minimal surfaces, finally specialised to rotational ones. He also considered the problem of passing a minimal surface through a space curve for which he got an honourable mention in vol. III of Darboux's great work *Théorie des surfaces* (1894).

Björling's chief interest was Cauchy's work and for a long time he served as Cauchy's propagator and interpreter in Sweden. One of the recurrent themes and the object of a correspondence with Cauchy was the definition of the functions x^y and $\log_\beta x$ when x and y are complex numbers and β is positive.

One result of the contact with Cauchy was Björling's interest in a number of essential but at the time often misunderstood concepts, in particular the convergence of series, continuity of functions and later also Cauchy's definition of the concept of an analytic function. The results of Björling's papers in these fields are less interesting than his view of convergence and continuity and his relation to Abel's famous paper on the binomial series, where these concepts are used correctly. Some of Björling's mistakes are due to a firm self-confidence that is very clear in his polemics with the younger Malmsten, but he was not alone in his mistakes and his papers show how difficult it was in his time to understand and accept strict analysis.

In a two-part paper (1844 b) Björling writes about the Euler-Maclaurin summation formula. Because Malmsten also wrote about the subject, a short, modern account follows here. The origin of the Euler-Maclaurin summation formula is an

---

[9]Head teacher requiring a doctor's degree.

2 The time 1800–1850

integration by parts as follows

$$\int_0^1 (t - \frac{1}{2})f'(t)dt = \frac{1}{2}(f(0) + f(1)) - \int_0^1 f(t)dt.$$

Here $B_1(t) = t - \frac{1}{2}$ is the Bernoulli polynomial of degree one. The other Bernoulli polynomials,

$$B_m(t) = \sum \binom{m}{k} t^k B_{m-k}, \quad m > 1,$$

where the B_k are the Bernoulli numbers, are defined recursively by the formula $B_n'(t) = nB_{n-1}(t)$ and the condition that $B_n = B_n(0) = B_n(1)$ with the value 0 when $n = 3, 5, \ldots$. Integration by parts on the left side above lead to the Euler-Maclaurin formula

$$\frac{1}{2}(f(0) + f(1)) - \int_0^1 f(t)dt = \sum_1^n B_{2k}(f^{(2k-1)}(1) - f^{(2k-1)}(0))/(2k)! - R_n$$

with Jacobi's remainder

$$R_n = \int_0^1 B_{2n}(t)f^{(2n)}(t)dt/(2n)!.$$

In the common version of the formula, the sum $\sum_0^{m-1} f(t+k)$ replaces $f(t)$. The result is an equality between the difference

$$\frac{1}{2}f(0) + f(1) + \ldots + f(m-1) + \frac{1}{2}f(m) - \int_0^m f(t)dt$$

and a sum

$$\sum_1^n B_{2k}(f^{(2k-1)}(m) - f^{(2k-1)}(0))/(2k)! - \int_0^m B_{2n}(t - [t])f^{(2n)}(t)dt/(2n)! \,,$$

where $[t]$ is the integral part of t. Björling deduces this formula without a remainder by working with the recursion formulas for the original Bernoulli numbers $B_k' = (-1)^{k+1}B_{2k}$.

The next volume of *Nova Acta* (1847) carries Björling's two-part study *Studies in the theory of infinite series*, written in Latin. The first theorem that Björling proves says in a not too precise formulation that a real or complex power series $\sum a_n x^n$ is convergent or divergent according as $|x|$ is larger or less than $1/|a_n|^{1/n}$ for all large n. Then follows a section about the problem whether or not a convergent sum $\sum f_n(x)$ of continuous functions in an interval is continuous. That the sum is continuous is proved in the following way (p.66,67): if $\omega > 0$ then there is to every x an $n = n(x)$ such that

(*) $$|f_n(x) + f_{n+1}(x) + \ldots| < \omega/2.$$

If $n(x)$ is maximal for $x = \xi$, the inequality follows for all x when $n = n(\xi)$. After this, Björling turns to the finite sum

$$g_n(x) = f_1(x) + \ldots + f_{n-1}(x).$$

He writes that
$$|g_n(x+\alpha) - g_n(x)| \leq n|f_m(x+\alpha) - f_m(x)|$$
for some $m < n$ and remarks that the right side tends to zero with α. In a footnote, the author writes that he has his theorem from Cauchy and discusses Abel's counterexample, namely the series
$$\sum_1^\infty n^{-1} \sin nx,$$
whose sum is $(\pi - x)/2$ in the interval $0 < x < 2\pi$ but zero when $x = 0, 2\pi$. One gets the impression that Björling was far from the solution of the paradox (the function $n(x)$ need not be bounded) and from the notion of uniform convergence. This is clear both from the paper in question and from what he wrote later (see below).

The origin of the remainder of the paper is a well-known paper by Abel from 1828 where he proves that the binomial series for $(1+x)^m$,
$$1 + mx + m(m-1)x^2/2 + ...,$$
where x and $m \neq 0$ are complex, converges when $|x| < 1$, when $|x| = 1$ and $\operatorname{Re} m > -1$, except when $x = -1$, in which case $\operatorname{Re} m > 0$ is the proper condition. In all other cases the series diverges. Finally, when the series converges, its sum is $(1+x)^m$.

Björling's paper is an adaptation of that of Abel and the author feels that this requires some excuses. The first part begins with a footnote (p. 62) where priority is conceded to Abel, but with two reservations. The first one says that Abel's general introduction is too long and involved in order to deal with what in Björling's opinion is just the case $|x| < 1$. The second one claims that Abel does not reap the full benefit of his work when he — unlike Björling— does not make a full separation of the complex numbers into real and imaginary ones. The first reservation means that Björling has not understood Abel's very clear general introduction. In particular, he has not grasped Abel's theorem about convergence at the boundary of the circle of convergence, i.e. that a series
$$\sum_1^\infty a_n x^n,$$
which converges when $0 \leq x \leq 1$, represents a continuous function in the same interval.

Abel's exposition of the binomial expansion is followed rather closely. The difference is that convergence and divergence for $x = -1$ is discussed case by case using Raabe's criterion. In a footnote in this part (p. 156) Björling says that it is impossible for him to understand how Abel can deduce the divergence of the series from the fact that the function $(1+x)^m$ does not have a limit as $x \to -1$. The explanation is of course that Björling has not understood Abel's theorem.

On page 158 Björling states that if a real power series $\sum a_n u^n$ converges for $u = U > 0$, then it converges for $0 \leq u \leq U$, but the proof works only when the terms are positive. From this he draws the wrong conclusion that if the corresponding function is continuous in the interval $0 < u \leq U$, then the series converges at the right endpoint. Björling also declares (p. 173) that the series
$$\sum f_n \sin nx$$

converges when f_n tends to zero at infinity. In a footnote he confirms this statement against Malmsten who was of another opinion. A simple counterexample, for instance $nf_n = \sin x, x = \pi/2$, would have settled the matter.

This long work by an intelligent and well-read mathematician shows that it was not only Abel's work on algebraic integrals which encountered incomprehension. What he wrote about convergence and divergence met with the same fate.

In (1841 d) Malmsten tried to prove that the series above converges when f_n decreases to zero. He writes

$$\sin nx = (\sin(n+\frac{1}{2})x - \sin(n-\frac{1}{2}))/\cos\frac{x}{2}$$

as a difference but fails to carry it over to f_n and the incorrect proof does not use the decrease of f_n. In a correct proof as by Abel, it is shown that the series converges at the same time as the series

$$\sum (f_{n+1} - f_n)\sin(n+\frac{1}{2})x,$$

where the convergence follows form the fact that the absolute value of the general term is at most $f_n - f_{n+1}$.

Malmsten's paper inspired Björling (1847) to carry over the differences to f_n, but his proof is wrong or at least incomplete because he does not say that the numbers $f_n - f_{n+1}$ must be zero or have the same sign for large values of n.

Björling returned several times to the question of the continuity of infinite sums of continuous functions, the last time in 1853 in *Öfversigt*. Here he writes about Cauchy's rectification of his erroneous theorem and states with some satisfaction that his own condition (the inequality (*) above for all x), in modern terminology uniform convergence, is also present in Cauchy's paper. He also mentions that Cauchy has checked that convergence is not uniform in Abel's example.

Against the background of papers of the time which have to do with uniform convergence, I. Grattan-Guinness (1986) and Yngve Domar (1987) have a more appreciative judgement of Björling's contribution than the one which is found here.

In his exegesis of the work of Cauchy, Björling could hardly avoid the fascinating subject of complex integration and Cauchy's theorem about integration of analytic differentials. In 1852, in *Förhandlingar*, Björling writes about the sensational piece of news that it is only necessary to know that a complex function of a complex variable is differentiable in order to be able to write it as a power series. He reproduces the proof including the formula

$$f(a) = \frac{1}{2\pi}\int_0^{2\pi}(e^{i\theta} - a)^{-1}f(e^{i\theta})e^{i\theta}d\theta,$$

but, to be absolutely certain, he considers it necessary to assume that f can be differentiated twice. Cauchy's important condition that the complex derivative must be independent of the direction in which it is taken, occurs in a footnote. On the other hand, the paper begins with a long and superfluous deliberation about derivatives of functions of a real variable.

Björling's mathematical frame of reference made it difficult for him to really understand Cauchy's theorem and other new, important mathematics. But we must give him credit for a number of serious efforts.

In school matters Björling used his energy and talent to achieve important reforms of the curriculum and he wrote a textbook of algebra which was used for a long time. It is remarkable that he could write his papers and at the same time carry a heavy teaching load. Björling's *In memoriam* was written by a colleague of his, C.F. Lindman (1816-1901), lektor in Strängnäs and the author of several well-known textbooks in mathematics. Lindman also wrote well about series and definite integrals in the Academy publications.

Malmsten

In competition with Björling, Carl Johan Malmsten was nominated professor of mathematics in Uppsala in 1841. Future developments proved that the choice was justified. During the 1840's Malmsten wrote many remarkable papers and he was the first Swedish mathematician after Klingenstierna who followed the mathematics of his time and was able to contribute to it. The young professor was a gifted public speaker, often in demand, and the papers he wrote in Swedish show his talent as a writer. In the beginning of his career he managed to make mathematics a fashionable subject in Uppsala.

Like Svanberg, Malmsten got his first inspiration from Legendre's *Exercises du calcul intégral*. His first printed papers deal with the consequences of certain basic formulas, used also by Svanberg, for instance

$$\int_0^\infty \frac{\cos mx}{1+x^2} dx = \frac{\pi}{2} e^{-|m|}$$

when m is real. Malmsten went much further than Svanberg and he was careful about convergence. He tried to show (1841'd, 1846) that the series

$$\sum_0^\infty f_n \cos nx$$

and the analogous one with $\sin nx$ converges when f_n decreases to zero and not all $\cos nx$ are equal to 1. His attempt has already been mentioned above in connection with Björling's work.

In a short paper, the value of the integral

$$\int_0^\infty \frac{\cos ax}{(b+x^2)^n} dx$$

is computed for arbitrary integers $n > 0$ using the property that the integral satisfies a differential equation as a function of b.

In 1847 Malmsten made his debut in *Crelle* 34 with a treatment of Abel's posthumus paper *Sur la résolution algébrique des équations* (1847 a), then available through Holmboe's edition of Abel's collected works.

Abel's paper starts with considerations about the divisibility and irreducibility of polynomials, things which are now general knowledge, and continues with the form of the roots of an equation $f(x) = 0$, where $f(x)$ is a polynomial of prime degree p which is irreducible over a field K containing the roots of unity. To a large extent, Malmsten's paper is a careful edition of the first part of Abel's, but it also contains a remark which gave him an answer to the question why general equations of degree > 4 are not solvable by root extractions. The present-day answer that the alternating group of more than four objects is not solvable, was not known at the time.

Let $x_1, ..., x_p$ with p a prime be the zeros of $f(x) = 0$. Malmsten could see that Abel had shown (see the appendix at the end of the chapter) that the Galois group $\text{Aut}(f/K)$ permutes Lagrange's resolvents

$$(\omega^k, x) = \sum_{j=1,...,p} \omega^{-jk} x_j, \quad k = 1, ..., p-1,$$

where $\omega^p = 1$, $\omega \neq 1$. Hence their product

$$\prod_{k=1,...,p-1} (\omega^k, x)$$

is invariant. For a general equation of degree p, we can consider the roots as independent quantities and the resolvents (ω^k, x) as linear forms of them. The decomposition of the product into factors (ω^j, x) is then unique, apart from multiplicative factors in K. An interchange of x_0 and x_1 gives

$$x_1 + x_0\omega + ... + x_{p-1}\omega^{p-1} = c(x_0 + x_1\omega^j + ...)$$

for some $c \neq 0$ in K and some number j, $1 < j < p$. A comparison between the coefficients of x_0 and x_1 then shows that $\omega^{j+1} = 1$ which is possible only when $p = 3$.

Here Malmsten gets an answer to his question why the impossibility (for equations of prime degree) starts with degree five. If he had succeeded in deciphering all of Abel's paper, he could have obtained a simpler proof: the affine maps $j \to aj + b$ mod p do not contain transpositions when $p > 3$ (see the Appendix).

In the next volume of *Crelle*, Malmsten wrote a paper (1847b) with the title *Sur la formule*

$$hu'(x) = \Delta u(x) - \frac{1}{2}\Delta u'(x) + \frac{B_1 h^2}{1.2}\Delta u''(x) + ...$$

Here $\Delta u(x) = u(x + h) - u(x)$ and the Bernoulli numbers are the original ones. One gets the formula by replacing $f(x)$ by $f'(x)$ in the Euler-Maclaurin formula (p. 14). Malmsten's contribution is that he (as did Jacobi) found the precise error term and that he proved that (in modern notation) $B_{2k}(x) - B_{2k}$ has the sign $(-1)^k$ when $0 < x < 1$ and, in addition, is symmetric around $x = 1/2$ (now this is well known). This gave him an excellent error term in applications, for instance to Stirling's formula. His paper was reprinted in *Acta Mathematica* 5 (1886), the editor Mittag-Leffler's motivation being to demonstrate that good mathematics had been done earlier in Sweden.

Functional equations

Crelle volume 38 (1849) begins with a paper in Latin by Malmsten entitled in English translation *About definite integrals and infinite series*. Some years earlier, when this paper, divided into parts, was defended by students in Uppsala as theses, the title was longer and more explicit: *New theorems about definite integrals, about summation of series and their transformation into other series*.

In this remarkable work Malmsten exhibits a complete mastery of the material in Legendre's *Exercices du calcul intégral* and uses it to prove new, important formulas. For certain functions f he wanted to deduce a connection between $f(s)$ and $f(1-s)$, analogous to the wellknown formula

$$\Gamma(s)\Gamma(1-s) = \pi/\sin \pi s.$$

His motives were a general curiosity and a feeling that such formulas could be important. In this he was right. His results are a great number of functional equations which are close to the functional equation

$$(3) \qquad 2\Gamma(s)\cos\frac{\pi s}{2}\zeta(s) = (2\pi)^s\zeta(1-s)$$

for Riemann's zeta function $\zeta(s)$, for $\operatorname{Re} s > 1$ defined by

$$\zeta(s) = \sum_{1}^{\infty} n^{-s}.$$

The functional equation shows that $\zeta(s)$ is entire analytic apart from a pole at $s = 1$. In his classical paper *Über die Anzahl der Primzahlen unter einer gegebener Grenze* from 1859 Riemann proved the formula (3) by noting that the integral

$$F(s) = \int_C (-x)^{s-1} dx/(e^x - 1)$$

is an entire function of s when C is a positively oriented loop close to the positive axis and $\arg(-x)$ grows from $-\pi$ to $+\pi$. When $\operatorname{Re} s > 1$, the loop can be drawn to the positive real axis which gives

$$(e^{-\pi i s} - e^{\pi i s}) \int_0^\infty x^{s-1} dx/(e^x - 1)$$

where the integral after a series expansion equals $\Gamma(s)\zeta(s)$.

When $\operatorname{Re} s < 0$, C can be replaced by a C' consisting of the circle $|x| = \varepsilon$ and the line $\operatorname{Re} x = \varepsilon$ with $0 < \varepsilon < 1$. When C' is drawn to $-\infty$, residues are produced at the points $x = 2\pi i n \neq 0$, so that

$$F(s) = 2\pi i \sum_{1}^{\infty} (2\pi n)^{s-1} (i^{s-1} - i^{-(s-1)}).$$

If the equality is made explicit, we get the functional equation by analytic continuation.

It is clear that these computations will produce similar results if we replace the function $1/(e^x - 1)$ in the formula for $F(s)$ with $f(e^x)/g(e^x)$ where f and g are polynomials. If $\operatorname{grad} f < \operatorname{grad} g$ and the zeros e^{ia_k} of the equation $g(z) = 0$ are simple with absolute values equal to 1, the function

$$\int_C f(e^x) x^{s-1} dx/g(e^x)$$

is entire analytic in s and equals

$$-2i \sin \pi s \int_0^\infty f(e^x) x^{s-1} dx/g(e^x)$$

when $\operatorname{Re} s > 0$ and a sum of residues

$$2\pi i \sum_{k,n} f(e^{ia_k})(g'(e^{ia_k}))^{-1} (a_k + 2\pi i n)^{s-1}$$

when Re $s < 0$. If this series converges for $s > 0$ and the series expansion of the integral converges when $s < 1$, one gets equality in the common interval.

Malmsten proves the equality in two cases, when

$$f(z)/g(z) = \frac{1}{z^2 + 2z\cos a + 1},$$

and when

$$f(z)/g(z) = \frac{1+z^2}{1 + 2\cos a z^2 + z^4}.$$

In the first case the result is the following identity where $|a| < \pi$ and $0 < s < 1$,

(4) $$\sum_{1}^{\infty}(-1)^{k-1}\left(\frac{1}{((2k+1)\pi + a)^s} - \frac{1}{((2k+1)\pi - a)^s}\right) =$$

$$\frac{1}{\Gamma(s)\sin\frac{\pi s}{2}}\sum(-1)^{k-1}\frac{\sin(2k+1)a}{(2k+1)^{1-s}}.$$

When $a = \pi/2$, it gives the functional equation

$$\Gamma(s)\sin\frac{\pi s}{2}\varphi(s) = (\pi/2)^s \varphi(1-s)$$

where

$$\varphi(s) = \sum_{1}^{\infty}(-1)^{k-1}(2k+1)^{-s}.$$

Malmsten did not know the calculus of residues. Instead, he used some known formulas, for instance

$$\int_0^\infty \frac{e^{au} - e^{-au}}{e^{\pi u} - e^{-\pi u}} u^{-s} du = \frac{\sin a}{\Gamma(s)\cos\frac{\pi s}{2}}\int_0^\infty \frac{e^{-x}x^{s-1}}{1 + 2e^{-x}\cos a + e^{-2x}}dx,$$

due essentially to Legendre. A natural series expansion of the two sides gives (4). The formula is proved when s is real, $0 < s < 1$, and the convergence is checked.

When Malmsten wrote his paper, he saw no connection with number theory and the theory of analytic functions was not so developed that he could see his formulas as identities between analytic functions. Therefore his paper did not play a part in later developments, but it shows Malmsten's analytical skill and his concentration on essential matters.

Continued fractions

Between his last two papers in *Crelle*, Malmsten wrote a paper by which his name remained in the mathematical literature, *About the convergence of continued fractions* in *Förhandlingar* (1848 c).

Continued fractions
$$b_0 + \cfrac{a_1}{b_1 + \cfrac{a_2}{b_2 + \cdots}}$$

are infinitely repeated fractions. They occur in the work by D. Bernoulli and Euler and play an important part in number theory. Every positive number x can be written as a continued fraction, namely by the rule

$$x = [x] + 1/x_1, \quad x_1 = [x_1] + 1/x_2, \ldots$$

where $[x]$ means the integral part of x. When x is rational the process ceases after a finite number of steps. If, for instance, all numerators and denominators are equal to 1, we get $x = 1 + 1/x$ so that $x = (1 + \sqrt{5})/2$.

In the sequel we shall put $b_0 = 0$ and write the continued fraction above as

$$a_1|b_1 + a_2|b_2 + ...$$

with the convergents a_1/b_1, $a_1b_2/(b_1b_2 + a_2), \ldots$, and

$$\frac{A_n}{B_n} = b_0 + a_1|b_1 + a_2|b_2 + \cdots + a_n|b_n.$$

when $n > 1$ where A_n, B_n are polynomials in $a_1, ..., a_n$ and $b_1, ..., b_n$.

Malmsten's paper ends with a theorem:

THEOREM. *A continued fraction with positive b_n and negative a_n converges if there are numbers $p_n \geq 1$ such that*

$$\left| \frac{a_n}{b_{n-1}b_n} \right| \leq \frac{p_n - 1}{p_{n-1}p_n}$$

for all $n > 0$.

Choosing $p_n = 1 + |a_n|$ we get a corollary: convergence if

(5) $$|b_n| \geq 1 + |a_n|.$$

Malmsten remarks that this was known earlier only when a_n, b_n are integers.

Malmsten's theorem follows from the special case since the denominators $a_1, \ldots$ and the numerators in a continued fraction are not unique. In fact, if $c_1, c_2, \ldots$ are positive numbers, the continued fraction above equals

$$b_0 + c_1a_1|c_1b_1 + c_1c_2a_2|c_2b_2 + c_2c_3a_3|c_3b_3 + ...$$

and the corollary above requires that $|c_nb_n| \geq 1 + |c_{n-1}c_na_n|$ which, with $c_k = p_k/b_k$, is Malmsten's theorem.

This theorem together with additional work by Malmsten's students Falk (1869) and Broman (1877) is mentioned in Perron's standard book (1929) about continued fractions. In this book it is also shown that (5) entails convergence when a_n, b_n are complex numbers (a theorem by Pringsheim). The proof is given below.

It is easy to verify the recursion formulas

$$A_n = b_nA_{n-1} + a_nA_{n-2}, \quad B_n = b_nB_{n-1} + a_nB_{n-2}$$

where $B_0 = 1$, $A_0 = 0$. In particular,

$$B_{n-1}A_n - A_{n-1}B_n = a_n(B_{n-2}A_{n-1} - A_{n-2}B_{n-1}).$$

We also get an inequality

$$|B_n| \geq |b_n||B_{n-1}| - |a_n||B_{n-2}| \geq |b_n||B_{n-1}| - (|b_n| - 1)|B_{n-2}|$$

whence

$$|B_n| - |B_{n-1}| \geq (|b_n| - 1)(|B_{n-1}| - |B_{n-2}|) \geq |a_n|...|a_1|$$

because $|b_n| - 1 \geq |a_n|$ and $B_1 = b_1, B_0 = 1$. This means that the numbers $|B_n|$ are positive and increase with n. With $C_n = A_n/B_n$ and $D_n = C_n - C_{n-1}$ we can write

$$\frac{A_n}{B_n} = C_1 + \sum_{2}^{n} D_k.$$

By the recursion formulas above, $D_k = -B_{k-2} a_k D_{k-1}/B_k$, which shows that the sequence C_n converges if and only if

$$a_1/B_0 B_1 - a_1 a_2/B_1 B_2 + a_1 a_2 a_3/B_2 B_3 - \ldots$$

converges. Hence, according to the above,

$$|a_1 \ldots a_n|/|B_n||B_{n-1}| \leq \frac{1}{|B_{n-1}|} - \frac{1}{|B_n|},$$

and this proves that the previous series converges absolutely and that the absolute value of the sum is at most 1.

Differential equations

Already in the early years of infinitesimal calculus it was found that not all ordinary differential equations could be solved explicitly by quadrature, i.e. by repeated integration. Although Newton found that all such equations could be solved by power series, the problem of quadrature lived on far into the nineteenth century and attracted famous mathematicians as for instance Kummer and Liouville. Malmsten was also interested and many of his papers deal with this problem which requires both imagination and a sharp mind.

An elementary formula says that if $y(x) \neq 0$ solves the differential equation $y'' + P(x)y' + Q(x)y = 0$ then

$$z = \int e^{-\int P dx} \frac{dx}{y(x)^2}$$

is another solution found by quadrature. In 1848 in Öfversigt, Malmsten proved an extension of this result where his delight in the new theory of determinants shines through. Let

$$y^{(n)} + P y^{(n-1)} + \ldots = 0$$

be the differential equation and let $y_1, \ldots, y_{n-1}$ be $n-1$ linearly independent solutions and

$$1/R = \det y_j^{(k)}$$

their Wronskian. A new solution is obtained from

$$y = z_1 y_1 + \ldots + z_{n-1} y_{n-1}$$

where

$$z_r = -\int e^{-\int P dx} \frac{\partial R}{\partial y_r^{(n-2)}} dx$$

The possibility to find a last solution by quadrature when the others are given is a special case of Jacobi's multiplier theory which caught Malmsten's interest in the

1850's. Jacobi's theory (*Crelle* 1844) deals with the connection between a system of differential equations

(6) $$dx_1/dt = X_1(x), \ldots, dx_n/dt = X_n(x), \quad x = (x_1, ..., x_n),$$

and the partial differential equation

(7) $$X_1 \frac{\partial u}{\partial x_1} + ... + X_n \frac{\partial u}{\partial x_n} = 0.$$

In classical mechanics there are plenty of systems (6) with t as time. The solutions $t \to x(t)$ are then orbits in n-dimensional real space.

A function $u(x)$ satisfying (7) is said to be an invariant of (6) since all functions $t \to u(x(t))$ are constant, i.e. u is constant on every orbit. It is obvious that functions of invariant functions are invariant and that an invariant function is uniquely determined by its values on a manifold of codimension 1 which meets the orbits transversally and that these values can be given arbitrarily. This can always be arranged locally, which means that there are $n-1$ invariant functions $z_1, \ldots z_{n-1}$ with the property that all invariant functions are functions of them. They have also the property that the equations

$$z_1(x) = c_1, \ldots, z_{n-1}(x) = c_{n-1},$$

where the right sides are constant, represent all orbits.

A differential form

(8) $$0 \neq \omega(x) = M(x) dx_1 \wedge ... \wedge dx_n$$

is said to be invariant if it is constant along the orbits of (6). A differentiation with respect to t of the form $\omega(x(t))$ gives the equivalent condition that

(9) $$\frac{\partial M X_1}{\partial x_1} + ... + \frac{\partial M X_n}{\partial x_n} = 0.$$

A solution $M(x)$ of this equation is said to be a multiplier of (6) or (7). According to (9), the product of an invariant and a multiplier is a multiplier. By the same equation, a multiplier solves the equation

$$d \log M / dt + \sum \partial_k X_k = 0,$$

and this shows that the quotient of two multipliers is an invariant. How a multiplier behaves under changes of variables x to another set y is a consequence of (8). The system (6) is then transformed very simply into a system (6′) with a multiplier $M'(y)$ such that, by (8),

$$M'(y) = M(x) D(x_1, ..., x_n) / D(y_1, ..., y_n)$$

where the quotient indicates the Jacobian.

When $n = 2$, M is the same as Euler's integrating factor because (9) means that

$$M X_2 dx_1 - M X_1 dx_2$$

is the differential $du(x)$ of an invariant function $u(x)$ which can be obtained by quadrature, here an integration along a path to x from a fixed point. The differential equation $X_2 dx_1 - X_1 dx_2 = 0$ then has the solutions $u(x) = $ const.

If the variables $x_1, ..., x_n$ are exchanged for other variables $y_1, ..., y_n$ where for instance $y_1, ..., y_i$ are invariant, then (6) is transformed into a system for y with right sides $Y_1, ..., Y_n$ where the first i ones vanish. This proves the following theorem by Jacobi, called the theorem of the last multiplier: For a system (6) with one multiplier and $n-2$ independent invariants, the remaining one can be computed by quadrature.

Due to the interest in the quadrature problem, Jacobi's paper attracted a lot of attention. Malmsten wrote a long paper with the title *About the integration of differential equations* (1860), later translated into French. In the introduction Malmsten writes, among other things,

> 'Now it is known of old that the mightiest general tool we possess in order to vanquish analytical difficulties consists in transformations, suitable for the purpose at hand. There is no part of analysis where they do not play the most important role— yes, one may say with good reason that the transformations form the backbone and purpose of the analytical method. The thought is therefore close at hand to change Jacobi's variables into other ones, which, although not *in general* more suitable than the others for the mastering of difficulties, yet may present *new* cases where this may be done. For such a transformation is in itself a generalisation, which, as it were, conducts us beyond the old borders and let us know, in addition to the old ones, new cases where quadrature is possible'.

This passage shows that Malmsten knew very well that, in modern terms, Jacobi's theory is invariant under transformation of variables, but that his main interest was to find new cases of quadrature. He begins his paper by carrying Jacobi's formulas into a system of general form,

$$d\varphi_k(x, x_1, ..., x_n)/dx = \psi_k(x, x_1, ..., x_n), \quad k = 1, ..., n.$$

A scalar differential equation may also be written in this form,

$$d\varphi(x, y, y', ..., y^{(n-1)})/dx = \psi(x, y, y', ..., y^{(n-1)}).$$

Malmsten proves among other things that the corresponding system has a multiplier when the derivative $y^{(n-2)}$ is missing in the left side and $y^{(n-1)}$ is missing on the right side. His main result, which must have cost him considerable work, is a list of twnety very varied and sometimes rather artificial equations which can be solved by quadrature, eventually when one or several solutions are known. The first example is an equation which expresses the curvature radius of a curve as a given function of the distance to a point.

The time after the professorship

In a paper (1865) after his time as a professor Malmsten wrote about Cauchy's integral theorem, i.e., that the integral of $f(z)dz$ between two points is (locally) independent of the path between them when $f(z) = f(x+iy)$ is an analytic function. Some years later, this theorem and the nature of analytic functions was to confuse the mathematicians at Uppsala and entice them to find their own interpretations. Malmsten begins his paper by observing unclear points in the two proofs by Cauchy.

Then he separates everything into real and imaginary parts and approximates the integrals by Riemann sums. The main ingredient of his reasoning is a proof that

$$\int_L pdx + qdy = \int_M pdx + qdy$$

when $p_y(x,y) = q_x(x,y)$ and L, M are separate paths from the corner of a rectangle to the opposite one. The remainder of the proof is a long computation with Riemann sums where the condition that $p_y = q_x$ is used in difference form. The paper grows into a kind of essay that shows Malmsten's ability to handle inequalities. His merit is to have brought out the condition that $p_y = q_x$, but he did not use it in Green's formula which probably was not known to him.

At the end of the 1850's Malmsten was on his way out of science. He had developed an interest in probability and insurance. In 1858 he joined a committee whose purpose was the construction of a pension system. In the following year he was appointed minister without portfolio and in 1866 Malmsten became governor of the province of Skaraborg, the first one to serve in this capacity who was not a member of the nobility. After retirement in 1879 he settled in Uppsala and died in 1886. In governmental circles, Malmsten was famous for a brilliant speech in 1859 about freedom of the press on the occasion of the fiftieth anniversary of the Swedish constitutional law and for his sumptuous dinners and excellent wines in the governor's residence.

The time came when Malmsten regretted that he had not returned to his chair after he left the government. One of the reasons was his concern for the economic position of his student and deputy H. Th. Daugh.

The library of the Mittag-Leffler Foundation has an exchange of letters between Malmsten and Mittag-Leffler. At first the subject was mathematical books that Mittag-Leffler, who at the time was in Berlin, procured for Malmsten. Later the correspondence turned more personal. One of Malmsten's daughters married the German mathematician Ernst Schering in Göttingen. After a visit to his daughter in connection with a scientific jubilee in Göttingen, Malmsten wrote to Mittag-Leffler: 'many people have reproached me that I spent too much money on the education of my daughters. But when I saw my daughter as a hostess speak with the same ease to Frenchmen, Germans and Englishmen in their own tongues, my heart swelled with affection and pride'.

Hjalmar Holmgren

Hjalmar Holmgren (1822-1885) was one of Malmsten's successful students. His thesis from 1847 is a routine computation of the nth derivative of a composite function, but as one of few students who wrote their own theses and showed talent and a genuine interest in mathematics, he was appointed docent at Uppsala University. After studies in Paris with Liouville and a short time as a deputy for Malmsten, Holmgren was appointed professor of mathematics and mechanics at the Technological Institute in Stockholm in 1857.

Holmgren wrote few papers, one about the change of variables in a multiple integral (1864a), one (1864b) with the title *About differential calculus with arbitrary indices*, which is an elaboration of the Riemann-Liouville integral, and, finally, a paper about the quadrature of a certain second order differential equation.

The impression that these papers, especially the one about the Riemann-Liouville integral, make on a modern reader calls for some comment. Holmgren has chosen important subjects and, broadly speaking, he has achieved his goals. But analysis at the time was not ripe for his material. In the first paper there is no adequate definition of an integral and the second one is marred by ungainly and unclear notation. When studying Holmgren's papers, the reader is caught in admiration of the concepts and notation of present-day analysis, developed and refined through many years.

During his travels in the 1850's, Holmgren studied the curricula and the teaching of the technological institutes on the European continent and this was the beginning of a long fight for the position of mathematics at the Technological Institute. Especially the newfangled notion of infinitesimal calculus met with resolute resistance. For a time Holmgren was forced to leave mechanics and mathematics to teach instead the construction of machines. His victory came in 1863 when the subject of his chair was changed to theoretical mechanics. Already before this the decried infinitesimal calculus had received a place in the curriculum, at first taught by special teacher and later represented by a professor.

At the end of the 1860's, Holmgren was the most prominent active Swedish mathematician and the natural successor to Hill in Lund. A petition from his students made Holmgren stay in Stockholm.

Already in 1841 Holmgren had published a note in *Botanical Notices* and botany remained his great hobby. His speciality was moss and his large collection went to the Academy of Sciences after his death.

Appendix: *Solvable equations according to Abel*

As a background to Malmsten's and later also Bendixson's and Wiman's interest in and work with solvable equations, follows here a review of the theory according to Abel (see Gårding and Skau (1994)).

Abel's proof (1824) that the general equation of degree five is not solvable by root extractions made it interesting to have a theory of all equations $f(x) = 0$ which are solvable in this way. The two founders of the theory, Abel and Galois, used different attacks, Abel investigating the structure of the roots and Galois using the Galois group, i.e., permutations of the roots which do not change the coefficients. Both wrote unfinished papers. Posterity followed Galois, which has made the theory strongly imbued with group theory, for instance as is done in Weber's Algebra (1899, 1912).

To begin with we must remind the reader of some definitions. Consider fields which are algebraic over a field K containing the rationals. Every number in such a field is a zero of a unique irreducible polynomial $f(x) = x^n + \cdots \in K[x]$ of least degree n. Its zeros $t_1, ..., t_n$ are said to be conjugate over K. The field $K(t_1, ..., t_n)$ of all rational functions of $t_1, ..., t_n$ with coefficients in K is the *splitting field* $K(f)$ of $f(x)$.

All automorphisms of a field $L \supset K$ leaving K fixed constitute a group, the Galois group $\mathrm{Aut}(L/K)$. Such a group is only interesting when L is a Galois field over K, i.e. L contains all conjugates over K of any one of its elements. The splitting field $K(f)$ of f has this property. Its Galois group $\mathrm{Aut}(K(f)/K)$ will also be written as $\mathrm{Aut}(f/K)$.

A *root extension* $K(t)$ of a field K is obtained by adjoining to K a root t of of an irreducible polynomial of the form $x^n - a \in K[x]$. In addition it is required that K contains a primitive n^{th} root of unity ω, which means that $K(t)$ contains all roots

$$t, \omega t, ..., \omega^{n-1} t$$

of $x^n - a = 0$. At the same time it is clear that $\mathrm{Aut}(K(t)/K)$ is cyclic order n and generated by $T : t \to \omega t$. If $n = kj$ is a product, then $K \subset K(t^j) \subset K(t)$ is a sequence of root extensions and it follows that it suffices to consider root extensions of prime degree.

We shall also consider *primary* extensions $M = L(t_1, ..., t_n)$ of primary degree p^n of a field L with the property that $t_1, ..., t_n$ are outside L but $t_k^p \in L$ for all k and no t_k is a polynomial in the others with coefficients in L. It is obvious that $\text{Aut}(M/L)$ has p^n elements and is generated by $T_1, ..., T_n$ where $T_k t_k = \omega t_k$ but $T_k t_j = t_j$ when $j \neq k$.

A polynomial $f(x) \in K[x]$ and the corresponding equation $f(x) = 0$ are said to be solvable if the splitting field $K(f)$ belongs to a field which is obtained from K by repeated root extensions. By not going outside the splitting field $K(f)$ and by adjoining all its conjugates together with every element, it is not difficult to see that an equation is solvable if and only if there is a sequence of Galois fields

(1) $$K \subset L_1 \subset ... \subset L_{m-1} \subset K(f)$$

where every field is a primary root extension of the preceding one (see e.g. Gårding and Skau (1994)). Neither is it difficult to see that every Galois field between K and $K(f)$ is a member of such a chain. Both Abel and Galois saw that polynomials $f \in K(f)$ which do not split in any Galois field between K and $K(f)$ are special. Galois called them primitive.

Remark. If we put $F_j = \text{Aut}(L_j/K)$, the sequence (1) has a mirror image in an analogous sequence of groups

(1') $$F_0 \subset F_1 \subset ... \subset F_m$$

where every group is a normal subgroup of the preceding one and the quotients F_{j+1}/F_j are abelian of primary degree. In the terminology of group theory this means that the group $F_m = \text{Aut}(f/K)$ is solvable or metacyclic.

If we assume that $f(x)$ is primitive and put $L = L_{m-1}$, then

$$K(f) = L(t_1, ..., t_n)$$

is a primary extension of L, the polynomial $f(x)$ is irreducible in L but it splits into linear factors in $K(f)$. The commutative group $\text{Aut}(K(f)/L)$ with p^n elements permuts these factors transitively and it follows that $\text{grad } f = p^n$.

We can now introduce multi-indices mod p, $j = (j_1, ..., j_n)$ and corresponding products

$$t^j = \prod_1^n t_k^{j_k}, \quad T^j = \prod_1^n T_k^{j_k},$$

where T^j runs through alla elements of $\text{Aut}(K(f)/L)$. Since

$$T^j t^k = \omega^{(j,k)} t^k, \quad (j,k) = \sum_1^n t_k j_k,$$

we can write the roots of $f(x) = 0$ as $x_j = T^j x_0$ where

(2) $$x_j = \sum a_k \omega^{(j,k)} t^k, \quad a_k \in L.$$

The terms $R(j) = a_j t^j$ can be written in terms of Lagrange's resolvents

$$(\omega^j, x) = \sum \omega^{(j,k)} x_k$$

where $(\omega^0, x) \in K$ is the sum of the roots. We get $R(j) = p^{-n}(\omega^j, x)$ with powers $S(j) = R(j)^p \in L$, which gives Abel's formula[10]

$$(3) \qquad x_k = a_0 + \sum_{j \neq 0} \sqrt[p]{S(j)}, \quad a_0 \in K$$

for some choice of the arguments. This reduces the solution of the equation $f(x) = 0$ of degree p^n to an equation of at most degree $p^n - 1$ with roots $S(j) \in L$.

The Galois group

It is now easy to see the structure of the Galois group $\mathrm{Aut}(f/K)$. For if $U \in \mathrm{Aut}(f/K)$ then Ut_k^p belongs to L since $t_k^p \in L$ and L is a Galois field. Hence $(T^j U t_k)^p = U t_k^p$ so that

$$T^j U t_k = \omega^h U t_k$$

for some h. It is easy to see that h is a linear function mod p of j and hence has the form $h = (j, a_k)$, which means that

$$U t_k = c t^{a_k}, \quad c \in L, \,, \quad a_k \in Z^n \mod p$$

so that

$$U t^j = c t^{Aj}, \quad Aj = \sum j_k a_k \quad c \in L,$$

where j is thought of as a column and A is a square matrix of order n with integral elements mod p. From this we see that if $Ux_0 = x_0$ then U permutes the terms of $R(j)$ in x_0 according to the formula

$$UR(j) = R(Aj).$$

We can also see how $U \in \mathrm{Aut}(f/K)$ operates on the roots x_j of $f(x) = 0$, namely as

$$(4) \qquad U x_j = x_{Bj+k}$$

where B is the transposed inverse of A. In other words: $\mathrm{Aut}(f/K)$ is isomorphic to a subgroup of the affine group modulo p in n dimensions and contains all translations $j \to j + k$. Galois guessed this result which is the culmination of Weber's exposition (1899 II) of the theory of metacyclic groups. It is clear from the general theorems which he states in the beginning of his paper that Abel had all essential aspects of the theory.

Remark. If $U \in \mathrm{Aut}(f/K)$ leaves $t_1, ..., t_n$ invariant, it is not difficult to see that $Ux_k = x_k$ for every k. Hence U must be the identity which means that

$$K(f) = K(t_1, ..., t_n), \quad L = K(t_1^p, ..., t_n^p).$$

In the simplest case when $f(x)$ is solvable of prime degree and then the formulas above are to be read with indices mod p. Abel's manuscript has one unclear point, namely that the field $L = L_{m-1}$ above is a Galois field. Abel has a proof, but it is not simple. Abel's and Galois's manuscripts were enigmas to their contemporaries. That of Galois was deciphered first and therefore he was credited with the discovery of the Galois group for solvable equations of prime degree.

---

[10] The right sides can be made symmetric in the $S(j)$ (see Gårding and Skau (1994) p. 100).

Bibliography

Hill.

HILL C.J.D.
1827. *Casum irreducibilem solvendi conatus*, Crelle 2,1827, 304-306.
1828. *Über die Integration logarithmisch-rationeler Differentiale*, Crelle 3,1828, 101-159.
1831. *Additamenta ad conatum irreducibilem solvendi*, Crelle 7,1931, 44-54.
1843. a. *Fragmenta theoriae aequationum lineariter differentialium*, Crelle 12,1843, 1-21.
1843. b. *De radicibus aequationis Riccatianae rationalibus*, ibid. 22-37.
1843. c. *Disquisitio qualis aequatio differentialis gaudeat integrali algebraico completo? qualisve primarie transcendenti? quaenamque forma integrali completa?*, ibid. 3873.
1844. *Prolegomena till hvarje blifvande almän storhetslära*, Lund 1844.
1848. *Deo favente Matheseos fundamenta nova analytica*, Lund 1848.
1849. *En åkers medelavstånd till hemmet*, KVA Öfversigt 1849.

Svanberg.

ABEL N. H.
1881. *Oeuvres complètes* I,II. Publiées par L. Sylow et S. Lie., Christiania 1881.
SVANBERG A.F.
1828. *De curvatura superficium annotationes*, Dissertation (m. Andreas Petrus Hellstedt) Uppsala 1828.
1832. *De integralibus definitis disquisitiones*, Acta Reg. Soc. Sc. Uppsala 1832, 231-288.
1838. *Mémoire de quelques intégrales définies*, Crelle 18,1838.
1847. *Observations sur la transformation des intégrales multiples*, Acta Reg. Soc. Sci. Uppsala bd XIII,1847, 1-13.
1854. *Observations sur l'intégration des équations différentielles*, Uppsala 1854, C.A. Leffler.

Björling.

BJÖRLING E.G.
1841. *Calculi Differentiarum finitarum inversi exercitationes. Pars I:ma*, Nova Acta Reg. Soc. Scient. Upps. vol. XII, 1841-44.
1844. a. *In integrationem aequationis derivatorum partialum superficiei, cujus in puncto unoquoque principales ambo radii curvedinis aequales sunt signoque con-trario*, Archiv d. Math. u. Physik 4,1844, 290-313.
1844. b. *Calculi differentiarum inversi exercitationes I,II*, Nova Acta Reg. Soc. Uppsala ser. 2 vol 12, 12.1, 299-344, 12.2, 14-60.
1845. *Om betydelsen af tecknen x^y, $\log_b(x)$, $\sin x$, $\cos x$, $\arcsin x$, $\arccos x$ i Analytisk mathematik*, KVA Handl. 1845.
1846. *Une classe remarquable de séries infinies*, Journal de Liouville t. 17,1852, 454-472, ursprungligen i KVA Handl. 13 maj 1846.
1847. *Doctrinae serierum infinitarum exercitationes Pars I:ma 61-86, II:da 141-186*, Acta Reg. Soc. Sc. Uppsala vol. XIII, 1847.
1852. a. *Om functionerna x^y och $\log_\beta(x)$*, KVA Handl.1852, 121-163.
1852. b. *Om det Cauchyska kriteriet på de fall, då functionerna af en variabel låta utveckla sig i serie, fortgående efter de stigande digniteterna af variabeln*, KVA Handl.1852, 169-228.
1852. c. *Om principal-potenser och principal-logarithmer*, l.c. 101-119.
1852. d. *Om det Cauchyska kriteriet på det fall, då functionerna af en variabel låta utveckla sig i serie, fortgående efter de stigande digniteterna af variabeln*, l.c. 151-155.
1852. e. *Om funktionerna x^y och $\log_\beta(xz)$*, KVA Handl.1852, 121-163.
1853. a. *Om oändliga serier hvilkas termer äro continuerliga functioner af en reell variabel mellan ett par gränser mellan hvilka serierna äro convergerande*, KVA Öfversigt 1853, 147-169.
1853. b. *Om den definita integralen $\int_0^\infty \frac{e^{-cx^n}}{a+x^n} dx$.*, KVA Öfversigt 1853, 1-5.
DOMAR Y.
1987. *E.G. Björling och seriesummans kontinuitet*, NORMAT 2,1987, 50-56.
GRATTAN-GUINNESS I.
1986. *The Cauchy-Stokes-Seidel story on uniform convergence: was there a fourth man?*, Bull. Soc. Math. Belg. t. 38,1986, 225-235.
LINDMAN C.F.

1878. *Emanuel Gabriel Björling*, Stockholm 1878.

Malmsten.

BROMAN K.E.
1877. *Om konvergensen och divergensen af kedjebråk*, Diss. Uppsala 1877.

FALK M.
1869. *Om konvergensen af Kedjebråk med blott negativa leder och Kedjebråk med omvexlande positiva och negativa leder.*, Diss. Uppsala 1969.

GÅRDING L., SKAU C.
1994. *Niels Henrik Abel and solvable equations*, Arch. Hist. Exact. Sc. 48 (1994) 81-103.

MALMSTEN C.J.
1841. a. *Mémoire sur les intégrales* $\int_0^m \frac{1-p\cos tx)\varphi(x)}{1-2p\cos tx+p^2}dx$, $\int_0^m \frac{p\sin tx\varphi(x)}{1-2p\cos tx+p^2}$, Nova Acta Reg. Soc. Sc. Upps. vol XII,1841-44, 155-176.

1841. b. *Note sur l'intégrale finie* $\sum e^x y$, ibid. 295-294.

1841. c. *Mémoire sur les intégrales définies entre* $x = 0$ *et* $x = \infty$, ibid. 177-254.

1841. d. *Note sur la convergence des séries*, ibid. 255-270.

1841. e. *Om integralen* $\int_0^\infty \frac{\cos ax}{(1+x^2)^n}$, KVA Förh. 1841.

1842. a. *Om integration af differentialekvationen* $y^{(n)} = ax^m y$, KVA Förh. 1842.

1842. b. *Theoremata nova de integralibus definitis, summatione serierum earumque in alias series transformatione*, Diss. Uppsala 1842. Samling av föregående avhandlingar.

1846. a. *Om seriers konvergens*, KVA Öfversigt 1846, 9-12, Grunerts Arkiv 6,1845, 28-41.

1846. b. *Om den generella expressionen på* $d^n u/d^n x$ *då* $u = f(\sin x)$, KVA Öfversigt 1846, 36-37.

1847. a. *In solutionem aequationum algebraicarum disquisitio*, Crelle 34,1847, 30-45.

1847. b. *Sur la formule* $hu'(x) = \Delta u(x) - \frac{1}{2}\Delta u'(x) + \frac{B_1 h^2}{1.2}\Delta u''(x) + ...$, Crelle 35 1847, 55-116.

1848. a. *Högre lineära differentialeqvationers integrering*, KVA Öfversigt 1848, 135-138.

1848. b. *Om Differentialequationen* $y''' + \frac{ry'}{x} = (Ax^m + \frac{s}{x^2})y$, KVA Öfversigt 1848, 189-191.

1848. c. *Om convergensen af continuerliga bråk*, KVA Förhandl.1848, 91-119.

1849. *De integralibus quibusdam definitis, seriebusque infinitis*, Crelle 38,1849, 1-38. (Originaltiteln var *Specimen analyticum Theoremata quaedam nova de integralibus definitis, summatione serierum earumque in alias series transformatione exhibens*, Dissertationer 1847 försvarade av diverse elever.

1858. *Om differentialequationers integrering*, KVA Öfversigt 1858, 387-389. Annons till samma titel i Förhandlingar 1859-60.

1859. *Om partiella Differential-equationer af första ordningen*, KVA Öfversigt 1859, 25767.

1860. *Om differentialequationers integrering*, KVA Förhandl. NF bd 3 1859, 1860, 1-94. Översatt till franska av Malmsten i Journal des mathématiques pures et appliquées Série 2,1862, 257-375.

1865. *Om definita integraler mellan imaginära gränsor*, KVA Handl. bd 6,1865, no 3 1-18.

PERRON O.
1929. *Die Lehre von den Kettenbrüchen*, B.G. Teubner 1929.

WEBER H.
1912. *Lehrbuch der Algebra* I,II, Braunschweig, Vieweg u. Sohn 1899,1912.

Holmgren.

HOLMGREN Hj.
1847. *Derivatorum n:ti ordinis functionum* $f(e^{ax}\cos mx)$ *et* $f(e^{ax}\sin mx)$ *expositio*, Diss. Uppsala 1847.

1864. *Om multipla integralers transformation*, KVA Handl. NF 1864, bd 5, no 6, 1-40.

1864. *Om differentialkalkylen med indices af hvad natur som helst*, ibid. no 11.

1867. *Integration af differentialeqvationen* $(a_2 + b_2 x + c_2 x^2)y'' + (a_1 + b_1 x)y' + a_0 y = 0.$, KVA Handl. NF 1867-68. bd 7,no 9.

CHAPTER 3

A New Time in Uppsala and Lund 1860 - 1890

Tidskrift för matematik och fysik[1]
The contest for a professorship in Lund
Björling as a mathematician and teacher

Modern Sweden was founded in the 1860's. The main new developments were a beginning industrialization and a constitutional reform in 1865 which replaced the old parliament by a modern one without a division into estates. Even the conservative universities had to submit to change. A bitterly resisted reform was the 1864 decision to take away the university entrance examination, the so-called student examination,[2] from the universities and give it to the gymnasiums, elite high schools run by the state and localized over all of Sweden. As a compromise, these examinations were supervised by travelling teams of university professors called censors. This feature, abolished a hundred years later, enhanced the status of the professors and added a new element of suspense to the student examination.

In 1870 the degree of *magister* was abolished and replaced at all faculties by three others, the degree of candidate, of licentiate and a doctoral degree, which required a public defense of a thesis written by the candidate himself. In a few years, this new order led to a great improvement in quality and remained unchanged for one hundred years.

For some time Uppsala University had a position as assistant professor in mathematics. As part of the general turn for the better, a similar position was created in Lund. It went to Edvard von Zeipel, a mathematician from Uppsala who started his teaching by instituting a mathematical seminar with problem solving and simple lectures by its members. Later, the seminar was divided into two, a higher seminar for the advanced students and a lower seminar for beginners. This order of things lasted till about 1950 when the lower part was replaced by classroom teaching.

The assistant professor in Uppsala, docent Göran Dillner, was to play a large part in mathematical life in Sweden around 1870. In his teaching he introduced the *synectic*, i.e., analytic, functions to Sweden. He founded a private gymnasium in Uppsala. Another result of his drive and enthusiasm was an elementary journal for physics and mathematics which lasted a few years. It gives a vivid picture of mathematics in Sweden 125 years ago and is therefore worth a few words.

Tidskrift för matematik och fysik

A journal with this name, 'dedicated to elementary teaching in Sweden', was published in five rather extensive volumes 1868 - 1874. The editors were Göran Dillner in Uppsala, lektor Frans Hultman in Stockholm and Robert Thalén, assistant professor

---

[1] Journal for mathematics and physics
[2] In Sweden the word 'student' meant university student

in physics in Uppsala. The journal carried articles on mathematical and physical subjects, often written by the editors themselves, and reviews of new literature. It also contained solutions of the problems of mathematics and physics given in the student examination. Economic difficulties forced the journal to cease publication.

The most frequent collaborator to *Tidskrift* was the main editor Dillner who wrote a long series of articles with the title *Grunddragen af den geometriska kalkylen*[3] The origin of this series was his displeasure with the algebraic view of complex numbers. Instead he wanted to follow Cauchy's paper *Mémoire sur les quantités géométriques*, and describe addition and multiplication of complex numbers in geometric terms. This line was followed right through till the definition of the complex derivative, analytic functions, expansion in power series and residue calculus, partly with personal terminology and formulas. As far as the author trusts his geometry, all goes well, e.g., in a proof of the fundamental theorem of algebra by the argument principle. But analysis is treated without insight. A continuous function $F(z)$ is assumed to have a derivative

$$\lim \frac{F(z+h) - F(z)}{h} = \frac{dF(z)}{dz} = F'(z)$$

(vol. III, 1870 s. 258), a formula that is commented on as follows: 'Because the contribution h does not appear in the derivative, this quantity is the same, in whatever direction h is drawn...'. As an explanation of the formula this is of course insufficient, also from a pedagogical point of view.

The other editors contributed each according to his interests. Hultman wrote very learned articles about the history of arithmetic in Sweden, and Thalén contributed notices about physicists and progress in physics. The physicist G. Lundquist wrote about the new spectral analysis that made it possible to identify the elements in the corona of the sun. Apart from these contributions, all of mathematical Sweden wrote in *Tidskrift*: Professor Daugh in Uppsala, hidden under the timid pseudonym D-g, Malmsten, the lektors E.G. Björling, C.F. Lindman, and Lars Phragmén; the docents Matths Falk and C.F.E. Björling, and Professor Hjalmar Holmgren at the Technical Institute. In the first volumes, we find the schoolboys Knut Wicksell, later professor of economics, and Gösta Leffler, who had not yet added his maternal grandfather's name to his own. Many university students contributed to the problem section. Geometrical constructions and the conic sections dominate.

Each volume before 1871 contained a prize problem with solutions a year later. The first one from 1868 was the following: 'draw a square whose sides (extended if necessary) pass each through four given points'. This excellent problem was solved by 23 people, including schoolboys, military people and one of the few women readers. The later prize problems did not have the charm of the first one and had less success. Once the sole solution came from an Italian reader.

Tidskrift för Matematik och Fysik hade an intense but short life and it presented things which were new for its time. It is a sign of an emergent interest in mathematics and natural science. At the same time, the journal gave the editors Dillner and Hultman an opportunity to write about their own interests. With time especially Dillner went too far to keep readers' interest alive. For the large majority of them, the theory of analytic functions was an incomprehensible novelty with no connection to the elementary mathematics of the time.

---

[3] The foundations of geometric calculus.

The contest for a professorship in Lund 1871 - 73

At about the same time as assistant professorships were created, Malmsten and Hill left their posts and new mathematicians took over. Malmsten's successor was Herman Theodor Daug, a differential geometer with moderate ambition. Hill's chair could be filled only after a protracted contest from 1871 to 1873. Its main outcome was the creation of a new system of appointing professors where, as a first step, a committee of experts examined the scientific merit of the applicants.[4] Because this system is still working, it may be interesting to review its origin. The contest which we shall describe has features which were to be repeated many times afterwards.

When the professorship of mathematics became vacant after Hill, the applicants were assistant professor E. von Zeipel born in 1823, active in Lund since 1861; lektor C.F.E. Björling in Halmstad born in 1839; associate professor G. Dillner in Uppsala born in 1832; docent A.V. Bäcklund in Lund born in 1845, and Sophus Lie from Kristiania (Oslo) born in 1842. Very soon Lie withdrew his application, but while his mathematical papers remained in the university office, Bäcklund got permission to read Lie's work on contact transformations, an event which determined his future as a mathematician. All remaining applicants except Bäcklund had studied in Uppsala. Viewed from the present, von Zeipel was a modest combinatorist with a recent paper about 'monomial and factorial coefficients' to his credit. Dillner and Björling had tried, each in his own way, to get acquainted with the use of complex numbers in analysis and Bäcklund had made an effort to establish himself in the new analytical geometry where algebraic curves and their singularities were classified. Lie made a career in Germany as the founder of the theory of continuous groups, now called Lie groups.

According to established procedure, it was the task of the philosophical faculty and thereafter the senate to rank the applicants with the purpose of advising His Majesty (i.e., the government) in the appointment of the worthiest of them to be professor of mathematics at Lund University. At every step except the last one it was possible to argue against decisions taken. Earlier, this procedure had yielded many written complaints but they were rare in comparison with things to come. The relevant documents, published by Björling in 1872, make a small book of 200 pages.

Part of the ordeal of applicants were trial lectures and, for Björling, who was not associated with a university, public defence ('disputation') of a recently written paper 'Theory of roots of algebraic equations' with Bäcklund as opponent, appointed by the faculty. In this paper the curves $\operatorname{Re} P(x + iy) = 0$ and $\operatorname{Im} P(x + iy) = 0$ are used to localize the zeros of a polynomial $P(z)$.

The faculty entrusted the astronomy professor Axel Möller with the examination of the applicants. The result is found in the faculty minutes of Nov. 30, 1871. Von Zeipel and Dillner were mentioned briefly and Bäcklund was critized for not checking his results against the literature. Björling, on the other hand, fared less well, partly because the discussion centered on his disputation. Möller quotes one passage after the other in Björling's paper where the author 'wrongly puts forward his own results as new although they can be found in work by Euler.' When this was revealed during the disputation, Björling answered 'that it was no less flattering for him to have found theorems which had been found before by Euler.' But Möller was not entirely negative: 'on the one hand algebraic research... which shows pretended

---

[4]In a letter to Marcel Riesz, March 1923, Hardy wrote 'You explained the system to me once, and I did not admire it. Of course it makes it impossible to make a scandalous appointment behind closed doors (as is too easy here).

or real ignorance of results by older authors, on the other hand shows detailed research of his own.'

The course of the disputation is made clear in a quote from Möller: 'in the defence of the thesis, the author exhibited a considerable superiority over his opponent, appointed by the faculty. With an unusual capacity to express his thoughts easily and correctly, the author directed his defence more with a purpose to dismiss the critical remarks from the opponent than to disprove them.'

Möller's summary says about Björling: 'I consider him and the other applicants competent for the post to which they aspire, but in respect to the originality of his results and their importance, I place him after the others.' Möller ranked the others in the order von Zeipel, Dillner, Bäcklund. The majority of the faculty followed Möller, but Björling's self-confidence had impressed the historian Odhner, who remarked that 'with respect to broad learning and many-sided formation, lektor Björling is vastly superior to the other applicants.'

After this, Björling had every reason to act. He did not feel inferior to his two most important rivals and he felt mistreated because his paper was the only one which had been criticized in detail.

To the session of the senate one year later, in January of 1872, Björling could present letters from the professors Adolph Steen in Copenhagen and Bjerknes in Oslo and the publisher Grunert of Grunerts Archiv, where a paper by Björling had appeared. Without contradicting Möller, they praised Björling's work for practically important results and methods and vouched for his suitability as a university teacher. Grunert in particular had no reason to criticize papers he himself had published. In the senate Möller defends his evaluation of Björling and finds support in a paper by Gauss 'in which we find most of the research contained in Björling's paper, which is only an exposition with details added ... '. After detailed 'vota' (explanations of votes) the senate voted as the faculty with the difference that Odhner now had Björling in the second place.

But Björling did not let himself be discouraged. It took him only a month to write and send a long letter to 'Your Ever Mighty, Ever Merciful Majesty' (i.e. the government). His main argument is the following: 'The decision of the Senate is based entirely on the vote of Prof. Möller. Should the reasons upon which this vote rests be found to be untenable in a more careful examination, then the aforesaid, almost unanimous decision would not have other significance than an expression of the trust that the members of the senate have for the competence and impartiality of Prof. Möller.' Björling knew that this argument would impress the lawyers at the goverment's office. He also calls for 'an examination by experts' which 'ought to give a result which to my advantage considerably deviates from that of Prof. M.'

In the sequel Björling attacks Möller's evaluation. The aim of his arguments, which are very unprejudiced when it comes to facts, is to show that Möller has concealed Björling's merit and that his criticism has been entirely misdirected. When Björling has made some obvious error, he concedes but defends himself by pointing out other minor matters which Möller has not mentioned. He accuses Möller of not having read the papers of the other applicants and proceeds to do this himself. Since von Zeipel is sometimes unclear and Dillner's paper are just rearrangements of the theorems of Cauchy, Björling has plenty of material for ironical exposition. After this protest, the temperature in the matter rose. Everbody's prestige was now at stake.

Björling's pamphlet caused the government to return the matter to the senate already in March of 1872. von Zeipel and Dillner defended themselves in letters to the governement via the senate. In addition von Zeipel accused Björling of having copied a well known textbook and Gauss's work on the separation of the zeros of a polynomial. Bäcklund gave his view of the course of the disputation. In a letter to the government, the senate defended its first decision. Möller answered at length Björling's attack, exposed some of his polemical tricks and reaffirmed his judgment.

Björling's last occasion officially to influence the matter was to write his final reminders to the King. This was done in a furiously polemical pamphlet where its author uses sixty pages to review all that has happened in the matter. He stresses again the dependence of the senate on Möller, the incompetence of his rivals is elucidated in long mathematical arguments, Möller's two reports are compared and his inability to understand Björling's work is demonstrated until fully evident. In his summary, Björling returns to his demand for experts: 'By a report of such quality, it should be clear that Prof. Möller has only verified the legitimacy of the protest I have submitted, namely that Your Majesty would gracefully in the interest of mathematical education admit that this important matter is worthy of examination by expert scientists.'

On Oct. 17, 1873, C.F.E. Björling was appointed professor of mathematics at Lund University. A year earlier his much criticized article had received a prize from the Academy of Science. But after all, it must be said that his success rested more on his polemical talent than scientific merit.

In 1876 the government decided that the procedure for the appointments of professors should start with an examination of the scientific merits of the applicants by a committee of three experts. Möller must have felt this as a reprimand for what was essentially a competent job.

Björling as a mathematician and teacher

The mathematics institute of Lund university has a magnificent portrait of Professor Björling, painted at the turn of the century when his retirement was near. Leaning back the professor sits at his desk, a thick watch chain adorns his waistcoat and his well-formed face has an expression of disillusion as if in front of a mediocre examinee. The picture recalls his pedagogical activity rather than the scientific one.

Carl Fabian Emanuel Björling (1839-1910), son of lektor E. G. Björling in Västerås, studied in Uppsala and got his degree with a prize-winning thesis (1863). The introduction is about the imaginary calculus and the principal branches of the inverse trigonometric functions, clearly a legacy from the father. The main result is a computation of

$$\tan \sum_0^n \arctan kc$$

as a rational function of c. His papers before 1871, one of them printed in *Archiv der Mathematik und Physik*, all contain rather unassuming computations and deliberations. The paper (1871) with which he competed in Lund in 1872, deals with the separation of a polynomial

$$P(z) = z^n + bz^{n-1} + ... + gz^2 + hz + k$$

of degree n with real coefficients into real and imaginary parts, $X(x,y) + iY(x,y)$. Both must vanish at a zero. The interest is concentrated to multiple zeros, also complex ones. The paper has only one reference to the literature, something which annoyed his critics.

Björling observes that $Y(x,y) = yK(x,y)$ is divisible by y. The curve $K(x,y) = 0$ is called the complex primary curve or the K-curve. If P does not have double zeros this curve has n separate branches and its asymptotes are given by $\sin n\theta = 0$. The curve meets the x-axis in the real zeros, the imaginary zeros lie on branches which do no meet the real axis. This theory is developed with much ado and many applications.

While in Lund Björling switched to algebraic geometry, probably inspired by the papers written by his rival Bäcklund. At that time algebraic geometry was a new and exciting part of mathematics. A general exposition of Björling's papers in this field, including his textbook *Nyare plan*[5], is given in the next chapter in the section *Algebraic geometry in Lund 1870 - 1900*.

Björling's teaching, his personality and his activity outside of mathematics are the subjects of a memorial article in *Svenskt biografiskt lexikon*.[6]. Some relevant passages including his work as a 'censor'[7] are quoted below.

When B. after an unusually sharp contest for promotion was appointed professor, he thought it imperative to arrange the teaching of mathematics in step with the times. To this end he worked out a higher course of lectures which took four years and gave a good initiation into certain function theoretic and geometric areas with a decided overweight for the latter. As a suggestion about the extent of these lectures it may be briefly stated that, as regards to methods and content, they followed well-known handbooks by Briot and Bouquet, Clebsch-Lindemann and Salmon-Fiedler, wherewith, as regards elliptic functions, the work by Weierstrass was taken into consideration. Björling repeated this course of lectures essentially unchanged during his time in office, yet at the end with some additional group theory, and it remained valuable for further studies. Already in the 1870's a mathematical seminar was instituted, later managed by paid assistants. In 1900 a post as assistant in mathematics was created at suggestion by Björling. After B.'s retirement it was changed into a professorship.

B.'s lectures were brilliant from a formal point of view. He had an unusual ability to express his thoughts easily, correctly and in a stimulating way. In contrast he was not to the same degree rigorous— neither orally nor in writing. Although equipped with a sharp intelligence and wit, it can be said that he was more gifted for reviewing and rounding off than for abstraction and strict reasoning. This feature is apparent in his remarkable textbooks. First of them is his *Lärobok i nyare plan geometri*, unique in its kind and used all over Scandinavia.

As an examiner B. was demanding and feared. On the other hand, he was always ready to even the way for gifted pupils by practical arrangements of examination and teaching. It was this feature which made him appear similarly as 'the most intelligent and the most ruthlessly severe among censors.' He knew everything, examined with the same penetrating expertise a mathematical specimen, as well as a Swedish, Latin, French one and was always watchful against teachers who were lax in their corrections. His extensive experience and great talent were used with advantage in the gymnasium committee of 1882 ...

B. was a person with a many-sided formation. But his main interest was natural science, which found its expression in his widely read popular books in this area. They gave a wide application to his formal talent and ability to stimulate. He was also till the end of his life sought after as a lecturer.

In all his activities B. was independent and indifferent to the views of others. He could appear cold and dismissive, but here the surface can betray. As regards

---

[5]Textbook of recent plane geometry, nicknamed Nyare plan.
[6]Swedish Biographical Encyclopedia.
[7]About censors see p. 33.

his relation to the students, it may be mentioned that he was elected chairman of the Academic Union in 1882 and remained in this post for several years. This means that the students with time saw in him a friend and benefactor.

BIBLIOGRAPHY

BJÖRLING C.F.E.
1863. *Om några arcus-tangens-summor och deras användade till definita integralers evaluering*, Westerås 1863.
1871. *Theori för algebraiska eqvationers rötter*, KVA Handl. bd 10 n:o 3, Stockholm 1871. 53 s.
1896. *Lärobok i Nyare plan geometri*, Lund 1896.

BÄCKLUND A.V.
1869. *Några satser om plana algebraiska kurvors normaler*, Lunds univ. Årsskrift 1869, 1-38.

von ZEIPEL E.
1962. *Undersökningar i Högre Algebran jemte några deraf beroende Theoremer i Determinanttheorien*, KVA Handl. NF bd 3 1859-1860. Stockholm 1862.
1869. *Om monomial- och fakultetskoefficienter*, Lunds Univ. Årsskrift VI 1869, 1-57.

CHAPTER 4

Algebraic Geometry in Lund before 1900

Bäcklund   Björling   Bergstedt   Wiman

By chance algebraic geometry came to dominate mathematics in Lund between 1870 and the end of the century. The incitement was probably the papers that Bäcklund wrote for the 1872 competition. The new professor Björling took up the subject and made it his own. At the time algebraic geometry went through a period of fast development and it had a strong representative in H. Zeuthen in neighboring Copenhagen. From the beginning Björling probably knew very little algebraic geometry but he studied energetically and, above all, taught forcefully which resulted in interested students, two good theses and a successful textbook.

This chapter starts with a review providing a background to the papers by Bäcklund, Björling, Bergstedt, and Wiman.

Projective and algebraic geometry in the nineteenth century

In the latter part of the nineteenth century algebraic geometry was a new and exciting part of mathematics, created in the first half of the century by the prominent mathematicians Cayley, Chasles, Cremona, Jacobi, Plücker, and Salmon and developed by Castelnuovo, Brill, Noether, and Riemann.

The result took the form of an almost coordinate-free classification of curves and surfaces in complex projective space, classification of singularities and a galaxy of geometrical maps and new structures.

The use of complex numbers and projective spaces gave a twofold completeness. Over the complex numbers a polynomial factors into linear factors and infinity has no priviliged position in projective space. The points in projective space are represented by rays through the origin in a linear space of one higher dimension. Hence the points in complex (real) projective plane are rays $t \to (tx, ty, tz)$ in $C^3(R^3)$ where $(x, y, z) \neq 0$. We speak of (x, y, z) as homogeneous coordinates. Close to a point where, e.g., $z \neq 0$, $x/z, y/z$ can be used as Cartesian coordinates in R^2 when x, y, z are real and as coordinates in the product of two complex planes when they are complex. Points where $z = 0$ have then no images in Cartesian coordinates, but represent what is called the line at infinity.

A projective point x, y, z determines a straight line $x\xi + y\eta + z\zeta = 0$ in a dual plane with the projective coordinates ξ, η, ζ. The duality point-line and, in one more dimension, point-plane is fundamental in projective geometry.

The permitted changes of coordinates in projective geometry are very simple: arbitrary linear bijections of the homogeneous coordinates. For the projective line with inhomogeneous coordinate x, this means that a complex transformation is rational,
$$x \to T(x) = (ax + b)/(cx + d), \quad ad - bc \neq 0,$$

where a, b, c, d are real or complex. Under such a transformation, three pairs of point-image can be prescribed and the double fraction

$$(x_1, x_2, x_3, x_4) = \frac{x_1 - x_3}{x_1 - x_4} \Big/ \frac{x_2 - x_3}{x_2 - x_4},$$

is invariant. If T^2 is the identity, T is said to be an involution.

Algebraic curves

When a polynomial $f = f(x, y, z) \neq 0$ is homogeneous of degree $n > 0$, i.e., $f(tx, ty, tz) = t^n f(x, y, z)$ for all t, the equation $f = 0$ is a projective invariant and represents an algebraic, plane, projective curve of degree n. In order to keep a bijection between polynomial and curve, we assume that the polynomial does not have multiple factors. If the polynomial is real, the real curve may be empty but when it exists, it is a smooth curve with singularities at points where f and grad f both vanish. When the polynomial is reducible, the curve splits correspondingly.

Close to a point on a complex algebraic curve where, for instance, $z \neq 0$ we can put $z = 1$ and use x, y as coordinates. We get a curve $f(x, y, 1) = 0$ and close to a point x_0, y_0 where for instance $\partial f / \partial x \neq 0$, we can solve with respect to the variable x and get a bijection $(x, y) \to x$ close to (x_0, y_0) onto an open part of the complex plane. In other words outside its singularities, the curve is a complex manifold of dimension one in a complex space of complex dimension two. Hence the curve has the real dimension two and it is situated in a space of real dimension four where we cannot use ordinary vision. Since the complex projective plane is compact, this means that a non-singular projective curve is a two-dimensional, compact and orientable manifold. One of the first serious theorems in topology says that there is a bijection from such a manifold to a sphere with a number of handles added. The number of handles, called the genus of the curve, also equals the number of independent closed curves which divide the curve into two parts.

This theorem appears for the first time in 1851 in Riemann's thesis. By projecting the complex curve into the complex plane including the point at infinity he could visualize the curve as a Riemann surface, i.e., a number of sheets lying on top of each other with branch points where the projection touches the curve. By means of closed curves avoiding the branch points the curve can be cut up into simply connected parts. The minimal number of such parts equals the genus g of the curve. An irreducible non-singular curve of degree 3 is topologically a torus whose genus is 1. A curve of degree n has genus at most $(n-1)(n-2)/2$. Genus is also the dimension of the linear space of holomorphic differentials on the curve (Abelian differentials of the first kind). Twice the genus is the dimension of the first homology class of the curve, the dimension of the two others is 1, which means that the Euler characteristic χ equals $2 - 2g$. It can be obtained from a triangulation of the curve as the sum of the number of corners and triangles diminished by the number of edges.

The fact that a polynomial has as many zeros as its degree extends to curves as Bézout's theorem: two algebraic curves of degrees m and n in general position meet in mn points. In special cases, the points must be counted with their multiplicities and if the number exceeds mn, the two curves must both contain the same algebraic curve. The case $n = 1$ says that an algebraic curve of degree m meets a line in general in m points and if this number is exceeded, the line must be part of the curve.

The class of a curve is the number of tangents in general which may be drawn from an outside point. If $f(x, y, z) = 0$ is the equation of the curve and the point is

(a, b, c), a point of tangency satisfies the equations $f = 0$, $af_x + bf_y + cf_z = 0$. Hence Bézout's theorem shows that a curve of degree n in general has the class $n(n-1)$.

An irreducible complex conic is an irreducible curve of degree two and class two. In suitable coordinates it has the equation $x^2 + y^2 + z^2 = 0$ and hence is a single geometric object. This illustrates the drastically reduced classification of geometric objects which may occur in complex, projective geometry. If only real projective coordinates are admitted, we still get one normal form $x^2 + y^2 - z^2 = 0$. If one of $z = 0, y = 0, y - z = 0$, represents the line at infinity, we get the three possibilities: ellipse, hyperbola, and parabola. If only orthogonal coordinates are permitted we get a classification by eccentricity and so on.

A point $P = (x_0, y_0, z_0)$ on a curve $f(x, y, z) = 0$ is said to be regular if the gradient of f does not vanish at P. Regular points have multiplicity one if the multiplicity $\nu(P)$ of P is defined as the degree of the first non-vanishing term in the Taylor series

$$f(x_0 + x, y_0 + y, z_0 + z) = h_0(x, y, z) + h_1(x, y, z) + h_2(x, y, z) + \ldots$$

To see what that means we can take $x_0 = y_0 = 0, z_0 = 1$ and $z = 0$. With multiplicity two, h_2 is a homogeneous polynomial in x, y which vanishes on two rays through the origin. If the rays do not coincide, the singularity is a double point where two branches cross. If the branches coincide, the singularity is a cusp, a name which comes from the real image of, e.g., $x^2 = y^3$. These singularities are said to be regular and are the only possible ones for an irreducible curve of degree three. Other interesting points on the curve are the points of inflexion where the Hessian vanishes, i.e. the determinant of the matrix of the second order derivatives.

The dual C' of a curve $C : f(x, y, z) = 0$ is the locus of points grad f when $f = 0$. It is easy to see that a point of inflexion of C produces a cusp in C'. In general, the degree of C' is $n(n-1)$. This degree decreases with $2d + 3s$ when C has d double points and s cusps and no higher singularities. This fact is one of Plücker's formulas. Another one says that the number of cusps of C' equals $3n(n-2)$ minus $6d + 8s$. The main term is the number of points common to $C : f = 0$ and Hess $f = 0$ with the degrees n and $3(n-2)$.

Clebsch proved that an irreducible curve of degree n with only regular singularities has the genus

$$g = \frac{(n-1)(n-2)}{2} - d - s.$$

A corresponding formula in the general case is due to M. Noether.

If $\pi : P \to P'$ is a projection of irreducible algebraic curves C and C' such that C covers C' a times, Hurwitz showed that

$$\chi(C) = a\chi(C') - \sum(\nu(P) - 1)$$

where $\nu(P)$ is the multiplicity of a point $P \in C$ over C' and the sum has only finitely many terms. The proof is very simple: in a triangulation of C' with corners in all P' with $\nu(P) > 1$, the number of triangles and sides in C are multiplied by a. The number of corners are also multiplied by a but the second term on the right must be subtracted from this number. The best-known result of the Dane H. Zeuthen is a similar formula for multiple correspondences between algebraic curves.

Higher singularities could be studied using Puiseux expansions. When an irreducible polynomial $f(x, y)$ in two variables vanishes of higher order in the origin, it splits into branches represented by fractional power series in one of the variables. These can be studied systematically according to order and branching order and characteristic numbers can be defined through which the formulas of Plücker and Clebsch extended to curves with arbitrary singularities (see Brieskorn-Knörrer (1981) for a modern exposition).

Algebraic surfaces

Next in complexity after algebraic curves come the algebraic surfaces. They are defined by equations $f(x,y,z,w) = 0$ where x,y,z,w are projective coordinates in space and $f(x,y,z,w)$ is a homogeneous polynomial whose degree determines the degree of the surface. We assume that f does not have multiple factors. A point in the surface where grad $f \neq 0$ is said to be regular, otherwise double or singular. The classification of singular points is much more involved for surfaces than for curves and play a large part in the classification of surfaces. Two hundred pages of the German edition from 1874 *Analytische Geometrie des Raumes* of a classical treatise by Salmon deals with surfaces of degrees three and four. All plane tangents of a general surface of degree n form the dual surface of degree $n(n-1)^2$. This is also the number of tangent planes to the surface which pass through a line. A curve in space is algebraic when its coordinates are algebraic functions of a complex parameter. Its degree is defined as the number of points it has in common with a plane in general position.

What has been said above, for instance the theorems by Plücker and Bézout, is just a small part of what in the second part of the nineteenth century was called enumerative geometry. The best-known result is perhaps the existence of twentyseven straight lines on a general cubic surface. The following simple argument shows that the number ought to be finite. A line is given by four parameters, for instance the intersections $P = (x,y,0,1)$ and $Q = (u,0,v,1)$ with the coordinate planes $z = 0$ and $y = 0$. That a homogeneous polynomial of degree three vanishes for all points $\lambda P + \mu Q$ gives four equations of degree three with four unknowns.

In the 1870's, algebraic geometry was ripe for large scale expositions, for instance Salmon-Fiedler (1874). Enumerative geometry obtained a classic in Schubert's book *Kalkül der abzählende Geometrie* (1879) with far-reaching results, partly obtained by intuitive methods. One of Hilbert's problems at the mathematical congress in Paris in 1900 was to find a strict foundation for this book whose most complicated results still challenge modern algebraic geometry.

Algebraic geometry in Lund: Bäcklund, Björling, Bergstedt, Wiman

Bäcklund

Bäcklund's geometric papers, written when he was under thirty, were motivated by an interest in a new attractive part of mathematics and perhaps also with a view to obtaining Hill's chair. For this occasion he was able to present five papers, all written in Swedish. Plücker and Chasles are quoted, but the references are sparse. The main part of the text consists of enumerations of geometric objects associated with algebraic curves. These papers show that Bäcklund was able to absorb new material in a short time and then do things of his own.

The most interesting paper (1869) has to do with normals of real algebraic curves. The objects of the last theorem are points $P_1, ..., P_n$ where a curve C of class n touches n lines through a fixed point Q and no point of tangency is infinity. If the lines are oriented from Q, if the radius of curvature at the point P_k is R_k and if its distance to Q is r_k, Bäcklund's formula says that

$$\sum_1^n R_k/r_k^3 = 0.$$

It is implicitly assumed that all P_k are real and that all tangencies have order one. By a projective map, Bäcklund could place Q at infinity. A short computation then shows that the formula now amounts to

$$\sum R_k = 0$$

for the curvature radii at all points of the curve with parallel tangents. Unluckily for Bäcklund, this theorem was found earlier by Liouville. Liouville's proof depends on the fact that a certain sum of residues vanishes. Using the same method we shall correct Bäcklund's result to

$$\sum R_k/r_k = 0.$$

The explanation is that Bäcklund's projective map did not properly preserve curvature.

We assume that the curve C is given by $f(x,y) = 0$ where $f(x,y)$ is a real and irreducible polynomial. The point outside the curve is taken to be the origin and we assume that the projective curve C^* is regular. In the points $P_1, P_2, \ldots$ of tangency, assumed to be real and different, the polynomial $N(x,y) = xf_x + yf_y$ vanishes and has simple zeros. The differential form

$$\omega = (-f_x dy + f_y dx)/N(x,y),$$

is analytic in C^* except for simple poles at the points P_k and the sum of its residues vanishes. The form is also invariant under rotation around the origin. Hence, at every point $P = P_k$ we may assume that its coordinates are $(r,0)$ with $r > 0$. In P both y and $f_x = 0$ vanish, but $f_{xx}(P) \neq 0$ so that $-f_y dx + f_x dy = -(f_y + f_x/f_y)dx = -(f_y(P) + O(x-r))dx$ and $N(x,y) = rf_{xx}(P)(x-r) + O((x-r)^2)$ close to P. Hence the residue of ω at P is $-2\pi i f_y/rf_{xx}$. In addition, $yf_y(P) = -(1+O(|x-r|))f_{xx}(P)(x-r)^2$ for $f(x,y) = 0$ and small $x-r$, which gives the radius of curvature $-f_y/2f_{xx}$. In the sum of residues the curvatures get proper signs.

Björling

Outside of his teaching, Björling worked with problems in algebraic geometry that can be seen as results of his own studies. The theme of his main papers is the classification of higher singularities of algebraic curves.

One example is the paper (1878) where theorems by Cayley and Zeuthen are proved anew. Björling studies Puiseux expansions at the origin of the equation $P(x,y) = 0$, $P(0,0) = 0$ and P has degree m, n, i.e., degree n in x and m in y. Under the assumption that m and n are coprime, the author computes the number of double and stationary points and tangents.

In a similar paper (1886) Björling considers a differential equation

$$(P(x,y) + R(x,y))dx + (Q(x,y) + S(x,y))dy = 0$$

where P, Q are homogeneous polynomials of degree m and R, S vanish of order $> m$ when $x = y = 0$. The author assumes that R, S are without influence, puts them equal to zero and gets an explicitly solvable equation where the solution orbits can be studied close to the origin. Neither Björling nor the author of a passive review in *Jahrbuch* seem to have understood that R, S cannot be neglected. The paper (1881) by Poincaré, where this problem is analyzed, is quoted, but only as a special case. It is probable that Björling never knew what deep waters he had tried to navigate.

In his late papers Björling went over to geometry in space. In a rather involved paper (1888) he considers curves

$$x = M\alpha^m + \ldots, \quad y = B\alpha^n + \ldots, \quad z = C + \ldots$$

with $A, B, C \neq 0$ and $n > m \geq 0$ and classifies them. By intersecting a ruled surface (generated by straight lines) with the coordinate planes, his paper gives a classification of the singular generators. At that time ruled surfaces was the subject of two Lund theses, Bergstedt (1886) and Wiman (1892), the latter a remarkable piece of work.

We shall mention yet another paper (1890) by Björling about Plücker's characters. For a plane curve with ordinary singularities these are the degree, the number of double points and the number of cusps of the curve itself and its dual. Plücker's characters for a surface can be said to be those of a plane section in general position.

Cayley introduced Plücker's characters for a space curve as those for the cone generated by lines through the curve from a point outside and those for the ruled surface generated by its tangents. Salmon-Fiedler's classical treatise (1874) carries an explicit example of these notions, which is generalized in Björling (1890).

A projective curve $C : t \to x(t)$ in spaces has, according to Björling, a singularity of type (l, m, n) in a point $x(0)$ if $l < m < n$ and

$$x_1(t) = Lt^l(1 + O(t)), \ldots, x_3 = Nt^n(1 + O(t)), \, x_4 = P(1 + O(t))$$

by a suitable choice of coordinates and $L, M, N, P \neq 0$. The dual projective curve $t \to x(t) \times x'(t) \times x''(t)$ then has a singularity of the type $(n - m, n - l, n)$. Björling gives a detailed geometric analysis of this situation, including local Cayley-Plücker characters as functions of l, m, n.

Björling's textbooks

Björling's textbook of analysis (1867),(1893) was the first systematic book of this kind written in Swedish. It is the result of a considerable pedagogical experience and the informal style makes a favorable impression on a present-day reader. The fact that Björling restricts himself to 'real' quantities mirrors the doubts that at the time surrounded complex numbers and analytic functions.

Björling's algebraic geometry studies resulted in a Swedish textbook in geometry (1894) *Lärobok i Nyare Plan Geometri*[1] known under the name of *Nyare Plan*. Now entirely superannuated, this book was a pioneering effort which was used also outside Sweden.

In the complex projective plane conics can be defined as curves which meet every straight line in two points, eventually coinciding. This simple definition produces exceedingly simple, computation-free proofs of the properties of conics. It was Björling's idea to expound this theory in simple way in a book which also was to give the elements of algebraic curves.

---

[1] Textbook of recent plane geometry

In this book, the author had to abandon his earlier cautious view of the complex numbers, but he was not able to give them their rightful place. The result was a collection of vague, operational definitions of the basic notions of projective geometry. It is true that they were current at the time, but the exposition makes an almost archaic impression on a present-day reader. A striking lack of systematics is apparent. Desargues's basic theorem that a pencil of conics through four points induces an involution on every straight line occurs as one of three equivalent definitions of the notion of involution.

For the naive reader these peculiarities give the book an air of sorcery. Coordinates, introduced as real numbers, suddenly change to complex ones. The author makes a show of his definition of a circle as a conic which passes through the imaginary circle points I and J with homogeneous coordinates $(1, \pm i, 0)$. Pascal's theorem that the continuations of opposite sides in a hexagon, inscribed in a conic, meet in three points on a straight line is proved in two ways. In the first one, the conic is transformed into a circle with an inscribed hexagon where two pairs of opposites sides are parallel. The problem is then to show that the sides of the third pair are parallel. Many exercises are supposed to be solved by the same method. The natural proof of Pascal's theorem via that of Desargues occurs much later than the first one.

The book was used in the teaching of mathematics in Lund until the middle of the 1920's and created a kind of I, J subculture. One must admire every student who was able to absorb this book on his own. But he was helped by an overflow of exercises, many of them rather difficult.

As is the case with many textbook writers, Björling could not resist the temptation to write more than it was reasonable to require of the readers. The second half of the book starts with the transformation theory of binary and ternary forms, i.e. homogeneous polynomials in two and three variables, and the corresponding theory of invariants, polynomials in the coefficients which remain invariant under linear changes of variables. There are also geometric interpretations. The last part of the book treats algebraic plane complex curves, defined by the property of meeting every straight line in a fixed number of points equal to the degree of the curve. Here the author proceeds to consider the number of points that determine a curve, the number of points common to two curves, the dual of a curve, regular singularities etc. and he proves Plücker's formulas.

In spite of its shortcomings Björling's book *Nyare Plan* is a remarkable attempt to give students some insight into the algebraic geometry of the nineteenth century.

Ruled surfaces. Bergstedt and Wiman

In algebraic geometry Björling was second to his students Jakob Bergstedt and Anders Wiman. Both devoted themselves to ruled surfaces, a popular subject at the time.

A ruled surface is a surface generated by straight lines, called generators, which is not a cone for which all generators pass through a point or a developable surface, generated by the tangents of a curve. A ruled surface can have directrices, straight lines or curves through which all generators pass.

A tangent plane at a point of a ruled surface must obviously contain a generator through the point. If the surface has degree n, it follows that a tangent plane meets the surface in a curve of degree n where the generator is a double line. The remainder of the intersection is a curve of degree $n-2$ which meets the generator in $n-2$ points on $n-2$ generators. The locus of these points, which are double points on the surface, is the *double curve* of the ruled surface which plays an important part in the classifications. Another invariant, introduced by H.A. Schwarz, is the genus of the surface defined as the genus of the irreducible curve of highest degree contained in the intersection of a tangent plane and the surface.

The tangent planes along a generator form a linear pencil. In fact, if, for instance, $x = 0, y = 0$ is a generator and the surface has degree n, its equation has the form

$$H(x, y, z, w) \equiv f(x, y, z, w)x + g(x, y, z, w)y = 0$$

where f, g are homogeneous of degree $n - 1$. We may assume that $w = 1$ and that z is a parameter on the generator. If $F(z) = f(0, 0, z, 1)$ and analogously for $G(z)$, the equation of the tangent plane at the point z on the generator has the equation $F(z)x + G(z)y = 0$. By the above, F and G vanish simultaneously in $n - 2$ points. Hence the equation $F(z) = mG(z)$ has only one solution z for every complex m. It also follows that there is a bijection between the tangent plane of a ruled surface and its points, i.e. a ruled surface is its own dual.

It is easy to see that every nondegenerate second degree surface has an equation of the form $xz = yw$. It is then a ruled surface with two families of generators, one where $x = cy, z = c^{-1}w$ with c constant and another one where y and w have changed places. Every generator in one family is a directrix for the other one.

To have an idea of ruled surfaces we shall now with Cayley classify those of degree three. The double curve of degree one must be a straight line. If its equation is $x = y = 0$ and that of the surface is $F(x, y, z, 1) = 0$, we must have $F = 0$, grad $F = 0$ when $x = y = 0$ which means that $F(x, y, z, 1)$ has the form $F_1x^2 + F_2xy + F_3y^2$. Hence

$$F = A(x, y)z + B(x, y)w + C(x, y)$$

must be a polynomial of degree one in z, w. Here $A(x, y)$ and $B(x, y)$ are nonvanishing polynomials of degree two while C is homogeneous of degree three. In the general case A and B may be diagonalized simultaneously, which means that all terms of F are divisible by x^2 or by y^2. If we collect the coefficient of x^2 to a new z and the coefficient of y^2 to a new w, we see that the equation can be written in the form

$$x^2z - y^2w = 0.$$

This is also the general form of a ruled surface of degree three. Every generator has the form $y = tx, z = t^2w$ with a complex parameter t. Except for the double straight line directrix there is another one, namely $z = w = 0$, i.e., the line at infinity in the plane $z = 0$. The corresponding real surface in x, y, z-space is generated by the two lines $y = \pm\sqrt{c}x$ in the plane $z = c \geq 0$. When $z = 0$ the lines come together on the y-axis and disappear out into the complex.

There is a remaining case which prevents the simultaneous diagonalization of A and B, namely when A is a square and has common factor with B. Then we may assume that $A = x^2$ and $B = xy$. With new coordinates x, y we may also assume that C has a term y^3. A small computation then shows that we may choose coordinates z, w in such a way that the surface gets the equation

$$y^3 + x(xz + yw) = 0,$$

a case fundamentally different from the earlier one. There is now only one directrix and the generators are given by $y = tx, t^3x + z + tw = 0$.

The classification of ruled surfaces of degree four was done by Cayley, Cremona, and Salmon. For the degree five, H. A. Schwarz (1865) introduced the genus of the surface which in this case is at most 2 and is attained when the surface has two directrices of multiplicities 2 and 3. The largest number of cases occurs when the genus equals zero in which case the double curve has maximal degree six. For genus 0,1,2, Schwarz found 10,4,1 possibilities for the double curve. Ruled surfaces of genus zero can be seen as intersection of two planes

$$a_m t^m + ... + a_0 = 0, \quad b_n t^n + ... + b_0 = 0,$$

where the coefficients are linear homogeneous functions of the coordinates and t is a complex parameter. The degree of the curve is $m + n$ since the resultant of the two polynomials is a polynomial in the coordinates of this degree. The genus of the surface is zero because the intersection with a plane is a curve whose coordinates are rational functions of the parameter t. Schwarz's classification of ruled surfaces of genus zero is based on a careful analysis of the formulas above with $n + m = 5$ and $n = 1, 2$. His article ends with a description of the ten kinds of double curves which may appear for genus zero.

When the mathematicians in Lund tried to classify ruled surfaces of degree six, there had been many before them and some established methods were available. The first one to try was Jakob Bergstedt (1886). He restricted himself to genus zero and followed Schwarz rather closely. One of his results is a list of the main possibilities for the double curve which in his case has degree at most ten.

Wiman (1892) used methods where the ruled surface is mapped onto a space curve in different ways. Every straight line which meets the line $x = 0$, $y = 0$ may, for instance, be written uniquely in the form $xx_1 + yy_1 = 0$, $yy_1 + zz_1 + ww_1 = 0$, and hence it is represented by the point x_1, y_1, z_1, w_1 in projective space. It follows that every ruled surface R with a straight line as a directrix can be mapped onto a curve C. That a generator meets a straight line means that a certain quadratic form in x_1, y_1, z_1, w_1 vanishes. Hence the degree of the surface is the number of points common to this quadratic surface and C. The connection between R and C is not simple, but in spite of this, the curve C turns out to be a simpler object than R. There are many details and technicalities in the 109 pages of this thesis which cannot be discussed here. The main result is an almost full classification with over sixty cases of the ruled surfaces of degree six.

Wiman's thesis is written in Swedish and was hardly ever quoted, but it is a fine achievement and shows the sharp intellect and strong character of its author. We can also see complete mastery of the algebraic geometry of the time, codified in Salmon-Fiedlers great treatise *Analytische Geometrie des Raumes*. The copy which is preserved at the mathematics institute in Lund bears the marks of a thoroughgoing study.

The nineteenth century was a time of expansion of algebraic geometry with many new results and new methods. The weaknesses are the same as those of the analysis explosion in the seventeenth century. Intuition and approximative proofs preceded a strict analysis. After 1950 a new time began for algebraic geometry with the introduction of topology and a new emphasis on the algebraic side.

Bibliography

BERGSTEDT J.
1886. *Om regelytor af sjette graden I. Unikursala ytor*, Diss. Lund 1886.
BJÖRLING C.F.E.
1863. *Om några arcus-tangens-summor och deras användade till definita integralers evaluering*, Westerås 1863.

1871. *Theori för algebraiska eqvationers rötter*, KVA Handl. bd 10 n:o 3. Stockholm 1871, 53 s. 76 figurer.
1878. *Om eqvivalenter till högre singulariteter i plana algebraiska kurvor*, KVA Öfversigt 1878 nr7, 33 s.
1886. *Ueber die singuläre Punkte der gewöhnlichen algebraischen Differentialgleichung erster Ordnung und ersten Grades*, Arch. der Math. u. Physik R. 2 Bd 4, 1886, 358-384.
1888. *Singuläre Generatricen in algebraischen Regelflächen*, KVA Öfversigt 1888,587-604.
1890. *Über Raumcurven-Singulariteten*, Arch. d. Math. u. Physik Ser. 2 bd 8, 1890, 83-91.
1867. *Elementerna af den algebraiska analysen och differential-kalkylen efter Cauchy, Bertrand, Todhunter m. fl. i korthet framställda* Del. 1. Reella quantiteter, Uppsala 1866-67, 291 s.
1893. *Lärobok i algebraisk analys och differentialkalkyl*, Lund 1893.
1896. *Lärobok i Nyare plan geometri*, Lund 1896.

BRIESKORN E., KNÖRRER H.
1981. *Ebene algebraische Kurven*, Birkhäuser 1981.

LIOUVILLE J.
1841. *Mémoires sur quelques propositions générales de géometrie*, J. Math. pures et appliquées VI, 1841, 345-411.

POINCARE H.
1881. *Mémoire sur les courbes définies par une équation différentielle*, J. des Math. pures et appliqués 3.7, 1881, 375-422.

SALMON G.
1874. *Analytische Geometrie des Raumes deutsch hergestellt von Dr Wilhelm Fiedler. Zweite Aufl.*, Leipzig 1874.

SCHUBERT H.
1879. *Kalkül der abzählende Geometrie*, Leipzig 1879.

SCHWARZ H.A.
1865. *Über die geradlinige Flächen fünften Grades*, Journal für die reine und angewandte Mathematik bd 67, 1865, 23-57.

WIMAN A.
1892. *Klassifikation af regelytorna af sjette graden*, Ak. Afh. Lund 1892.

CHAPTER 5

Bäcklund

The most gifted mathematician among those who competed with Björling for the professorship in Lund was Albert Viktor Bäcklund (1845-1922). He entered the university at the age of sixteen and devoted himself to science: astronomy, mechanics, physics, and mathematics. The astronomer Möller encouraged him and employed him at the astronomical observatory. In his thesis (1868) Bäcklund determined the precise latitude of the new observatory. He was nominated docent of geometry a year later after having written a paper in the subject. It was von Zeipel's seminars rather than Hill's lectures that made him interested in mathematics. Fate ruled that Bäcklund was too young in 1872 for a professorship in mathematics and he chose another subject with better prospects of promotion. He was nominated assistant professor in mechanics in 1878, professor of physics in 1900 and he was rector of the university 1907-1909. Before 1872 Bäcklund wrote about algebraic geometry, but later switched to partial differential equations where Bäcklund transformations are named after him.

One of the applicants for the 1872 professorship was the Norwegian Sophus Lie, at the time working on the theory of partial differential equations of the first order. Lie found a class of transformations, contact transformations, which map solutions of two such equations to each other. As mentioned in the previous chapter, Lie's application arrived too late, but it was decided that his papers could remain at the university office during the competition. By permission of the chief administrator Bäcklund was able to read these papers. This decided his future. Bäcklund's generalizations of contact transformations, written between 1873 and 1883, are known today as Bäcklund transformations and continue to be of interest.

Bäcklund's work must be seen against the theory of partial differential operators current at the time and, of course, Lie's contact transformations. Thus this chapter will deal with the mathematics of an epoch and not only the work of one mathematician.

Differential equations at the end of the nineteenth century

The article *Partielle Differentialgleichungen* in *Enzyklopädie der Mathematischen Wissenschaften* (see under *Enzyklopädie* in the bibliography) reviews the theory of partial differential equations with roots in the eighteenth century that was developed in the nineteenth and is codified in two books by Goursat (1893, 1896).

In this theory, a partial differential equation of the first order is written in the form

$$f(z, x, p) = 0,$$

where $z = z(x)$ is the unknown function, $x = (x_1, ..., x_n)$ the independent variables, and $p = (p_1, ..., p_n)$ with $p_k = \partial z/\partial x_k$ the partial derivatives. The triple z, x, p is seen as coordinates in a phase space where, in order to keep the connection between

z and p it is required that the Pfaffian form, i.e., first order differential form,

$$dz - p\,dx, \quad p\,dx = \sum_1^n p_k dx_k$$

vanishes. This point of view gives the theory two faces, one turned to the differential equation, the other one to phase space. The theory in the way it was expressed a hundred years ago assumed tacitly that all functions were sufficiently differentiable or even analytic. We shall choose the latter category.

In order to simplify notations we shall in the sequel denote derivatives by an index in such a way that z_x means the gradient of z with respect to x when z is a function of $x = (x_1, ..., x_n)$. In the same way, z_{xx} means the second order derivatives etc.

For a differential equation of the first order,

$$f(z, x, z_x) = 0,$$

the notion of phase space proved very successful. Through work by Lagrange and Cauchy the theory could be reduced to the characteristics of the equation, i.e., orbits of the differential system

$$dx = f_p dt, \quad dp = -(f_x + pf_z)dt, \quad dz = pf_p dt,$$

for Pfaffian forms. This system annihilates both df and the form $dz - p\,dx$ and hence it is invariant under changes of coordinates $z, x, p \to z', x', p'$ if $z'(x') = z(x)$, $p' = \partial z'/\partial x'$ and $f'(z', x', p') = f(z, x, p)$.

Since the system annihilates df, f is constant on all characteristics. A surface $F : z = h(x)$ is said to be noncharacteristic at a point (z, x, z_x) if no characteristic touches the surface there, i.e., if $f_p(z, x, z_x)h_x \neq 0$. Under these circumstances we can solve Cauchy's problem locally: find a solution z of the equation $f(z, x, z_x) = 0$ when $z = u$ is given on F. In fact, the tangential derivatives of z along F are then given and the equation $f(u, y, u_y) = 0$ on F defines the entire gradient u_y on F. Finally, the characteristic flow z, x, p from the triple u, y, u_y when $t = 0$ defines a local function $z = z(x)$ such that $f(z, x, p) = 0$ and $dz = p\,dx$, i.e., $f(z, x, z_x) = 0$.

In special cases it is possible to find a linear combination of the elements of the differential system which is the differential of some function $V(z, x)$. It is then invariant and has the property that $V(z, x) = $ const implicitly defines a solution $z(x)$ of the differential equation. In the early stages of the theory one frequently looked for such invariants and tried to construct a general solution from them. Later, boundary problems, in particular Cauchy's problem, played a larger part. For analytic equations with analytic data, the Cauchy-Kovalevski theorem gives a complete insight locally.

That two functions $f(z, x, p)$, $g(z, x, p)$ are constant on each other's characteristics is expressed by the vanishing of the Jacobi bracket

$$[f, g] = f_p(g_x + pg_z) - g_p(f_x + pf_z).$$

Two functions with this property are said to be in involution. The condition $[f, g](z, x, z_x) = 0$ is necessary for the two equations $f = 0$, $g = 0$ to have a common solution $z(x)$. One gets the condition by differentiating both equations after x and then eliminating z_{xx}.

A system of differential equations for the same function z,

(1) $$f_1(z, x, p) = 0, \quad \ldots, \quad f_k(z, x, p) = 0,$$

raises new problems. If the system (1) has a solution it is necessary that all $[f_i, f_j]$ vanish for this solution. When the condition is satisfied identically and the functions $f_1, ..., k_k$ are independent, the system is said to be involutive. Such a system has $n + 1 - k$ independent solutions but no more.

In the 1870's Sophus Lie radicalized the old theory by redefining a solution as a manifold in phase space which solves the system and where the Pfaffian form, also called *contact form*,

$$\omega = dz - p\,dx,$$

vanishes. In this way z, x, p appear as variables of the same kind. The term contact is geometrical: the variables p represent the tangent plane of a surface $z = z(x)$ at the point x.

Contact transformations

In his new theory Lie introduced the notion of contact transformation which is a local map $z, x, p \to Z, X, P$ in phase space such that

$$dZ - P\,dX = \varrho(dz - p\,dx),$$

where ϱ is a nonvanishing factor. This implies in particular that

$$dZ \wedge dX_1 \wedge ... \wedge dX_n \wedge dP_1 \wedge ... \wedge dP_n = \varrho^{n+1} dz \wedge dx_1 \wedge ... \wedge dx_n \wedge dp_1 \wedge ... \wedge dp_n$$

and hence that the map has a local inverse. In one of his papers Lie brags that contact transformations map any given differential equation to any other one.

The contact transformations form a local group where the elements may depend on arbitrary functions. For example, every ordinary transformation of variables $X = X(z, x)$, $Z = Z(z, x)$ induces a contact transformation where every P_j is a linear function of $p_1, ..., p_n$. But there are many others. If. e.g., $n = 1$ then

$$X = p, \quad Z = f(z - px), \quad P = -xf'(z - px)$$

defines a contact transformation when $f' \neq 0$.

Equations of the second order

Already in 1784 Monge found a way to reduce a second order partial differential in two variables to a first order differential system. The condition is that the equation is quasilinear, i.e., linear in the highest derivatives. Such an equation has the form

(2) $$Hr + 2Ks + Lt + M = 0$$

where the coefficients are arbitrary functions of z, x, y, $p = \partial z/\partial x$, $q = \partial z/\partial y$ and r, s, t are the second order derivatives of z, defined by

(3) $$dp = r\,dx + s\,dy, \quad dq = s\,dx + t\,dy.$$

Monge's theory depends basically on some simple linear algebra. The proof of the following lemma is left to the reader.

LEMMA. If $dy = \lambda dx$ solves the characteristic equation

(4) $$H dy^2 - 2K dx dy + L dx^2 = 0,$$

assumed to have separate zeros, then (2) and (3), considered as a linear system of equations for r, s, t, is simply degenerate and has a solution if and only if dp, dq, dx satisfy the linear condition

(5) $$H\lambda dp + L dq + M\lambda dx = 0.$$

Remark. The two quadratic forms in (2) and (4) are dual to each other.
If $z = f(x, y)$ solves (1), the system

(6) $$dz = p\,dx + q\,dy, \quad H dy^2 - 2K dx dy + L dx^2 = 0, \quad H\lambda dp + L dq + M\lambda dx = 0$$

describes how the coefficients p, q of the tangent plane vary along the characteristics of the solution, i.e., the two families of curves that satisfy the characteristic equation (4).

To get from the system (6) to solutions of the equation (2) is much more difficult. When $H\lambda^2 - 2K\lambda + L = 0$, it is sometimes possible to find a linear nontrivial combination

(7) $$a(dz - p\,dx - q\,dy) + b(H\lambda dp + L dq + M\lambda dx)$$

which is integrable, i.e., equals the differential of some function

$$V = V(z, x, y, p, q).$$

If $z = z(x, y)$, $p = z_x$, $q = z_y$ solve the differential equation $V = \text{const}$, then, by virtue of (7), the condition (5) is satisfied. Hence r, s, t as given by (3) satisfy the equations (2) when $dy - \lambda dx = 0$. But this is an equation that we can satisfy in some open part of R^2, and hence we have found a local solution of (2).

The theory has two weaknesses. It is restricted to two variables and the last step supposes that some differential form (7) is exact. The same weakness is also present in various generalizations, for instance an analogous theory for the equation of Monge and Ampère

$$Hr + 2Ks + Lt + M + N(rt - s^2) = 0,$$

and in the theory of Darboux where the reasoning above is transferred to systems obtained by differentiating the equation several times and replacing the highest derivatives by a system of differential forms (Goursat (1896)).

Bäcklund on partial differential equations

In the article about partial differential equations in *Enzyklopädie d. Math. Wiss* (see the bibliography) Bäcklund's papers are mentioned in several places. Many have to do with general theory, for instance (1877b). In this paper the author wants to study manifolds in R^{2n+1} of maximal dimension on which the Pfaffian form

$$dz - p_1 dx_1 - \ldots - p_n dx_n$$

vanishes together with m independent functions

$$f_1(z,x,p), \ldots, f_m(z,x,p).$$

When $m = n + 1$ and all $[f_j, f_k]$ vanish, this maximal dimension is n and the manifolds are given by

$$f_1(z,x,p) = c_1, \ldots, f_{n+1}(z,x,p) = c_{n+1}.$$

Bäcklund finds that the maximal dimension is $n - \varrho$ where 2ϱ is the rank of the skew matrix $([f_j, f_k])$. This result is contained in later work by von Weber and Cartan about the rank of Pfaffian forms restricted to submanifolds.

In his paper (1878) Bäcklund computes the characteristic form of a differential operator of high order. The main part of the paper is the deduction of the equation of a characteristic surface $\varphi(x) = 0$ of a second order partial differential equation in n variables, $f(x, z, z_x, z_{xx}) = 0$. The equation is

$$\sum_{i,k} f_{ik}(x, z(x), z_x(x), z_{xx}(x))\varphi_{x_i}\varphi_{x_k} = 0$$

where the f_{ik} are the partial derivatives of f with respect to $\partial^2 z/\partial x_i \partial x_k$. It depends on the solution z of the equation $f = 0$ and expresses the property that the surface $\varphi(x) = 0$ cannot carry Cauchy data: if one knows z and z_x on the surface, it is not possible to compute the higher derivatives on the surface. The simplest example is the equation $z_{xy} = 0$ and the x-axis. This result, which is immediately transferrable to equations of any order is nowadays general knowledge. In a rather messy paper (1879) the author treats among other things second order differential equations which are invariant under an infinitesimal contact transformation.

Bäcklund transformations

In his first paper (1875) in an international journal Bäcklund answers a question put to him by Lie. He gives two proofs that an osculation transformation in the plane, i.e., a map from curves to curves which preserves second order contact, must be a contact transformation. To show the sense of this result it is proved in precise form below. Bäcklund's rather sketchy geometrical proofs are not accessible to a modern reader and were probably accessible at the time only to a limited circle of initiated readers.

In analytical terms an osculation transformation is a map

(1) $$z, x, p, p' \to Z, X, P, P'$$

between two phase spaces of dimension 4 such that

(2) $\quad dz - p\,dx = 0, \quad dp - p'dx = 0 \Rightarrow dZ - PdX = 0, \quad dP - P'dX = 0.$

THEOREM. *An osculation transformation which is not a contact transformation cannot be locally surjective.*

Remark. In Bäcklund's geometrical language the theorem says that an osculation transformation cannot map a three parameter family of curves to another such family. What he means is that an osculation transformation cannot depend on three independent functions. In the degenerate case below, the transformation depends on two independent functions.

PROOF: If an index denotes derivatives in the coordinate system x, z, p, p' we get

$$(3) \quad dZ - PdX = \\ (Z_x - PX_x)dx + (Z_z - PX_z)dz + (Z_p - PX_p)dp + (Z_{p'} - PX_{p'})dp'$$

and an analogous (3') for $dP - P'dX$. These expressions vanish for $dz = p\,dx$, $dp = p'dx$ if and only if

$$(4) \quad Z_{p'} - PX_{p'} = 0, \quad P_{p'} - P'X_{p'} = 0$$

and

$$(5) \quad LZ - PLX = 0, \quad LP - P'LX = 0$$

where $L = \partial_x + p\partial_z + p'\partial_p$. Let us first consider the main case when $X_{p'} \neq 0$ and hence x, z, p, X can be chosen as coordinates. By (4) and (3),(3') Z, P, P' are functions of z, x, p, X such that

$$P = f_X, \quad P' = f_{XX}, \quad Z = f(z, x, p, X).$$

Insertion into (5) and a passage to the variables z, x, p, X shows that

$$LZ - Z_X LX = 0, \quad LZ_X - Z_{XX} LX = 0.$$

In the new variables we have $L = \partial_x + p\partial_z + p'\partial_p + (LX)\partial_X$ so that $[\partial_X, L] = (LX)_X \partial_X + p'_X \partial_p$. If we differentiate the first equation according to X and subtract the other one, the result is that

$$p'_X Z_p = 0.$$

If $Z_p = 0$ the functions X, Z, $P = Z_X$, $P' = Z_{XX}$ depend only on X, x, z and cannot be independent. If $p'_X = 0$ then dz, dx, dp, dp' are not independent against the hypothesis.

In the degenerate case $X_{p'} = 0$ (4) shows that the functions X, Z, P are independent of p'. The first equality (5), i.e., $P = LZ/LX$, proves that $X_p = 0$, $Z_p = 0$ and that P is a certain piece-wise linear function of p. The second equality (5) shows, finally, that P' is a certain affine function of p'. With a suitable choice of X, Z as functions of x, z the mapping (5) is an osculation transformation and simultaneously a contact transformation. It is surjective in phase space, but does not depend on three independent functions. This is probably what Bäcklund means when he says that an invertible osculation transformation is a contact transformation.

Remark. The proof above can easily be extended to cover all higher orders of contact and several variables. This is at least what Bäcklund suggests in a later paper (1877 a). The result is the following: a surface map

$$(6) \quad Z = f(z, x_1, ..., x_n, p_\alpha), \quad X_1 = f_1(\ldots), \quad \ldots, \quad X_n = f_n(\ldots), \quad \ldots$$

from a phase space of derivative symbols p_α with $|\alpha| \leq m$ corresponding to the partial derivatives

$$\partial^{|\alpha|} z / \partial x_1^{\alpha_1} ... \partial x_n^{\alpha_n}$$

is uniquely determined by

(7) $$Z = g(z, x_1, ..., x_n, p_\beta, X_1, ..., X_n)$$

as a function of
$$z, x_1, .., x_n, p_\beta, X_1, ..., X_n$$

where $|\beta| < m$. In addition, the functions P_α are equal to the corresponding derivatives of g with respect to $X_1, ..., X_n$. Since their number when $|\alpha| = m > 1$ is bigger than n when $n > 1$, it follows that the map (8) cannot be surjective in phase space.

In his first paper Bäcklund writes that it was conceived on a journey to Germany during which he discussed osculation transformmations with Lie and with Felix Klein. But he is careful to stress that the proof is his own. His paper also contains some preliminary notions in the subject of surface transformations and closes with a transformation which maps solutions of a differential equation to solutions of another one without being a contact transformation. This is a comment on Lie's statement that only contact transformations could have this property.

After the result that bijective surface transformations must be contact transformations, Bäcklund turned to general surface transformations. A first step was a paper in two parts, (1877 a) and (1878 a) about intermediary integrals which means differential equations having solutions in common with differential equations of lower order.

In the simplest case $F(x, z, z') = 0$ is an ordinary differential equation of order one and $G(z, x) = 0$ an equation which defines a solution. This case can be arranged so that $G(z, x) = G(z, x, c)$ depends on a constant c and has the property that every solution of $F = 0$ satisfies $G(x, z, c) = 0$ for some value of c. By differentiating $G = 0$ with respect to x and elimination of c, the original differential equation is retrieved.

The very rare analogue of this advantageous situation in n variables occurs when the function G and the common solutions depend on n independent constants. By differentiation the constants may be eliminated and the differential equation recovered.

In his first paper Bäcklund discovered an example of precisely this situation. Namely, the equations (6),(7) show that the equalities

(8) $$f = c, \quad f_1 = c_1, \quad ..., \quad f_n = c_n,$$

where $z(x)$ is a given function and $c, c_1, ...$ are constants have the consequence that

$$c = F(z, x_1, ..., x_n, p_\beta, c_1, ..., c_n).$$

If, as Bäcklund assumes, there is a function $z(x)$ for every choice of constants, this means that the equation above is an intermediary first integral of every differential equation in (8).

This observation causes Bäcklund to investigate situations that lead to first integrals with many constants. His results are too complicated for a review here.

In the second part (1878 a) Bäcklund takes the opposite path by investigating the structure of differential equations with intermediary first integrals. One of his results is the following one. A second order differential equation with solutions $z = z(x, y)$ for which there are two independent functions $f(z, x, y, p, q)$ and $g(z, x, y, p, q)$ such that df and dg expressed in dx, dy are linearly dependent for all (x, y) must be of the Monge-Ampére type

$$Ar + Bs + Ct + D(rt - s^2) + E = 0$$

with the classical notation $r = z_{xx}$, $s = z_{xy}$, $t = z_{yy}$ and $A, ..., E$ functions of z, x, y, p, q. The proof is simply an elimination of the parameter λ between the two equations
$$df/dx + \lambda dg/dx = 0, \quad df/dy + \lambda dg/dy = 0.$$
This gives $(df/dx)(dg/dy) - (df/dy)(dg/dx) = 0$ which is the equations above. Bäcklund completes the proof with geometric considerations which say that the equation represents a ruled surface, considered as an equation in the coordinates r, s, t.

With this example in mind, Bäcklund wants to construct partial differential equations with given intermediary first integrals of order one. The source of these constructions is what Bäcklund calls surface transformations.

Let us write the coordinates of a phase space $F^{(1)}$ of order one over R^n as
$$z, \quad x = (x_1, ..., x_n), \quad p = (p_1, ..., p_n)$$
and the contact form as $p\,dx = \sum p_k dx_k$. With the notations we can now state what Bäcklund more or less explicitly meant by a surface transformation, a notion which later was to carry his name.

DEFINITION. *A Bäcklund transformation is a relation in the product $F^{(1)} \times F^{(1)}$ where the equations*
$$dz - pdx = 0, \quad dz' - p'dx' = 0$$
are combined with at most $n+1$ equations

(9) $\qquad f_1(z, x, p, z', x', p') = 0, \quad ..., \quad f_k(z, x, p, z',' x,' p') = 0.$

In his paper (1878 a) Bäcklund gives his future research program: to investigate systems (9) with the property that an insertion of certain functions $z = z(x)$ produces one or more functions $z' = z'(x')$. As in the definition above, this means $dz - p\,dx = 0, dz' - p'\,dx' = 0$ is added to the system. We can also say that a Bäcklund transformation is a contact transformation with side conditions.

In practice the contructions are performed so that an insertion $z = z(x)$ gives conditions that $dz' = p'dx'$, which in turn can entail conditions on the function $z(x)$.

In this paper Bäcklund starts with $n = 2$ and a single equation (9) where p' is missing. A function $z(x)$ then results in a surface $z' = z'(x')$. The rest of the text is deliberations about what diverse situations in one phase space, e.g. surfaces, differential equations etc. produce in the other phase space. There are also remarks about the theory of partial differential equations of order 2 and Darboux is mentioned.

After his papers 1877-1879 about general theory, Bäcklund returns to surface transformations in (1880) and now for two variables x, y. He combines the basic condition
$$dz - p\,dx - q\,dy = 0, \quad dz' - p'dx' - q'dx' = 0$$
with three equations $f_k(z, x, p, q, z', x', p', q') = 0$, $k = 1, 2, 3$. Inserting a function $z = z(x, y)$ and eliminating x, y from two equations, the result is a first order differential equation for z' whose nature is investigated.

Bäcklund proceeds to the same problem now with four equations. If the equations are differentiated after an insertion of $z = z(x, y)$, the result is a system where all differentials can be expressed in just two of them, e.g. dx', dy', so that
$$Ndp' = Hdx' + Kdy', \quad Ndq' = Ldy' + Mdy'.$$

If p', q' are derivatives with respect to x', y' of a function, this means that $K = L$. After some computation this condition produces a fifth equation

(10) $$\sum (f_{1x}f_{2y} - f_{1y}f_{2x})[f_3, f_4]_{z', x', y'} = 0$$

where the sum runs over all even permutations of $1, 2, 3, 4$. With five equations the functions z', x', y', p', q' may in general be expressed in terms of z, x, y, p, q, and the condition produces two differential equations of order three such that the transformation maps solutions of one to solutions of the other.

But other situations may also arise. If the equations are solved with respect to x', y', p', q' and do not contain z' the condition (10) reduces to a second order equation for z of the Monge-Ampère type. One solution of this equations gives several functions z'.

In his paper (1882) Bäcklund continues with two equations. At the end there is an explicit example that connects with a known result to which his name has become attached. Let F be a surface and C, C' the corresponding central surfaces, i.e., loci of the two principal radii of curvature of F. The surfaces C and C' have orthogonal tangent planes which meet in the normal of F. The theorem of Bianchi and Lie says that if the difference between the two radii is constant $= a$, then C and C' both have the constant curvature $-a^2$.

This theorem fitted into Bäcklund's general scheme as follows. If the point x, y, z is on C and x', y', z' on C' and $p = z_x$, $q = z_y$ and analogously for p', q' we have

(11a) $$(x' - x)p + (y' - y)q - (z' - z) = 0$$

(11b) $$(x' - x)p' + (y' - y)q' - (z' - z) = 0$$

(11c) $$1 + pp' + qq' = 0.$$

When these equations are completed with the constant distance

(11d) $$(x - x')^2 + (y - y')^2 + (z - z')^2 = a^2,$$

and the equations $dz - p\,dx - q\,dy = 0$, $dz' - p'dx' - q'dy' = 0$, the situation fits into Bäcklund's scheme. The involution condition is here

$$z_{xx}z_{yy} - z_{yx}^2 + a^2(1 + p^2 + q^2)^2 = 0,$$

i.e., the differential equation of a surface with constant curvature $-a^2$. This example was generalized by Bäcklund and later by Darboux ((1894), III s. 412) by replacing the right sides of (11 a,b,c) with, respectively,

$$aK \sin \alpha, \quad aK \sin \beta, \quad a \cos \gamma$$

where $K = \sqrt{1 + z_x^2 + z_y^2}$. The geometrical meaning is that the two tangent planes form the angles α and β with the line from x, y, z to x', y', z' and the angle γ among themselves. The involution condition is then a relation

$$ARR' + B(R + R') + C = 0$$

between the principal radii of curvature. This means that the surface belongs to a Weingarten class, defined as the class of surfaces for which there is a fixed relation between the principal radii of curvature.

With C, C' as in the first case with curvature -1, let

$$ds^2 = \cos^2 \omega \, du^2 + \sin^2 \omega \, dv^2, \quad \omega_\omega(u,v),$$

with u, v coordinates along the curvature lines be the line element of C and let $\theta = \theta(u, v)$ describe a corresponding line element of C'. As proved in Darboux (1896), III p. 427, the relation between C and C' is given by the following systems of equations

(12) $$\theta_u - \omega_v = \cos \omega \sin \theta, \quad \theta_v - \omega_u = -\sin \omega \cos \theta,$$

where, as a consequence, ω, θ solve the equation

(13) $$\omega_{uu} - \omega_{vv} = \cos \omega \sin \omega$$

which expresses that the curvatures are -1. If a function $\theta = \theta(u, v)$ solves this equation, there is a solution ω of (12) which can be obtained by quadrature and depends on an arbitrary parameter (Darboux l.c. Ch. XIII). This situation is often given as an example of a Bäcklund transformation, but it cannot be found in Bäcklund's papers.

In his last paper (1883) from this period Bäcklund generalizes Bianchi's theorem according to Darboux as above. A footnote gives three simple examples of surface transformations. The first one is the hodograph map $x' = u_x$, $y' = u_y$ which maps harmonic functions to harmonic functions, the second one is a transformation of minimal surfaces which is a consequence of Weierstrass's representation formula and the third is a transformation of surfaces with constant curvature.

After his retirement as professor of mechanics and physics Bäcklund went back to mathematics and wrote about surface transformations in (1913,1918,1920) without adding anything essential.

Bäcklund transformations after Bäcklund

In his paper (1915) Elie Cartan proved that if a transformation $z, x, p \to z', x', p'$ in R^{2n+1} is invertible in phase space and maps surfaces in a manifold E of dimension $n+1$ in phase space to surfaces in another such manifold, then the contact forms $dz - p\,dx$ and $dz' - p'dx'$ are proportional. In this sense many Bäcklund transformations are also contact transformations.

Goursat took an interest in Bäcklund transformations and treated them in a long article (1925). It deals with Bäcklund's problem in two independent variables with four equations

$$f_k(z, x, y, p, q, z', x', y', p', q') = 0, \quad k = 1, ..., 4,$$

and the conditions

$$dz - p\,dx - q\,dy = 0, \quad dz' - p'dx' - q'dy' = 0.$$

This framework is generalized to a study of manifolds of dimension two which annihilate two Pfaffian forms in six variables,

$$\omega_1 = \sum_{1}^{6} a_k dx_k, \quad \omega_2 = \sum_{1}^{6} b_k dx_k.$$

It is then no restriction to assume that

$$\omega_1 = dx_5 + ..., \quad \omega_2 = dx_6 + ...$$

When $\omega_1 = 0$, $\omega_2 = 0$ then

$$d\omega_1 = \sum_{i,k=1}^{4} A_{ik} dx_i \wedge dx_k, \quad d\omega_2 = \sum_{i,k=1}^{4} B_{ik} dx_i \wedge dx_k$$

with antisymmetric matrices $A = (A_{ik})$, $B = (B_{ik})$. The classification of the pair ω_1, ω_2 is based on those nonvanishing pairs λ, μ for which

$$\det(\lambda A + \mu B) = 0$$

and on the rank of the corresponding Pfaffian forms. There are many cases and only part of them have been observed by Bäcklund.

The interest in Bäcklund transformations has increased somewhat since 1950 in connection with their use in gas dynamics and other parts of physics. A typical example is Loewner (1952) which treats the equations for a compressible irrotational flow in a gas with variable density. It is shown that these equations are close to the Cauchy-Riemann equations after an infinitesimal Bäcklund transformation. Springer Lecture Notes in Mathematics 515 (1976) is a collection of loosely connected papers about the use of Bäcklund transformation for the Korteweg-de Vries equation and other nonlinear equations with solutions in the form of progressing waves, the so-called solitons.

Oseen's obituary of Bäcklund

After his nomination for professor of mechanics in 1878, Bäcklund abandoned mathematics and devoted himself to the nature of electrostatic forces. In this he based his theory on an analogy between these forces and pulsating spheres in an incompressible fluid suggested by the Norwegian physicist Bjerknes. For Bäcklund the fluid was the hypothetical ether. He now found a new outlet for his powers, and during twentyfive years he devised a comprehensive theory of electricity and magnetism.

The Swedish theoretical physicist Carl Wilhelm Oseen (1879-1944), who had studied in Lund and knew Bäcklund well has written an obituary of Bäcklund (Oseen (1924)). One section describes his teaching.

> Bäcklund devoted himself to his teaching duties with ardour and delight and he often talked about the pleasure his lectures had given him. During the time that the author heard him (the 1890's) Bäcklund carried a double burden. Outside of the regular four lectures a week he gave a beginner's course in mechanics with exercises, also four hours a week. His interest in his audience was manifested by, among other things, invitations to a yearly party where lobster was a standing ingredient.

In his article Oseen comments on Bäcklund's entire scientific production. The reviews of the mathematical papers now seem to be a bit passive, but about the papers in mathematical physics Oseen wrote as follows:

> There are many kinds of physical theories. For experimental research most useful, but in the long run least tenable, are those which, on the basis of accepted

knowledge, examine recent experiments and their consequences for a systematic theory. For theoretical physics most important is the theoretical work, which, from contradictions between reality and established theory, carves out a new basic hypothesis for the theory. As an example we may mention PLANCK's quantum hypothesis. Yet another kind is the theoretical work which, by a deeper mathematical treatment of an existing theory, attempts to eliminate a contradiction between theory and experiment or be useful to experimental research. Bäcklund's physical theory falls into none of these categories. With contemporary experimental work it had no connection. Neither had it any ties with the theoretical work which at the same time was done in England or on the continent. The aim of his theory was neither to be useful to experimental work nor to eliminate some contradiction between theory and experiment. Then, what was the aim of this theory? To reduce physics to mechanics! To prove that all physical phenomena can be deduced from various forms of motion! This is the object, more philosophical than physical, which Bäcklund in his activity as a mathematical physicist tried to realize. In his youth this was the established aim of theoretical physics. He remained true to this conception while around him physics went through a development which for the majority of physicists radically changed the circle of problems. For this reason his theory was never tested in experiment. At the time when it was so developed that experimental tests were possible, the center of gravity in physics had moved from mechanics to electromagnetism. The foundations of his theory had failed before the theory could be tested. This was the tragic element in Bäcklund's life as a theoretical physicist. Or perhaps, this was just one side of his tragedy. There was another one. In the extensive world he had created he was the only living being. During his twenty-five years of work in theoretical physics he was completely alone. There are things he has said where it is clear that he felt this loneliness as a tragedy.

Bäcklund's mathematical work was more successful. His discovery and characterization of generalized contact transformations is a permanent piece of mathematics which remains of interest one hundred years after it was done. But also in the mathematical papers it is possible to see the combination of power and insensitivity which Oseen describes. It is perhaps regrettable that Bäcklund was not able to devote himself to mathematics all his life.

Bibliography

AMPERE A.M.
1820. *Mémoire contenant l'application de la théorie exposée dans le XVIIe cahier etc.*, Journal de l'Ecole polytechnique XVIIe cahier 1820.

BÄCKLUND A.V.
1869. *Några satser on plana algebraiska kurvors normaler*, Lunds Univ. Årsskr. 1869, 1-38.
1875. *Über Flächentransformationen*, Math. Ann. IX, 1875, 298-320.
1877. a. *Über partielle Differentialgleichungen höherer Ordnung die intermediäre erste Integrale besitzen*, Math. Annalen XI, 1877, 199-241.
1877. b. *Über Systeme partieller Differentialgleichungen erster Ordnung*, Math. Ann. XI, 1877, 412-433.
1878. a. *Über partielle Differentialgleichungen höherer Ordnung die intermediäre erste Integrale besitzen. Zweite Abhandlung*, Math. Ann. XIII, 1878, 69-108.
1878. b. *Zur Theorie der Charakteristiken der partiellen Differentialgleichungen zweiter Ordnung*, Math. Ann. XIII, 1878, 411-428.
1879. *Zur Theorie der partiellen Differentialgleichungen zweiter Ordnung*, Math. Ann. XV, 1879, 39-88.
1880. *Zur Theorie der partiellen Differentialgleichungen erster Ordnung*, Math. Ann. XVII, 1880, 285-328.

1882. *Zur Theorie der Flächentransformationen*, Math. Ann. XIX, 1882, 387-422.
1883. *Om ytor med konstant negativ krökning*, Lunds Univ. Årsskrift vol. XIX, 1883.
1913. *Einiges über Kugelkomplexe*, Annali di Matematica. Ser. III vol XX, 1913, 65-107.
1918. *Ein Satz von Weingarten über auf einander abwickelbare Flächen*, Lunds Univ. Årsskrift N.F. 1918, Avd. 2 bd 14 nr 13.
1920. *Zur Transformationstheorie partieller Differentialgleichungen zweiter Ordnung*, Lunds univ. Årsskr. XVI, 1920, 1-31.
  ENZYKLOPÄDIE der mathematischen Wissenschaften bd II 1.1 A5, *Partielle Differentialgleichungen* 296-396, Leipzig 1899-1916.
1976. *Bäcklund Transformations*, Lecture Notes in Mathematics 515, 1976.
  CARTAN E.
1915. *Sur les transformations de Bäcklund*, Bull. Soc. Math. de France t. 43, 1915, 6-24.
  DARBOUX G.
1894. *Leçons sur la théorie générale des surfaces vol III*, Paris 1894.
  GOURSAT E.
1893. *Vorlesungen über die Integration der Partiellen Differentialgleichungen erster Ordnung...*, Leipzig 1893.
1896. *Leçons sur l'intégration des équations aux dérivées partielles du second ordre*, Paris 1896.
1925. *Le problème de Bäcklund*, Mémorial des Sciences mathématiques. Fasc VI, 1925.
  LOEWNER C.
1952. *Infinitesimal Bäcklund transformations*
  Journal d'analyse mathématique 2, 1952/53, 219-242.
  MONGE G.
1784. *Mémoire sur le calcul intégral des équations aux différences finies*, Histoire de l'Académie des Sciences 1784.
  OSEEN C.W.
1924. *Albert Viktor Bäcklund*, KVA Årsbok 22, 1924.

CHAPTER 6

Uppsala 1860 - 1900

Daug   Dillner   Falk   Berger

At the end of the 1850's Malmsten left Uppsala and academic life to join the government. His successor Daug did not have the energy and initiative of his predecessor. Instead, teaching and research were dominated by the assistant professor Dillner whose main interest was Cauchy's work on analytic functions and his own version of this theory.

Daug

Herman Theodor Daug (1828-1888) studied in Uppsala under Malmsten and got his degree in 1856 with a thesis in Swedish, in English translation *Contributions to the theory of the radius of curvature, the radius of torsion and the osculating sphere*. It has only two references, to Cauchy and to a certain professor Schellbach. One natural reference is missing, namely the paper (1847) by Frenet, printed in *Journal de Liouville*, which contains the classical Frenet formulas. The missing reference is explained by the slow communications at the time. In any case, Daug wrote his thesis himself which was somewhat unusual at the time. When Malmsten was away from Uppsala, Daug took over his teaching and in 1867 he was nominated professor.

Differential geometry dominates Daug's sparse scientific production, almost entirely in Swedish. His lectures in Swedish, in English translation *The use of differential calculus for the investigation of lines in space and curved surfaces*, (1877-94), were published as a textbook after his death. The exposition is careful and clear, but the book was used only for a short time.

The locus of the center of the osculating sphere of a space curve is another space curve. One of Daug's results (*Öfversigt* 1858) says that the quotient of the curvature and torsion of the first curve is reversed for the second. Since I have not found this result in the literature, the simple proof follows below.

If $s \to x(s)$ is a C^3-curve in space with arc length s and x', x'', x''' are linearly independent, Frenet's formulas

$$e'_1 = e_2/r, \quad e'_2 = -e_1/r + e_3/t, \quad e'_3 = -e_2/t$$

describe the dependence on s of a certain triple of orthonormal vectors e_1, e_2, e_3. Here $e_1(s) = x'(s)$ is the tangent vector, e_2 is the normal vector and e_3 is orthogonal to e_1 and e_2. The parameter $r = r(s)$ is the radius of curvature and $t = t(s)$ the radius of torsion.

The center z of the osculating sphere and its radius a at the point s are determined by the condition that the $\| x(s) - z \|^2 - a^2$ vanishes and is stationary of order three at s, i.e., $(x', z - x) = 0$, $(x'', z - x) = 1$, $(x''', z - x) = 0$. After some computation with Frenet's formulas we find that

$$z = x(s) + re_2 + r'te_3,$$

as a function of s. From this is follows that the tangent vector $z'(s)$ is

$$z'(s) = f(s)e_3, \quad f(s) = (r't)' + r/t.$$

If the right side is not zero, i.e., if the curve $s \to x(s)$ is not spherical, $d\sigma = f(s)ds$ defines the arc length σ of the curve $\sigma \to z(\sigma)$ so that its tangent vector is simply $E_1 = e_3$. Hence

$$dE_1/d\sigma = -e_1/tf(s) = E_2/R, \quad E_2 = e_1, \quad R = -f(s)t,$$

where R is the new radius of curvature. Hence $E_3 = e_2$ so that, finally, $dE_3/d\sigma = -E_3/R + E_1/T$ where $T = -f(s)r$ is the new torsion.

Dillner

Göran Dillner (1832-1906), assistant professor from 1865 to 1876 and full professor from 1877 to 1897, did many things. He started the short-lived Matematisk Tidskrift (1868-1874) (see Chapter 4) and was one of the founders of a private gymnasium in Uppsala which still exists and was nicknamed the Scrape because it accepted students who had failed at the state schools. Dillner advocated a university reform and gave research in Uppsala a new direction. In these both cases his initiatives ended in disappointment.

In Dillner's time the old tradition with many subjects required for every degree still prevailed. Dillner critized this order of things. He wanted to modernize the curriculum and thought that Sweden was backward in comparison with France and Germany. He received a grant from the government to study the universities on the continent and came back with a suggestion for a radical reform whose main ingredient was a scientific degree without a compulsory earlier schooling. This reform did not materialize, but the university curricula were eventually changed in the direction he had advocated.

Dillner's many mathematical papers were not as successful as he had expected. His thesis (1861), *Geometric Calculus or the laws of computation with geometric entities* in English translation, has very few references to the the literature and the text can be described as a compromise between linear algebra and coordinate geometry. The author defines addition and multiplication of complex numbers in geometrical form and thinks that this reveals their true nature. There are also remarks about analytic functions.

The same ideas appear in the long paper (1873). In the preface the author's geometrical calculus is presented in magisterial style. Dillner derides Riemann surfaces whose global nature has eluded him. According to Dillner, a simple residue formula ('circular integral') replaces Abel's famous theorem about the form of rational integrals from a point of a complex curve to its points of intersection with another complex curve. Dillner has a vague idea of convergence but he is fond of new terms and replaces the word monodrome (one way) with eichodrome (way home). With a term from Cauchy (1821), analytic functions are called 'synectic' with the following vague definition: 'continuous, single-valued and finite at a point, at a line etc.' The complex derivative has a correspondingly unclear definition from which it is claimed to follow that it is direction independent. Cauchy's integral theorem has a long and very complicated proof. At the end Dillner treats the elliptic functions whose subtle theory he has not grasped.

Dillner's many and long mathematical papers are marked by his self-confidence and inability to understand the work of others. It is however possible that Dillner was a better algebraist than analyst. He wrote a nice paper (1881) which is a modern and demystified exposition of quaternions. Here, for once, the rules of computation come first and the geometry afterwards.

Maybe it is too easy to criticize Dillner's treatment of analytic functions. Before his time only the best mathematicians had the combination of intuition and good taste that permitted them really to understand what the theory was about. The mathematician who finally made the theory of analytic functions understandable to many was Weierstrass. He based the theory on local expansions in power series, a definition that is not easy to vary or to misunderstand.

Dillner's geometrical program had the charm of novelty and attracted many students, but they did not adapt the views and terminology of their teacher. His most successful student was Gösta Leffler who wrote his thesis under Dillner, went abroad and came under the influence of Weierstrass.

Falk

Malmsten's paper about continued fractions resulted in two theses, one of them in 1861 by Matths Falk (1841-1920), docent in 1869, lektor in Uppsala in 1877. From time to time he substituted for Dillner and was nominated professor in 1890.

Falk, who wrote few papers, used English at a time when German and French were the dominating languages for scientific publications. His speciality was differential equations. In two papers (1872, 1875) he treats the case of two independent variables. His implicit assumption is that every equations has a general solution of the form

$$f(x, y, \varphi_1(u_1), ..., \varphi_r(u_r)),$$

where $u_1, ..., u_r$ are given functions of x, y and $\varphi_1, ..., \varphi_r$ are arbitrary. In the beginning Falk observes that t differentiations of f gives $a = t(t+3)/2$ equations of for $b = rt$ derivatives of the unknown functions. Hence they can be eliminated when $a > b$. An example: if $z = f(y+x)g(y-x)$ differentiation and elimination produces the equation $z_y^2 - z_x^2 = z(z_{yy} - z_{xx})$.[1]. Other examples are also given. The author also considers characteristics and the Monge-Ampère theory (see p. 51). His text shows that he believes that the Pfaffian forms of the theory are all exact.

Falk also considers linear partial differential equations whose coefficients are functions of x and y and proves the following result.

Theorem 1: Every solution is the sum of a particular solution and a solution to the homogeneous equation. Theorem 2: Linear combinations of the solutions to a homogeneous equations are solutions. Theorem 3: Homogeneous equations with constant coefficients have solutions of the form $z = \varphi(y - mx)$.

The reason why these results are stated is that linear algebra was only implicit knowledge in Falk's time and that he was happy to state some precise results in a paper where they are otherwise scarce. He ends his paper with the following statement

'We do not suppose that we have overcome all the difficulties which exist in the delicate problem of obtaining a theory of partial differential equations of the form treated in this memoir. As to our method of deduction, it is essentially the same as that, which *Boole* in his *Treatise on differential equations* applies to Monge's linear partial differential equation

$$Rr + Ss + Tt = V.$$

We hope in a future memoir to extend the theory, now given, to any number of independent variables.'

This paper from 1872 and a later one from 1875 got long reviews in *Jahrbuch über die Fortschritte der Mathematik*, the *Mathematical Reviews* of its time, which

---
[1] This is the wave equation for $\log z$ in disguise

started publication in 1867. But the referee had difficulties understanding them and their relation to Darboux's theory of 1870 where a given differential equation is differentiated many times in the hope of finding integrable equations for the characteristics (see Goursat (1896, Ch. VII). The explanation is that there is no connection. Falk had his knowledge from Boole's textbook where this theory does not occur.

In a volume of *Nova Acta*, issued on the occasion of the 400th anniversary in 1877 of the founding of Uppsala University, Falk wrote about *imaginary functions*. To avoid complex numbers he splits a complex function into a real and an imaginary part,
$$F(z) = f(x,y) + ig(x,y).$$

By replacing on the right side x by $z - iy$, differentiating with respect to y and putting the result equal to zero, Falk gets the Cauchy-Riemann equations $f_x = g_y, f_y = -g_x$. This absurd reasoning illustrates the vague ideas about the nature of analytic functions which were current in Uppsala at the time.

Falk's other papers deal with determinants (1878) and elimination according to Bézout (1879). He also wrote a simple paper in *Acta Mathematica* vol. 7 (1885-86) about elliptic functions.

Falk worked for almost twenty years as a full-time gymnasium teacher and that is perhaps one reason why his papers are shallow. In 1899 he was nominated professor at Uppsala in competition with a better candidate (Berger). The proud young mathematicians of Stockholm University considered Falk as an example of the inferiority of mathematics in Uppsala.

Berger

Alexander Berger (1844-1901) got his degree in 1873 and worked as docent and acting professor in Uppsala. His dissertation, *Om periodiska funktioner*, reflects the new interest in analytic functions in Uppsala. Mittag-Leffler wrote his thesis one year before Berger and Berger's list of references, literally, 'Books used', quotes his friend's thesis.

Berger has a correct definition of complex derivative but he assumes that existence implies continuity. The main subject is periodic analytic functions, for instance those with the property that $f(z+1) = f(z)$. The author almost knows that such a function has the form $g(e^{2\pi i z})$ where g is analytic. He considers the case when $f(x+iy)$ is regular and has limits for $y \to \pm\infty$ and calls them functions of the first order. This means that g is a linear function. When g is rational with an n-valued inverse, we get what the author calls functions of order n. Examples of the theory are taken from the trigonometric functions.

The next item of Berger's bibliography is a paper in Swedish with the title *About the occurrence of primes among the integers*, printed in Uppsala in 1875. This paper is remarkable because it is essentially identical to Chebyshev's classical *Mémoire sur les nombres premiers* from 1852 although this fact is not mentioned in the paper.

Until the middle of the nineteenth century, all great mathematicians knew the classics and the work of their colleagues, but a beginning mathematician could have difficulties to find his way in the literature. With the periodicals published by Crelle, Liouville, and Borchardt the situation improved and with the new review journal *Jahrbuch* a complete view of the literature became possible. At the same time, it was easier to decide priorities.

For a long time Swedish universities considered theses as exercises where priority was not important. Before the university reform in the 1860's, it was sufficient to defend something a professor had written and later only a few 'Books used' were listed as a bibliography. With time requirements sharpened but as late as in 1875

6 Uppsala 1860-1900

Alexander Berger could print a paper about the distribution of primes that has no references and is essentially identical to a classical paper by Chebyshev from 1852, printed in the most important French journal. Presented as a study in Swedish of Chebyshev's work, Berger's paper could be acceptable, but as it now stands without a reference, it gives the reader a bad feeling.

Chebyshev's paper is the first one which comes near to the prime number theorem, i.e., the statement that the number $\pi(x)$ of primes p which are $\leq x$ obeys the asymptotic law

$$\pi(x) = (1 + o(1)) \int_2^x dx/\log x$$

for large x. The integral may be replaced by $x/\log x$. This formula was first proved by Hadamard and then by de la Vallée Poussin in the 1890's. Chebyshev shows, rather elementarily, that the upper and lower limits of $\pi(x)\log x/x$ lie in a small interval around 1.

The proof introduces three functions,

$$T(x) = \sum_{n \leq x} \log n,$$

$$\theta(x) = \sum_{p < x} \log p,$$

$$\psi(x) = \sum \theta(x^{1/n}),$$

connected by the fundamental identity

$$T(x) = \sum_1^\infty \psi(x/n).$$

The order of magnitude of $T(x)$ is $x \log x$. We see that

$$T(x) - 2T(x/2) = \sum (-1)^{n-1} \psi(x/n)$$

and, more precisely, that

$$0 \leq \psi(x) - \psi(x/2) \leq T(x) - 2T(x/2).$$

Here the right side is $O(x)$ which is also shown to be the bound for $\psi(x)$. The author can then show that the limit points of $\theta(x)/x$ belong to a small interval around 1. Finally a partial summation in the formula

$$\pi(x) = \sum_2^x (\psi(n+1) - \psi(n))/\log n$$

is used to show that $\pi(x)\log x/x$ has the same property (see Landau (1909), pp. 71-85). Chebyshev's paper follows this line of reasoning completed with careful estimates. It is clear that Berger follows Chebyshev very closely, but while Chebyshev is interested in bounds of the limits, Berger constructs upper and lower bounds for all primes. His summary has the formula

$$\frac{3}{5}\frac{x}{\log x} < \pi(x) \leq 2\frac{x}{\log x}, \quad x \geq 3.$$

By using the ζ-function and Dirichlet's lectures in number theory, he verifies that the quotient $\pi(x)\log x/x$ comes arbitrarily close to 1 for certain large numbers x.

Berger's papers after 1875 deal with analytic number theory in a general way. He deduces for instance the main term in sums like

$$\sum_{n\leq x} n^{-z}, \quad \sum_{n\leq x} \log n,$$

and occupies himself with Euler's summation formula, Bernoulli's numbers and the Gamma function (1880, 1881). In a long and careful paper without a news value (*Nova Acta* 1880) Berger reworks papers by Liouville and proves, among other things, a result by Dirichlet which says that if $\sigma(n)$ is the sum of the divisors of an integer n and $T(n) = (\sigma(1)+...+\sigma(n))/n$ is its mean value, then $\lim T(n)/n = \pi^2/6$.

In the paper (1886) Berger wants to compute the asymptotics of the number of solutions of

$$g_1 x_1^m + g_2 x_2^m + ... + g_s x_s^m \leq n$$

where $m > 0$ and the coefficients of the left side are positive and fixed. If $g(x)$ is the left side, the main term is

$$\int_{g(x)\leq n, x\geq 0} dx,$$

an integral which may be computed explicitly. It follows that every sufficiently large number may in the mean be written in the form $x_1^m + ... + x_s^m$ in $\Gamma(1+1/m)^m$ different ways.

Without naming the originator Berger uses in (1887) Kronecker's function $\Delta = \Delta(n)$, defined as $n, 4n, 8n$ according as $n \equiv 1, -1, 2 \mod 4$, in order to sum various number theoretic functions. A peculiar paper (1887) uses a formula by Dirichlet for the number of integral solutions of $x^2 + y^2 = s$, s odd, to compute the double integral.

$$\int f(x^2 + y^2) dx dy.$$

After 1887 Berger continued to write long pedagogical articles about number theory, still with almost no references.

Berger was the first Swedish mathematician with number theory as his first interest. His papers have very dubious originality but he gives the impression of being a serious mathematician.

The contest for a professorship in Uppsala 1889

Falk and Berger applied for Daug's chair, vacant after Daug's retirement. The experts were Mittag-Leffler, the Dane H. Zeuthen and the German professor Schering. Mittag-Leffler considered Falk a shallow mathematician and recommended Berger. Schering starts with a list of important properties for advancement in science: suitability; scientific production; achievements as teacher, practical and theoretical. He finds that Falk is more versatile and Berger deeper and ends with a diplomatic judgement: both are equal in merit. Berger's prime number paper, which, if new, would have given him a prominent place in the history of number theory, is not mentioned. Zeuthen remarks that Falk's papers do have references while Berger's do not. He excuses this with Berger's originality but does not mention the paper about prime numbers.

The experts recommended Berger with some emphasis, but the faculty voted unanimously for Falk. The oldest professor D. (astronomy) summed up the opinion of the faculty: Mittag-Leffler's statement is too short to give real information,

Schering is neutral and Zeuthen recommends Berger on scientific grounds, but he is not absolutely positive and may be interpreted otherwise. The long review in *Jahrbuch* of Falk's papers about differential equations speaks in his favour .... .

Since D. brings up the passive review, he had probably some hidden reason to favour Falk. However, the main feature of the contest was certainly the dislike felt by the representatives of the old Uppsala University for the upstart university in Stockholm and its cocky representative.

Berger would have been Dillner's obvious successor had he not died in 1891. But when Dillner later retired, the name of the professorship was changed from mathematics to algebra and number theory and in this way Berger was remembered.

BIBLIOGRAPHY

FRENET M.F.
1847. *Sur les courbes à double courbure*, Thèse, Fac. Sc. Toulouse 1847, utdrag i J. de Liouville 17, 1852, 437-447.

DAUG H.Th.
1858. *Framställning af några formler som angifva förhållandet mellan curvaturen hos en curva hvilken som helst och curvaturen hos den curva som innehåller alla mot den förra svarande osculerande sferers centra*, KVA Öfversigt bd 15, 1858.
1877. *Differential- och integralkalkylens användning vid undersökning af linier i rymden och bugtiga ytor*, Uppsala 1877-94.

CAUCHY A.L.
1821. *Cours d'Analyse de l'Ecole polytechnique de l'année 1821*, Paris 1821.

DILLNER G.
1861. *Geometrisk kalkyl eller geometriska qvantiteters räknelagar*, Vet. Soc. Ups. Årsbok årg. 2, 1861.
1873. *Traité de calcul géométrique supérieur*, Nova Acta Reg. Soc. Sc. Ups. Ser. III, vol. VIII Uppsala 1873, 1-136.
1877. *Versuch einer neuen Entwicklung der Hamiltonschen Methode, genannt "Calculus of quaternions"*, Math. Ann. XI, 1877.
1893. *Mémoire sur la solution analytique du problème des N corps*, Nova Acta Reg. Soc. Sc. Uppsala 1893.

FALK M.
1869. *Om konvergensen af Kedjebråk med blott negativa och Kedjebråk med omväxlande positiva och negativa leder*, Diss. Uppsala 1869.
1872. *On the integration of partial differential equations of the nth order with one dependent and two independent variables*, Nova Acta Reg. Soc. Sc. Uppsala 1872.
1875. *Om partiella differentialeqvationer af högre ordning än första*, Uppsala 1875, Edquist.
1877. *Sur les fonctions imaginaires à l'égard spécial du calcul des résidus*, Nova Acta vol e.o. 1877.
1878. *Sur une propriété des déterminants nul*, Nouvelle correspondence de mathématiques publiée par E. Catalan et P. Mansion, Mons Manceaux, Paris, GauthiersVillars.
1879. *Sur la méthode d'élimination de Bézout et Cauchy*, Ups. Årsskrift 1879.
1885. *Beweis eines Satzes aus der Theorie der elliptischen Funktionen*, Acta Math. 7, 1885-86, 197-200.

BERGER A.
1873. *Om periodiska funktioner*, Diss. Uppsala 1873.
1875. *Om primtalens förekomst i talserien*, Uppsala 1875.
1880. a. *Elementära bevis för några formler i differenskalkylen*, KVA Öfversigt 1880 nr 10, 39-53.
1880. b. *Sur quelques applications de la fonction gamma à la théorie des nombres*, Nova Acta Upsala vol. XI, 1880, 87 s.
1881. *En generalisering af några formler i Gammafunktionens teori*, KVA Öfversigt 1881 nr 9, 13-30.
1886. *Om antalet lösningar till en viss indeterminerad eqvation med flera obekanta*, KVA Öfversigt 1886, 355-366.
1887. a. *Sur une application de la théorie des équations binômes à la sommation des séries*, Nova Acta Ups. vol. XIII, 1887, 1-36.
1887. b. *Om rötternas antal till kongruenser av andra graden*, KVA Öfversigt bd 44, 1887-88.
1888. *De Bernoulli'ska talens och funktionernas teori baserad på ett system af funktionalekvationer*, KVA Öfversigt 1888, 433-461.

1891. a. *Recherches sur les valeurs moyennes dans la théorie des nombres*, Nova Acta vol. XIV, 130 s.

1891. b. *Déduction des propriétés principales de la fonction elliptique générale du second ordre*, Nova Acta vol XIV, 1891, 50 s.

LANDAU E.

1909. *Handbuch der Lehre von der Verteilung der Primzahlen* I,II, Leipzig 1909.

CHEBYSHEV P.

1852. *Mémoire sur les nombres premiers*, Journ. des math. pures et appl., Sér. 1 vol. 17, 1852, 366-390.

CHAPTER 7

Gösta Mittag-Leffler — A Biographpy

At the turn of the century Gösta Mittag-Leffler (1846-1927) was the leading and most influential Swedish mathematician. He was the first professor of the new Stockholm university, he founded the international journal *Acta Mathematica* and, by donating his house and library to the Swedish Academy of Sciences, he laid the foundations of the present Mittag-Leffler Institute. In his lifetime Mittag-Leffler was well known to the general public. Together with Alfred Nobel, Adolf Nordenskiöld, Sven Hedin[1] and other celebrities he was one of the leading Swedes of his time. A picture of mathematics in Sweden would be incomplete without a short biography of Mittag-Leffler. There is ample material for a complete biography in his diaries and newspaper clippings and, above all, in his enormous collection of letters.

Family and growing up

Mittag-Leffler was born on March 16, 1846, first child to Johan Olof Leffler (1813-1884) and Gustava Wilhelmina Leffler born Mittag (1817-1903). At this time his parents lived in the old Klara school where his father was a teacher. Later, when the family had moved to a house of their own, the other three children were born. At a mature age they became known as the writer Anne Charlotte Edgren-Leffler, the linguist Fritz Läffler and the civil engineer Artur Leffler.

The Swedish Lefflers came originally from Breslau in Silesia. One member emigrated to Sweden in 1655 and founded a large family. Mittag-Leffler's grandfather was a sailmaker in Gothenburg and a member of parliament. His mother's family was also of German origin. Her father, the dean Mittag from a rural part of Sweden, was a well-known member of the clergy.

Mittag-Leffler's father was promoted to rector of a school in Stockholm and during one period he was elected member of parliament. He had a large circle of friends who were often invited to his house and provided a favorable atmosphere for the children's intellectual development. The young Gösta Leffler admired his mother and spent the summer vacations with her parents, Dean Mittag and his wife. Later in life he talked warmly of his grandparents and in his twenties he added the name of Mittag to his own.

In elementary school and later in the Stockholm gymnasium, Gösta Leffler showed talent for and interest in mathematics. The last three last years of school he was excused from the regular teaching of the subject. From 1865 to 1872 he studied in Uppsala and became one of the more active problem solvers in Dillner's and Hultman's periodical *Tidskrift för Matematik*.

From an early age Gösta Leffler kept an irregular diary. His early entries show an interest in literature and later, when he was around twenty, the diary gives the picture of a well-behaved and well-educated young man with general interests.

---

[1] Nordenskjöld sailed through the Northwest passage and Hedin explored Central Asia

Studies in Uppsala, Paris and Berlin

Like many others in the same situation, Gösta Leffler supported himself by private teaching. Many of his pupils were young sons of the landed gentry.

Studies in Uppsala, Paris and Berlin

Gösta Mittag-Leffler studied in Uppsala at a time when Göran Dillner was the driving force in mathematical teaching and research. The big novelty was the theory of *synectic*, i.e. analytic, functions. But Dillner had his own ideas about analysis and his efforts were, on the whole, rather amateurish. Mittag-Leffler defended a less than remarkable thesis in 1872 on applications of the argument principle. It earned him a position as docent.[2]

His life's turning point came one year later when he received the so-called Byzantine stipend that had been donated by a Swede living in Constantinople. The donor had attached a provident condition: the receiver must promise to live for three years outside Sweden. Inflation had made this promise an adventurous wager, but Mittag-Leffler succeeded in keeping it. At this time he was twentyseven years old, an attractive, well-educated young man with a burning desire to learn mathematics and assert himself in the subject.

On his voyage Mittag-Leffler carried a letter of recommendation from Malmsten, the result of a letter from Hultman to Malmsten, in which the young Leffler was mentioned in warm appreciation. His thesis is praised and it is said that 'Leffler is also a pleasant companion who can talk about almost any subject. His ability to speak French is rather good.' The greater part of the letter describes Leffler's strict education and praises his good character. As proof of this Malmsten tells that when Leffler was teaching the younger brothers of the chief royal equerry, he once refused to take part in a party at which 'many members of the diplomatic corps arrived in elegant coaches ..... each one with his camelia lady'.

Mittag-Leffler came to Paris in October of 1873, equipped with a beautifully bound diary where, until New Year of 1874, he carefully noted his observations and daily life. Darboux seems to have been the first mathematician that he met. Afterwards there were Chasles, Liouville, Briot, Bouquet, and others. His main contact was Hermite whose lectures on elliptic functions he attended. At their first meeting in the family pension where Mittag-Leffler lived, Hermite spoke enthusiastically and with admiration about Riemann, Weierstrass, Fuchs, and other German mathematicians. He regretted that he had not been able to follow their lectures as one in the audience. This was the first time that Mittag-Leffler heard his future idol Weierstrass mentioned by name. The diary does not support later statements by Mittag-Leffler himself that Hermite told him directly to go to Berlin and Weierstrass, but it is clear that Hermite's words were enticing. The fact that Hermite's lectures were difficult to understand was another reason why Mittag-Leffler moved to Berlin in the spring of 1874. Hermite's old-fashioned version of the theory of elliptic functions did not have the clarity that Weierstrass was to give to it.

Dillner had sent his papers to Paris and Hermite discussed them with his young Swedish friend. In his opinion they were weak and contained nothing new. Hermite was hurt on Riemann's behalf when Dillner claimed that his theory replaced Riemann's idea of imagining the values of an algebraic function as forming several layers over the complex plane, the construction which was later called a Riemann surface. Not without satisfaction, Mittag-Leffler wrote to Dillner about Hermite's reaction.

Most of the diary notes the happenings of daily life, a visit to the theatre (Sarah Bernhardt), discussions on political and religious matters, lessons in French and English, a visit to the workers' quarter, reflections on people met and so on. Sometimes

---

[2] Unpaid unversity teacher with prospects of a paid position, limited in time.

there is a sigh over Hermite's lectures, so difficult to understand.

One gets a vivid picture of the diarist, curious and reflecting. The many entries from Swedish social life in Paris show that he is anxious to cultivate his contacts. Now and then the diary speaks of insomnia and stomach ache, afflictions that were to follow Mittag-Leffler all his life.

The diary says nothing about the time in Berlin 1874-1875 but by that time the novelty of living abroad had worn off. With letters of recommendation from Malmsten and Hermite, his vivid interest in mathematics and his charm, Mittag-Leffler had no difficulty in gaining the confidence of Weierstrass and Kronecker.

Mittag-Leffler's stays in Paris and Berlin occurred just after the end of the Franco-Prussian war, 1870-1871, which had left deep resentments. Patriotism and loathing of the victor were strong in French academic circles. Hermite's admiration for German mathematicians was an exception. In Berlin, the capital of Prussia, a blend of patriotism, adoration of the Kaiser, and an arrogant view of lesser nations was the rule. But all this was foreign to Weierstrass.

The liberal positions of Hermite and Weierstrass made a great impression on Mittag-Leffler. In a speech to the Scandinavian mathematical congress in 1925 he says, among other things, the following:

> 'I got a vivid impression of the sharp tension between academic circles in Paris and Berlin during my visits to the two capitals. I was therefore struck by my experience that Hermite and Weierstrass were absolutely free of nationalistic feelings or leanings. Both were born catholics and Hermite, as Cauchy before him, was a warm believer. Weierstrass was interested in, or rather amused by, talking to learned prelates about the finest points of the church dogmas.'

The difference between Hermite and Weierstrass was also the difference between the intuitive view of analytic functions current in the early nineteenth century and a systematic theory with local power series, locally uniform convergence and analytic continuation as its main building blocks. In Weierstrass's papers and lectures, the so-called strict analysis with the notions of upper and lower bounds, upper and lower limits, if necessary in explicit form with the well-known letters epsilon, delta and omega, is an integral part of the text. The precision obtained in this way permitted him among other things to construct the famous example of a continuous function which is nowhere differentiable.

Weierstrass's view of analysis was foreign to Hermite who definitely refused to deal with his colleague's sick functions. For Mittag-Leffler on the other hand, Weierstrass's methods and his view of mathematics became the gospel which was to lead him through his entire life. At the end of the Paris diary there is a note about the well-known example of a variational problem without a solution which Weierstrass had devised in polemic with Riemann's unfounded but later vindicated use of a variational principle to solve Dirichlet's problem.

In 1875, when Mittag-Leffler was in Berlin, he got wind of a vacant professorship in Helsingfors.[3] The professor there, Lorenz Lindelöf (1827-1909), had studied in Germany and worked with minimal surfaces, but now he was going to leave the university for a high administrative post. About this we have Mittag-Leffler's own words (1914).

> '.... In 1875 during my stay in Berlin, I was informed in a roundabout way through Uppsala that the professorship in mathematics in Helsingfors was open. It was L. Lindelöf, an important mathematician and rector of the university, who was going to the National School Board... and I decided to apply for his post. When

---

[3] The Swedish name for Helsinki, the capital of Finland. Between 1809 and 1918 Finland was under Russian domination but institutions and daily life changed little.

I payed a visit to my teacher, the famous mathematician Weierstrass, and told him about my plans, he exclaimed: No, please, do not do that! I have written to the minister of culture and asked him to institute an extraordinary professorship for you here in Berlin and I just got the message that my application has been granted!

I was not blind to the great advantages, mathematically speaking, of such a position compared to the one in Helsingfors. Hardly ever was there such a brilliant collection of distinguished mathematicians: Weierstrass, Kummer, Kronecker, Helmholtz, Kirchhoff, Borchardt etc. But the conditions were not endurable for a foreigner. It was not so long after Germany's victorious war against France and German arrogance was at a high point. Foreigners were treated with haughty condescension, der grosse Kaiser, Bismarck, Moltke etc. were words heard everywhere. It was taken as a matter of course that Holland, Sweden etc. would be members of a German Bund. For those who were not born Germans, it was impossible to live under such conditions. At least this is what I thought. Now things are different, the brilliant victory has not born the fruits that the Germans imagined and they have taken a more realistic view. However, my application to Helsingfors was not sent immediately...'

At the time he is talking about, Mittag-Leffler had only his thesis, two or three insignificant papers and a first sketch (1876) of what was to become Mittag-Leffler's theorem. It is therefore not probable that the Prussian state was ready to offer him an extraordinary professorship. However, it is possible that there had been some loose talk about the matter. In addition to the reasons he mentions, it is almost certain that Mittag-Leffler had another one. With his experience of the high standards of German mathematics, he must have realized at the age of 30 that his mathematical talent was hardly sufficient for a distinguished position in Germany. A position with less competition probably seemed preferable.

The time in Helsingfors

Before Mittag-Leffler, Finland had two professors of mathematics in Helsingfors, N.G. af Schultén (1794-1860) and Lorenz Lindelöf (1827-1908).[4] Af Schultén had made a journey to Paris in 1824 and in Lindelöf did the same in 1860. Lindelöf, who started out as an astronomer, worked from time to time at the large Russian observatory in Pulkova. Both were distinguished mathematicians with good knowledge of the classics. Lindelöf's speciality was the theory of minimal surfaces, parametrized by analytic functions.

In the competition for the professorship in Helsingfors, the applicants submitted original and newly written papers that had to be defended in disputations. Apart from Mittag-Leffler, there were Matths Falk from Sweden and Ernst Bonsdorff, Skari Levänen, and Edvard Melberg from Finland. The most serious competitor was Bornsdorff who had written about elliptic functions and the theory of invariants of the general linear group in two and three variables. Mittag-Leffler arrived in Helsingfors in February of 1876, armed with letters of recommendation from Hermite, Schering, Weierstrass, and Kronecker and with a newly won insight into the theory of elliptic functions. He had prepared a paper, *En metod att komma i besittning af de elliptiska funktionerna*,[5] in Berlin, but his manuscript had disappeared on the way and he had to start again. His paper follows Weierstrass's ideas but to prove the addition theorem, the author uses Abel's theorem. The paper has a

---

[4]For the history of mathematics in Finland see Elfving (1981), where there is also a full account of Mittag-Leffler's time in Helsingfors.
[5]A method to acquire the elliptic functions

quiet, factual and pedagogical tone, but sometimes the author is overwhelmed by the genius of Abel and Weierstrass. It is clear that Mittag-Leffler in four years had come a long way from his thesis. The opponent Lindelöf was appreciative.

Lindelöf ranked Mittag-Leffler first without hesitation, but with this the matter was not finished. After 1865 there was a rule in force which said that teachers of the philosophical faculty had to understand written Finnish. But exceptions could be made for specially qualified candidates. With the votes eleven for and nine against the senate determined that Mittag-Leffler satisfied the stipulations.

In Helsingfors, Mittag-Leffler encountered the language dispute[6] and the friction with the Russian administration. The young professor succeeded well on both fronts. Once, when receiving an examination paper written in Finnish he gave it a pass for mathematics but not for its language (after consulting a language expert). On another occasion, he ejected the police chief from Mittag-Leffler's rightful seat in the theatre, not knowing the identity of the interloper. We may also assume that he was successful in social life. Mittag-Leffler was soon to marry the young Signe af Lindfors, daughter of a distinguished and wealthy Finland Swede.

During his time in Helsingfors Mittag-Leffler lectured on basic analysis and elliptic functions in the spirit of Weierstrass. He had many students many of whom wrote theses, some after studying also in Stockholm where their teacher moved in 1881. Elfving (1981) mentions almost ten such cases, among them Hjalmar Mellin, the originator of the Mellin transform, who became professor at the recently founded Institute of Technology in Helsingfors.

The eventful decade 1880-1890

Sweden's first university was founded in Uppsala in 1477. The southern part of the country was then part of Denmark but after a successful war it became Swedish in 1658. Ten years later Lund university was established in the recently conquered province. For the next two hundred years the Swedish state used these two universities to educate servants of the Lutheran state church and the Judiciary. Living languages and the natural sciences were also taught. In the expansive second part of the nineteenth century, leading circles in Stockholm felt the need for a third university and already in 1865, the town council decided to create a foundation whose purpose was to institute a center for higher education payed for by private means.[7] The plans materialized in 1880 and Mittag-Leffler became the first professor. This was the beginning of the great decade of his life with three big events: *Acta Mathematica* was founded, the new university hired the Russian mathematician Sonya Kovalevski and an international mathematical competition on the highest level was announced.

Acta Mathematica

How this journal came about has been described by Yngve Domar (1982). In his speech (1925) the aging Mittag-Leffler credits his two heroes Hermite and Weierstrass with the idea of starting an international mathematical journal in Sweden. But the initiative came in 1881 from the Norwegian Sophus Lie, then professor in Leipzig, Germany. According to him, the editorial board and the greater part of the material ought to come from Finland and Scandinavia. It was not difficult to

---

[6]Finland was under Swedish domination for several centuries and Swedish used to be the language of the administration. In the nineteenth century, the Finnish majority began asking for a place for their own langauge.
[7]The Swedish name till 1950 was 'Stockholms Högskola'. Because 'Högskola' means literally 'high-school', the institute in question is referred to here by its present name 'Stockholm University'.

compose a reasonable editorial board, but doubts began to be heard about the economic basis. A nordic journal would have difficulties in finding enough subscribers. But Mittag-Leffler began visualizing an international journal. There were several reasons for optimism.

In 1880, Kronecker and Weierstrass had been chosen editors of *Crelle's Journal*, founded in 1828. Because they were enemies and, in Mittag-Leffler's eyes, bad administrators, it was very probable that this potential rival would not be a serious competitor in the market. In addition, by his frequent contacts with Hermite, Mittag-Leffler knew that his correspondent had recently acquired three promising and extraordinarily gifted students, Paul Appell, Emile Picard, and, above all, Henri Poincaré. Mittag-Leffler hastened to start a correspondence with the young genius Poincaré to find out his plans. In this situation he saw the analogy with the importance that Abel once had for *Crelle's Journal* and he wrote to his mentor Malmsten:

'It is my firm belief that we are now at a time similar to the period when the elliptic functions were discovered. Then Abel made a success of Crelle's *German* Journal. In the same way Poincaré will make our *Swedish* journal a success.'

After this flash of insight Mittag-Leffler put everything on one card. He described the situation to Poincaré and asked in the first place for the big manuscript on Fuchsian groups, intended for *Journal de l'Ecole Polytechnique*, and for other manuscripts. In exchange Poincaré was promised fast publication and maximal editorial service. Already before he had Poincaré's answer, Mittag-Leffler revealed his plans to his confidents in the project, Lie and Malmsten, the astronomer Gyldén and the Danish geometer Zeuthen. All of them agreed.

Now an intense campaign started in order to secure subscriptions and contributions from various sources. An appeal to interested parties by Malmsten and Mittag-Leffler stressed that there was a need for an international mathematical journal, important also for Scandinavian mathematics. The appeal was met with success which accelerated when it was heard that the Swedish king, Oscar II, had given 1500 crowns.[8] The total sum collected was 2600 crowns, almost four times a professor's salary and this was considered enough. The international distribution was secured by contracts.

French mathematicians would probably contribute manuscripts, but it was possible that the patriotic and self-sufficing Germans, which so far had been kept ignorant of the project, might prefer their own great journals, *Crelle's Journal* and *Mathematische Annalen*. In order to prepare the German market, Mittag-Leffler got the king's permission for Malmsten to write to selected German mathematicians, Kummer, Kronecker, Schering and Weierstrass with a wish, on behalf of the king, for contributions to the new journal. Through Malmsten's providence, all these had earlier received Swedish orders from the king. In the summer of 1882 Mittag-Leffler embarked on a honeymoon trip through Europe with his Signe. This gave him many opportunities to pay visits to distinguished mathematicians in many countries. The subject of conversation was given: *Acta Mathematica*.

The name had not been without competitors. Already from the beginning there were many suggestions, Disquisitiones Mathematicae, Arkimedes, Analecta Mathematica, Museum Mathematicum. At the end of October, when the journal started to be typeset, the name was *Acta Mathematica Eruditorum*, but in the last minute, the agent in Berlin suggested that the third word should be cancelled. The professor of Latin in Uppsala gave his permission but would have prefered the more precise *Acta Mathematicorum*. On December 12, Mittag-Leffler could deliver the first copy to the journal's first subscriber, King Oscar.

The title page had German and French text in parallel and the editor stressed in his preface that the journal was devoted to *pure* mathematics. For Mittag-Leffler,

---

[8] Approximately 20.000 dollars today.

this meant something special. His encounters with the great mathematicians had convinced him that mathematics was the science of pure thought and the first of all sciences.

From the start there was a good flow of manuscripts. The first volume is graced by Poincaré's paper on Fuchsian groups and other important papers. In the second volume Mittag-Leffler started printing Cantor's controversial papers on the structure of point sets, translated into French. The manuscripts that Weierstrass and Kronecker had promised never appeared. In a letter to Sonya Kovalevski in 1884 (Kochina (1987) p. 30) Kronecker explains his reason: the papers of his antagonist Cantor had been printed in *Acta*. At the same time he takes the opportunity to be ironic about Mittag-Leffler's pure mathematics: the concept ought to fit Kronecker's rather than Cantor's work.

The editorial committee of *Acta Mathematica* was a collection of representative mathematicians and mathematical physicists from Denmark, Norway, Sweden and Finland. They were carefully weighted and discussed, there was some hesitation about Bäcklund in Lund but not about Daug in Uppsala. In the eyes of posterity, the choice should perhaps have been the opposite.

For a couple of years *Acta Mathematica* contained a supplement, *Bibliotheca Mathematica*, edited by the librarian Gustaf Eneström (1852-1923), from the start a proof-reader, but later a bibliophil and historian of mathematics. After an argument with Mittag-Leffler in 1888, Bibliotheca Mathematica vanished from *Acta Mathematica* but Eneström continued to publish his journal until 1914. *Acta Mathematica* still publishes two volumes a year. Up till now (1996) there are 250 volumes. Except for grants during the first years, the journal is self-supporting. The credit for this goes to Mittag-Leffler. Without his powerful marketing effort in the beginning, Acta Mathematica together with most Swedish journals devoted to art and culture would now stand in the long line of applicants for state aid.

Sonya Kovalevski

In Berlin, Mittag-Leffler could hardly avoid hearing about a brilliant student of Weierstrass, Sonya Kovalevski[9] (1850-1891), who had been Weierstrass's private student from 1870 to 1874, written a wellknown paper, and had received the degree of doctor of philosophy in Göttingen. Now she lived with her husband and her little daughter in St. Petersburg.

To be able to go abroad and study she had broken with her family and, eighteen years old, formally married Vladimir Kovalevski (1842-1883) in 1881. The couple lived for one year in Heidelberg and then moved to Berlin.

With Mittag-Leffler as an intermediary Sonya Kovalevski was employed by Stockholm University, first as a docent and later as professor. As a woman mathematician and professor she was a sensation both in Sweden and abroad. There is a large literature about her with a biography in Russian by P. Kochina (1981) and in English by Koblitz (1983). In her biography (1987) of Mittag-Leffler, Kochina has a detailed chapter about the contacts between Sonya Kovalevski and Mittag-Leffler. They are also discussed by Hörmander (1991) through excerpts from Mittag-Leffler's diary.

The first encounter was in St. Petersburg in 1876 when Mittag-Leffler followed Weierstrass's recommendation to see his former student. Sonya Kovalevski's spirituality, charm and mathematical knowledge made a stunning impression on the young Swede. The next encounter followed in 1880 in the same town at the sixth Russian

---

[9] Her full Russian name as married can be transliterated as Sonya Vasilievna Kovalevskaya. When she wrote in German she signed herself Sophie Kowalevsky. In French this was changed to Sophie Kovalevski. I shall write Sonya Kovalevski which was her name in Sweden at the time except that Sonya was written Sonja.

congress for the natural sciences. Both gave talks, Sonya about ellipic integrals of the third kind and Mittag-Leffler about differential equations with doubly periodic coefficients. This was followed by a correspondence that lasted to her death in 1891 (published in Russian in 1954 by Palubarina-Kochina). Mathematics and matters of general and personal interests were the subjects, both correspondents expressing themselves very well. In addition to portraits of the two writers, the letters give a good idea of the mathematical world of Europe, especially of what Weierstrass had been doing. Sometimes one gets the impression that the two friends orbit Weierstrass as two planets orbiting the sun.

Sonya Kovalevski wanted to work as a mathematician, but as a woman she had no prospects in Russia or at the old European universities. She asked about the possibilities in Helsingfors and Mittag-Leffler did some enquiries. He found that she would have been welcome if her nationality had not been Russian. The professors feared that she could perhaps bring with her some young Russian revolutionaries.

In June of 1881, already before Mittag-Leffler was nominated to Stockholm University, he mentions the possibility that Sonya Kovalevski could be employed as a docent or professor, without salary to start with. In earlier letters he had described the attractions of Stockholm and his closest friends, among them the astronomer Gyldén. Sonya Kovalevski answered that she was willing but could foresee difficulties. In her letters she also wrote about her work on the propagation of light in doubly refracting crystals and on the movement of a rigid body around a point.

In 1884, after her husband had died, Sonya Kovalevski was employed in Stockholm for a five-year period. The economy of the university was shaky and her employment was a diplomatic triumph for Mittag-Leffler.

Her work was to begin in September of 1883, but Sonya Kovelevski asked for a postponement since she thought that she needed a couple of months with Weierstrass in Berlin. She arrived in Sweden in November of 1883 and was received with open arms by Mittag-Leffler and his friends. Sonya Kovaelvski's little daughter found a home with the Gyldén family who had a child of the same age. Her work started the first of January and at the end of the month she gave her first lecture.

In Stockholm Sonya Kovalevski was a sensation and was soon to play an important part as subject of conversation and star in the social life of the higher bourgeoisie, marked by a never-ending sequence of dinner parties. She made friends with Mittag-Leffler's sister, the writer Anne Charlotte Edgren-Leffler, and they wrote a play together. The subject was of burning interest: The position of women in men's society. Sonya Kovalevski was also active as a writer. In 1889 she published the novel 'The sisters Rajevski' and posthumously, 'Vera Vorontzoff' 1902.

Mittag-Leffler introduced Sonya Kovalevski into Swedish academic life and advised her about her lectures. Another task was to get the unpredictable Russian to appear in time at the sittings of the senate of the university. Friction was then unavoidable, but it never disturbed their basically friendly connections. When her term of employment came to an end in 1889, Sonya Kovalevski had plans to leave Sweden but her position was renewed and she stayed in Stockholm.

Scientifically, Sonya Kovalevski's time in Stockholm was successful with four mathematical papers (see the next chapter). The most spectacular was a discovery in the theory of the movement of a rigid body around a point. The principles of classical mechanics give a system of ordinary differential equations. It is rather simple, but the movements it describes may be very complicated. The simplest cases, for instance the motion of a top around a slowly varying axis, had been found by Euler and Lagrange. In her last paper, which received the Bourdin prize from the French Academy of Sciences, Kovalevski found yet another so-called integrable motion. Her unexpected death in a pulmonary infection ended her scientific and literary career. Her friends in Stockholm were stunned by the news of her death.

Sonya Kovalevski challenged a male dominated society and this has coloured her posthumous fame, both negatively and positively. For Mittag-Leffler she was a triumph. He had proved himself the leading professor at Stockholm University and capable of a feat which has become history.

The royal prize

The founding of *Acta Mathematica* and the employment of Sonya Kovalevski at Stockholm University were both proofs of Mittag-Leffler's unique energy and initiative. They were soon followed by a third example. King Oscar II was very interested in art and science and Mittag-Leffler used this in a spectacular way. In 1885, *Acta Mathematica* announced a competition about the best mathematical paper within four given fields. The papers were to be anonymous except for mottos for identification and had to be delivered before the first of June 1888. The judges were Hermite, Weierstrass and Mittag-Leffler. The prize, a gold medal with a picture of the king and and 2500 crowns was paid for by the king himself and was to be given by him in a ceremony on his sixtieth birthday, January 21 1889. By putting himself on the same level with Hermite and Weierstrass and appearing as a friend of the king, Mittag-Leffler took the final step to a position in Sweden which he kept all his life: a magnate of mathematics and a member of the king's inner circle.

The competition had four questions. The first one asked for series which describe, for all future times, the motions of bodies which attract each other according to Newton's law and do not collide with each other. The three other questions were mere reflections of their time and dealt with questions about elliptic and hyperelliptic functions which could be asked in connection with new papers by Fuchs, Briot, and Bouquet.

As intended, the competition caused a sensation already from the beginning. Kronecker felt insulted because he was not on the prize committee and he was also critical of the questions. The first question was a generalization of the three body problem, i.e. the motions of the sun, the moon and the earth which had kept astronomers occupied for centuries. This was also the speciality of the astronomer of the Academy of Sciences, Hugo Gyldén. He did not doubt for a moment that his series had solved the problem and hence that he was the obvious choice for the prize.

The prize committee had hoped that Poincaré would participate and their hope came true. His contribution *Sur le problème des trois corps et les équations de la dynamique* was deemed superior and got the prize. The problems began when the paper was to be printed. Mittag-Leffler's assistant, Lars Edvard Phragmén, was also employed as proofreader of Acta Mathematica. Poincaré's paper was probably hastily written and Phragmén found one dubious point after another. In December of 1888 he writes: 'If the author were not what he is, I would not for a moment hesitate to say that he has made a great mistake here.' By correspondence with Poincaré, everything was arranged to the author's satisfaction and the printing started. Then Phragmén made a new check and found new doubtful points. Poincaré admitted his mistakes and thought they could be corrected in an additional note. There is now a gap in the letters dealing with the matter, but on the fifth of December 1888 Mittag-Leffler wrote to those prominent mathematicians who had already received their copies and had asked to get them back. The volume had to be printed again. Three subscribers, among them Gyldén, refused. There are two copies at the Mittag-Leffler Institute and one in the main Swedish library.[10]

Poincaré rewrote his manuscript and it was printed again. He was full of remorse and offered to decline the prize but Mittag-Leffler prevented this. Later and at

---

[10]Poincaré's mistake is analysed in Andersson (1994).

Mittag-Leffler's suggestion, Poincaré paid for a large part of the reprinting.

Poincaré's paper states that the series which have been suggested by Gyldén and the Swedish mathematician Lindstedt are useful but unable to represent the motion for all time. This made Gyldén furious and led to a schism between him and Mittag-Leffler. But it is not surprising that the temperamental Gyldén, who was more of a physicist than a mathematician, was unable to understand Poincaré's subtle arguments. His and Lindstedt's work will be analyzed in Chapter 9.

The royal competition, begun with such fanfare, had three winners, Mittag-Leffler, Poincaré, and Phragmén. His remarkable proofreading feats gave Phragmén a great reputation. It was maybe overstated, but his penetrating critical ability became clear to everyone. In 1892 Phragmén succeeded Sonya Kovalevski as professor of mathematics at Stockholm University.

The time after the age of fifty

In the 1890's honorary doctorates and memberships in foreign academies and learned societies started to shower on Mittag-Leffler. He had become an international celebrity. The reason for this was his mathematics, his position as chief editor of a famous journal, an impressive appearance, and energetic contacts with important mathematicians all over Europe. The experiences of his youth of the rivalry of France and Germany made him an indefatigable proponent of international contacts, and he was one of those who initiated the international congresses of mathematicians, the most famous one in Paris in 1900.

As many of his equals Mittag-Leffler built himself a house in Djursholm, the new garden suburb in Stockholm, and moved there in the early 1890's. Here he lived in state with a secretary and servants, here he managed his professorial duties, his correspondence and business transactions, edited *Acta Mathematica*, and built his library. He ordered portraits of himself and the family from famous Swedish painters from the turn of the century and was frequently interviewed in the better weeklies. His secretary kept track of the many times his activities were reported on in the newspapers. The house was enlarged to make room for the growing library.

Some years after the turn of the century Mittag-Leffler's house got the form it now has. Inside the main door we are met by the master himself in the form of a portrait painted for the fiftieth birthday by the Finnish painter Edelfeldt. Dressed in a big, red gown, the sign of an honorary doctorate from Oxford, the object of celebration holds an issue of *Acta Mathematica* in his right hand.

Mittag-Leffler's view of mathematics as the main expression of pure thought is expressed as a somewhat awkward maxim, carved in stone over the fire-place in the Mittag-Leffer Institute:

Talet är tankens början och slut, Med talet föddes tanken, Utöfver talet når tanken icke.

The literal meaning is 'Number is the beginning and the end of thought. Beyond number thought does not reach.' In Swedish the word 'tal' can also mean 'speech' and, taken in this sense, the motto makes perhaps better sense.

In a small paper (1924-25) with the title *What is number? Infinity? Continuity?* the author has explained himself. He says that man is able to imagine things *in a row* and *in a heap*, he quotes Gauss's words about number theory being the queen of mathematics, and then continues with the integers, the rationals, and the real numbers according to Weierstrass.

Mittag-Leffler's activity, the large household, the long and frequent journeys, consultations with famous doctors about his abdominal trouble, a summer house in Tällberg, etc., had not been possible without a fortune. His marriage brought him money, but this is not a sufficient explanation. Mittag-Leffler's talents as an entrepreneur extended to the world of business and he took good care of his money. The details cannot now be reconstructed, but at least part of Mittag-Leffler's business life is probably recorded in his large personal archive, now with the main Swedish library in Stockholm.

From the beginning Stockholm University was intended to be an institute for free studies without examinations. The economical basis was donations from wealthy citizens and grants from Stockholm city. The city, represented by the chairman of the city council Albert Lindhagen, wanted the university to offer young people of the city educational opportunities equivalent to those of the two state universities, but the council of professors lead by Mittag-Leffler wished to make the university a place only for an elite and to control the nomination of professors. The diminishing flow of money decided the outcome of this fight. In spite of the efforts of Mittag-Leffler and his partisans, the university was finally granted the right to organize examinations and at the beginning of this century the state took over the funding.

Mittag-Leffler's name also appears in a *cause célèbre* immediately before the first world war. In a meeting at the university he accused the liberal prime minister Karl Staaff of negligent handling of secret documents. Staaff then accused Mittag-Leffler of injury to his honour and won in court.

The will and the Gösta and Signe Mittag-Leffler foundation

In 1916, when he was seventy years old, Gösta and Signe Mittag-Leffler bequeathed their fortune, house and library to the Academy of Sciences. The purpose of Gösta and Signe Mittag-Leffler's mathematical foundation was to further the development of mathematics under the leadership of a director, in the beginning Mittag-Leffler himself. Because of the economic crisis after the first world war all the stipulations of the will could not be realized. In the end the money was only sufficient for the upkeep of the house and the management of the library but not for a director's full salary. This problem was solved when one of the professors of mathematics at Stockholm university was paid an allowance to act as director. The situation changed in 1968 through Lennart Carleson's initiative. The foundation, now called the Mittag-Leffler Institute, enlarged its scope, houses for guests were built and the activities of the institute are now paid for by money from the various research agencies.

Mittag-Leffler had a good nose for young talent. In 1909 he had a mathematical visit from the young Hungarian Marcel Riesz. Two years later this visit led to Riesz's employment as editor of a supplementary volume of Acta Mathematica containing a register of the first thirty volumes including biographical data and portraits of the authors. This feature, unique for mathematical journals, is still another expression of *Acta Mathematica*'s character as a grandiose enterprise. Riesz stayed in Sweden and had a successful career as docent at Stockholm University and as a professor (1926-1952) at Lund University.

In the 1982 anniversary volume of *Acta Mathematica*, André Weil tells about a visit to Mittag-Leffler shortly before his death in 1927. True to his instinct, Mittag-Leffler took the young Frenchman's thesis unseen into *Acta*.

Mittag-Leffler was driven by his passion for pure mathematics and his admiration of the great mathematicians. Above all, he was a successful mathematical entrepreneur. His favorite child, *Acta Mathematica*, lives on in full health after more than a hundred years and the institute he founded is very active. Sonya Kovalevski's professorship and the great competition are legendary. In addition, Mittag-Leffler was a successful teacher and his importance for mathematics in Sweden is so great that we may call him the father of Swedish mathematics. The criticism that he met focussed on his somewhat superior manners, his business methods, and the fact that his mathematical work did not quite measure up to the rest.

Mittag-Leffler's students

At the beginning of his stay in Stockholm, Mittag-Leffler acquired two gifted students, Ivar Bendixson (1861-1935) and Lars Edvard Phragmén (1863-1937), who immediately got involved with the editing of *Acta Mathematica*. One result of this was that both of them could make their debuts with papers inspired by the work of Cantor. Their careers and the mathematical activity at Stockholm University, directly or indirectly inspired by Mittag-Leffler, are the object of the Chapters 10 and 11.

BIBLIOGRAPHY

ANDERSSON K.G.
1994. *Poincaré's discovery of homoclinic points*, Archive for History of Exact Sciences 48.2 (1994) 133-147.
CRAMÉR H.
1958. *Lars Edvard Phragmén*, KVA Årsbok 1958.
DOMAR Y.
1982. *On the foundation of Acta Mathematica*, Acta M. 148, 1982, 3-8.
ELFVING G.
1981. *The History of Mathematics in Finland 1828-1918*, Helsinki 1981.
HÖRMANDER L.
1991. *The first woman professor and her male colleague*, Miscellanea Mathematica, Springer 1991.
KOBLITZ A.H.
1983. *A convergence of lives*, Birkhäuser 1983.
KOCHINA P. Ja.
1987. *Gösta Mittag-Leffler*, Nauka Moskva 1987 (Russian).
1981. *Sofia Vasiljevna Kovalevskaja*, Nauka 1981 (Russian).
KOCHINA P. Ja. OZCHIGOVA E. P.
1984. *The exchange of letters between Gösta Mittag-Leffler and Sonja Kovalevski*, Nauka 1984 (Russian).
MITTAG-LEFFLER G.
1914. *Minnen från Finland (Memories from Finland)*, Manus 1914, Mittag-Lefflerstiftelsen.
1925. *Tale af Prof. G. Mittag-Leffler*, 6. Skand. Mat.-Kongr., Köbenhavn 1925.
1925. *Vad är tal? Oändlighet? Kontinuitet? (What is number? Infinity? Continuity?)*, Ark. Mat. Fys. Astr. 18, 1925.

CHAPTER 8

Mittag-Leffler's and Sonya Kovalevski's Mathematical Papers

The two professors at Stockholm University in the 1880's worked in different fields but both bore the imprint of their teacher Karl Weierstrass. Mittag-Leffler devoted himself to the technical part of the theory of analytic functions, Sonya Kovalevski worked with partial differential equations. Mittag-Leffler had a long life and wrote more than a hundred papers, Sonya Kovalevski died young and wrote almost all her papers within a span of ten years.

Gösta Mittag-Leffler

Gösta Mittag-Leffler began as a mathematical writer in 1868 with notices in *Tidskrift för matematik*. The last item from 1927 is a paper in *Acta Mathematica* about Georg Cantor. Mittag-Leffler made important contributions to the theory of analytic functions, but the majority of his papers consists of reviews, speeches and remarks that came with his position as a famous person.

When Mittag-Leffler studied in Uppsala there was a great interest in Cauchy's intriguing theorem that the integral of an analytic function along a closed path is locally independent of the path. Malmsten gave a proof in 1865 using Riemann sums and surface integrals. Mittag-Leffler wanted a direct proof without a surface integral and produced the first such proof in 1873. His paper has an estimate of the difference between the Riemann sums

$$A = \sum F(x_i)(x_{i+1} - x_i), \quad B = \sum F(y_i)(y_{i+1} - y_i)$$

of two complex line integrals of an analytic function F along concentric circles, chosen in such a way that $|x_i|$ and $|y_i|$ are constant and $x_i = (1+\varepsilon)y_i$ for all i. The end result is that the integral of F along a circle is independent of the radius. Mittag-Leffler's rather opaque computations can be rewritten so that $A - B$ is a sum of terms

$$F(x_i)(x_i - y_i) + F(x_i)(x_{i+1} - x_i) + F(x_{i+1})(y_{i+1} - x_{i+1}) + F(y_i)(y_i - y_{i+1})$$

where the sum over i of the first plus the third term vanishes since the index is circular. The sum of the four terms above vanishes when F is constant. If F is uniformly differentiable, its values can be referred to the points y_i with the result that the sum of the four terms has a majorant

$$\varepsilon(\sum_i |y_{i+1} - y_i|(|y_{i+1} - y_i| + \varepsilon)).$$

Hence, for small ε and fine divisions, the difference $A - B$ is arbitrarily small compared to the area between the concentric circles and this shows that the integral of $F(z)dz$ over $|z| = r$ is independent of r.

The insight gotten from this involved proof is easier to obtain from the fact that the integral of $F(z)dz$ over a small rectangle with the sides a, b has the majorant $o(ab)$. Goursat's well-known proof uses this fact at a single point and requires only differentiability without uniformity.

Mittag-Leffler was very insistent about his priority. When the paper was reprinted in the *Oxford Quarterly Journal* in 1925, the author complained that neither his nor Malmsten's proof had been mentioned in *Enzyklopädie der Mathematischen Wissenschaften*.

Mittag-Leffler's theorem about meromorphic functions

In 1876, when Mittag-Leffler was in Berlin, Weierstrass published his theorem about the existence of an entire function with prescribed zeros $z_1, ..., z_n, ...$ where each zero z_n has a given multiplicity m_n. In order for such a function to exist it is necessary that $z_n \to \infty$ when $n \to \infty$. We may assume that no z_n vanishes. Weierstrass's solution is an infinite product of factors

$$E_n(z) = F_n(z)e^{-G_n(z/z_n)},$$

where
$$F_n(z) = (1 - \frac{z}{z_n})^{m_n}$$

has the zero z_n with the multiplicity m_n and the functions $G_n(z/z_n)$ are chosen such that $E_n(z)$ is so close to 1 for bounded z and large n and z_n that $\prod E_n(z)$ converges. To succeed it suffices to choose $G_n(u)$ as a sufficiently large partial sum of the series for $\log(1 - u)$. This gives an estimate

$$\log F_n(z) - G_n(z/z_n) = O(m_n|z/z_n|^{k_n}).$$

If the sequence k_n grows sufficiently fast, the corresponding series converges uniformly in every bounded part of the plane.

During his time in Berlin, Mittag-Leffler tried to find something corresponding to Weierstrass's theorem for meromorphic functions. Such a function $f(z)$ has a countable number of poles z_n tending to infinity. The singular part $M_n(z)$ of the Laurent expansion of $f(z)$ around z_n is a polynomial $M_n(z)$ in $1/(z - z_n)$ such that $f(z) - M_n(z)$ is regular analytic close to z_n.

The problem was to construct a meromorphic function with prescribed poles and singular parts. If the sum

$$\sum M_n(z)$$

converges outside the poles, it is a solution. In the general case the terms have to be modified to give the same result. In his first try in Swedish in *Öfversigt*, Mittag-Leffler used Weierstrass's construction and managed to construct the desired series. He sent the proof to Weierstrass who found the result interesting but the proof so clumsy that he published a better version under the title *Über einen Satz des Herrn Mittag-Leffler*. The correct proof is rather easy. It suffices to approximate every $M_n(z)$ with a polynomial $Q_n(z)$ such that

$$|M_n(z) - Q_n(z)| < 2^{-n} \quad \text{for} \quad |z|/|z_n| < 1/2.$$

Then $\sum(M_n(z) - Q_n(z))$ converges locally uniformly and gives a solution to the problem. It is not necessary to find explicit forms of the polynomials $Q_n(z)$.

Two years after the foundation of *Acta Mathematica*, in 1884, Mittag-Leffler was ready to lay claim to his theorem, extended from the plane to an open part of the plane. The problem is now to construct an analytic function which is meromorphic in an open region Ω where it has prescribed singular parts

$$M_n(z) = P_n(1/(z - z_n))$$

such that z_n tends to the boundary of Ω when $n \to \infty$. After some experimenting Mittag-Leffler found compensating terms $Q_n(z)$ which are polynomials in $1/(z-w_n)$, where w_n is a point on the boundary of Ω which has least distance to z_n. The notions of open set and boundary were then not current and are expressed in terms of notions from Cantor's work on point sets.

With this paper in the fourth volume of his own journal, Mittag-Leffler became the sole proprietor of a theorem that later became widely known and with this he took his place in the circle of internationally known mathematicians.

In simply connected regions, in particular in all of the complex plane, and for singular parts of the form $m_n/(z-z_n)$ with m_n positive integers, Weierstrass's theorem follows from that of Mittag-Leffler and conversely. In fact, if $f(z)$ is meromorphic with the singular parts above, then

$$e^{\int^z f(w)dw}$$

is an entire function with the zeros z_n, each with the multiplicity m_n. Conversely, the logarithmic derivative of Weierstrass's product with these zeros and multiplicities gives Mittag-Lefffler's theorem. The German mathematicians could refer to Mittag-Leffler as Weierstrass's logarithmic derivative (in more than one sense) while in an old Swedish encyclopedia Mittag-Leffler is the author of a theorem 'of which Weierstrass's famous theorem is just a special case.'

Mittag-Leffler's star

Mittag-Leffler's most ambitious work apart from his theorem about meromorphic functions is a series of papers between 1900 and 1905 and a sixth one from 1920. He called them *Notes* and published them under the heading of *On the analytical representation of a single-valued branch of a monogeneous function*.[1] The ideological background comes from Weierstrass. Every power series is a germ of the analytic function it represents. In the *Notes*, the problem is to construct the analytic continuation of a power series outside the circle of convergence using the coefficients of the series. This succeeded in what is known as Mittag-Leffler's star. This is the part of the region of analyticity which can be seen from the center of the power series when the singularities obstruct the view. Later, the poles could be deleted from the obstructions.

The *Notes* contain in all 200 pages of very detailed mathematical text, interspersed by magisterial expositions and references to the classics. It is the father of Swedish mathematics who is speaking.

After three notes, Mittag-Leffler found himself out-distanced by, among others, Emile Borel, but in the fifth note he took a certain revenge. Mittag-Leffler's star has not been as important as the originator thought. But approximation of analytic functions by polynomials or rational functions has become important. In a basic paper by Carl Runge (1895) it is shown that an analytic function can be approximated locally uniformly by rational functions in every open, bounded region where it is defined.

---

[1]The word monogeneous means that the graph of the function is connected.

Summation of series

All of Mittag-Leffler's notes belong to the theory of summation of divergent series, a subject much in vogue in the decade after the turn of the century. In general terms, the problem is the following one.

A summation method is a family of weights

$$c_0(\omega), ..., c_n(\omega), ...$$

depending on a parameter ω with a limiting value ω_0. A real or complex series $\sum a_n$ is then summed by this family when the sums

$$\sum_0^\infty c_n(\omega) a_n$$

converge and tend to a limit when $\omega \to \omega_0$. It is natural to let the weights increase towards 1 when $\omega \to \omega_0$ because then the method gives the correct sum of a convergent series. We can also exchange the terms a_n against the partial sums

$$s_n = \sum_0^n a_k$$

and prescribe that the weights are non-negative with their center of gravity tending to infinity as $\omega \to \omega_0$.

A simple example: the partial sums $s_n = (-1)^n$ are summed to $1/2$ through $c_n(\omega) = \omega^n$ when ω increases to 1. Another example is Borel summation from 1896,

$$\lim e^{-\omega} \sum_0^\infty s_n \omega^n / n!$$

where $\omega \to \infty$. That this method sums the series $1 + z + z^2 + ...$ in the half-plane $\operatorname{Re} z < 1$ follows easily from the formula $s_n = (1 - z^{n+1})/(1 - z)$.

Borel summation of power series

In order to sum a power series by Borel summation (Borel 1901) considered the integral

(1a) $$\frac{1}{2\pi i} \int_S f(u) e^{-\omega(1 - \frac{z}{u})} du/(u - z) = f(z) - e^{-\omega} \sum_0^\infty f_n(z) \omega^{n+1}/(n+1)!$$

where

$$f_n(z) = f(0) + ... + f^{(n)}(0) z^n/n!$$

are the partial sums of the power series. Here the function $f(u)$ is assumed to be analytic on a straight line from 0 to z and S is a path around this line. The right side of (1a) is the sum of the residues of the residues of the left side, the first one for $u = z$ and the second one for $u = 0$. If $\operatorname{Re} z/u < 1$, the left side tends to zero when $\omega \to \infty$. Now z/u has the form $1 + is$ with s real precisely when $u = z(1 + is)^{-1}$ where $1/(1+is)$ runs through the boundary of the circular disk $D : |u - 1/2| \leq 1/2$. Hence, for the right side to tend to zero, it suffices that $f(u)$ is analytic in the disk zD and that S encloses this disk. A passage to the limit gives Borel's formula

(1b) $$f(z) = \lim_{\omega \to \infty} e^{-\omega} \sum_0^\infty f_n(z) \omega^{n+1}/(n+1)!$$

when $f(u)$ is analytic in zD.

A variant of this method starts from the formula

$$\text{(2a)} \quad f(z) = \frac{1}{2\pi i} \int_0^\infty (\int_S e^{-\omega(1-\frac{z}{u})} f(u) du/u) d\omega$$

$$= \int_0^\infty e^{-\omega} (\sum a_n z^n \omega^n / n!) d\omega$$

where $a_n = f^{(n)}(0)/n!$ and the function f and the path S around the origin and zD are as before. The integration with respect to ω in the middle formula gives the left side and the right side is a residue development at $u = 0$ which can be rewritten as

$$\text{(2b)} \quad f(z) = \lim_{\omega \to \infty} (\sum a_n z^n (\int_0^\omega e^{-t} t^n dt/n!)).$$

The set of all zD where f is analytic constitute a star in the domain of analyticity of $f(z)$ where Borel's formula sums the corresponding power series. When the singularities are poles z_0, Borel's star is the intersection of all half-planes $\operatorname{Re} z/z_0 < 1$. As the figure below shows, it is much less than Mittag-Leffler's star.

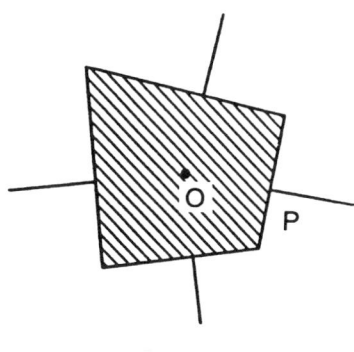

Fig. 1

Borel's star (the interior of the quadrangle) and Mittag-Leffler's star (everything outside the shadow of the four poles P from the center O of the power series).

Mittag-Leffler's notes

In the beginning Mittag-Leffler worked independently of others. The first note is a systematization of analytical continuation by way of overlapping disks. If $n > 1$ is an integer and the centers are

$$0, \ z/n, \ 2z/n, \ \ldots, \ (n-1)z/n$$

and the radii $|z|/n$, then $f(z)$ can be written as a Taylor series with its center at $(n-1)z/n$, the derivatives at the same point can be expressed by Taylor series with centers at $(n-2)z/n$ etc. Every time more and more factorials appear in the denominators and the end formula is

$$\text{(3)} \quad f(z) = \sum_{k_1=0}^\infty \cdots \sum_{k_n=0}^\infty \frac{1}{k!} f^{(|k|)}(0)(z/n)^{|k|}$$

where $k = (k_1, ..., k_n)$, $k!$ is the product of the factorials of the components and the summations are to be performed in succession. For convergence it suffices that the function $f(z)$ is analytic in a neighborhood of the closed set $D_n(z)$ whose largest distance to the interval from zero to $(1-1/n)z$ is $|z/n|$. For large n this region is close to the interval from 0 to z. Hence the formula represents the function $f(z)$ in Mittag-Leffler's star $S(f)$. It is shown that the function is approximated locally uniformly in $S(f)$ by those polynomials which one gets by replacing the upper limits in the summation by, for instance, in order $n^2, n^4, ..., n^{2n}$.

The second note is a detailed continuation of the first one with examples of the necessity that the summations are performed in the correct order. It is also proved that the series (3) do not converge outside the star.

The author likes his summations to have the property that convergence at a point z implies convergence at all points tz with $0 \le t \le 1$. Let us call this property ray convergence. Now the summation (3) holds when the function $f(z)$ is analytic in a neighborhood of the set $C(z,n)$ of disks with the radius $|z/n|$ with center in the points $0,, z/n, ..., (1-n^{-1})z$. Hence this method does not have ray convergence. The author circumvents this difficulty by introducing a star-shaped variant of is $C(z,n)$. This gives a modified series (3) where the coefficients get factors c_k independent of the function f.

In a footnote to the introduction of the third note, the author lists work by Borel, Painlevé, and Phragmén which are connected with the first note. All are about summation of series, not necessarily power series.

The third note gives a completely new method of summation in Mittag-Leffler's star. It employs a family of analytic functions $u \to h(\alpha, u)$, each one mapping the unit disk to a heart-shaped region $\Omega(\alpha)$ with its point at 1 and cusp at $-a(\alpha < 0)$ just below zero. When $\alpha > 0$ tends to zero, the heart-shaped region tends to the interval $0, 1$. The generating family $h(\alpha, u)$ (Mittag-Leffler's generating function) determines the method of summation.

If $f(z)$ is a given function with a star S, the condition that $z\Omega(\alpha) \subset S$ defines another star $S(\alpha) \subset S$. If z belongs to $S(\alpha)$ the functions

$$f(zh_\alpha(u))$$

are analytic in the unit disk with power series

$$\sum G_{\alpha,k}(z)u^k.$$

If the power series of $h_\alpha(u)$ converges to 1 for $u=1$ we get a series

$$f(z) = \sum G_{\alpha,k}(z)$$

which converges when z belongs to the open star $S(\alpha)$. It is also shown that the Taylor coefficients of the functions $G_{\alpha,k}(z)$ are multiples of the corresponding coefficients of $f(z)$. The main part of this note is an explicit construction of a generating family $h_\alpha(u)$ with the desired properties and of the corresponding coefficients of the summation formula. That the members of this family have power series which converge for $u-1$ is proved by Phragmén in a paper which is quoted in extenso.

The note concludes with a reference to Borel's formulas (1b, 2b). Mittag-Leffler also comments on the difficulties of getting a star-shaped convergence region with summations of the type

$$f(z) = \lim g_n(z).$$

On the contrary, this is possible with a double limit

$$f(z) = \lim_\alpha \lim_n \sum_0^n G_{k,\alpha}(z).$$

8 Mittag-Leffler's and Sonya Kovalevski's mathematical papers 91

In the fourth note, Mittag-Leffler combines his generating families with Cauchy's formula and Borel summation. The simplest results are in the second part of the note. For instance, by exchanging the integrand of (1a) by

$$f(zh(\alpha,w))e^{-\omega(1-w/u)}du/(u-w),$$

integrating around a circle which contains w, taking residues as before and putting $w = 1$ in the result, we get a summation of the function $f(z)$ in the part of Mittag-Leffler's star which contains zD_α. The coefficients of this summation are very complicated.

The position after the fourth note was now as follows. All of Mittag-Leffler's summations so far were much more complicated than Borel's summation which, in spite of its attractive simplicity, only worked in part of Mittag-Leffler's star.

The introduction of the fifth note comments on the papers that had simplified and generalized the earlier results by using Cauchy's formula. It is clear that Mittag-Leffler feels himself unjustly criticized. In a footnote he says that he had wanted to be elementary, and therefore it was not possible for these authors to understand the breadth of his thought.

He was referring in the first place of Borel, who in his book (1901) about the summation of divergent series had devoted an entire chapter to Mittag-Leffler. There he points out that this author performs analytical continuation in only one way and that the really interesting area is the summation of divergent series in general. Borel shows that many methods of summation can be obtained very simply by an expansion of $1/(1-z)$ in a series

$$\frac{1}{1-z} = \sum f_n(z)$$

of polynomials $f_n(z)$ which converges when $\operatorname{Re} z > 1$. By Cauchy's theorem,

$$2\pi i f(z) = \int_C f(u)du/(u-z) = \sum \int_C f(u)f_n(z/u)du/u,$$

a suitable choice of the contour C and a computation of the residues on the right side, we can sum the Taylor series in all of Mittag-Leffler's star.

In this situation, Mittag-Leffler thought of improving Borel's construction by replacing the disk C by a smaller region. His solution was to replace the function e^z of (1a) by an entire function $G(z)$ which is exponentially increasing in a small neighborhood V of the real axis but tends to zero at infinity outside. Under these circumstances the integral

$$(4) \qquad \frac{1}{2\pi i}\int_S G(\omega)^{-1}G(\omega z/u)f(u)du/(u-z)$$

has residues as before and tends to zero when $\omega \to \infty$ and z/u avoids V, which in turn means that the region of convergence can be chosen closer to the interval $(0,1)$. The choice of G was

$$E_\alpha(z) = \sum_0^\infty z^n/\Gamma(1+\alpha n), \quad \alpha > 1,$$

later called Mittag-Leffler's E-function. When $\alpha = 1$ it reduces to the exponential function. To see the properties of E_α we shall express it as an integral. Let $L(\varepsilon)$

denote the positively oriented polygonal path $z = 1 - \varepsilon|s| + si$ where s is real and $0 < \varepsilon < 1$. Then, by a classical formula,

$$1/\Gamma(1+n\alpha) = \frac{1}{2\pi i} \int_{L(\varepsilon)} e^u u^{-n\alpha-1} du.$$

Multiplying by z^n, summing and changing the integration variable to u^α, gives

$$E_\alpha(z) = \frac{1}{2\pi i \alpha} \int_{L(\varepsilon)^\alpha} e^{u^{1/\alpha}} du/(u-z)$$

when z is situated to the left of the path of integration $L(\varepsilon)^\alpha$. When z is to the right of this path, we must add a residue $e^{z^{1/\alpha}}/\alpha$ which is the dominating term of $E_\alpha(z)$.

When the path $L(\varepsilon)^\alpha$ is inverted, it bounds a region D_α around the interval $[0,1]$ which tends to this interval when $\alpha \to 0$.

Using the function $E_\alpha(z)$ Mittag-Leffler could prove new summation formulas for the partial sums $f_n(z)$ of a Taylor series, namely

$$f(z) = \lim_{\omega \to \infty} \frac{1}{E_\alpha(\omega)} \sum f_n(z) \omega^{n+1}/\Gamma(1+\alpha(n+1))$$

and

$$f(z) = \lim_{\omega \to \infty} (\sum a_n z^n (\int_0^\omega e^{-t} t^{\alpha n} dt/\Gamma(1+\alpha n)).$$

Both are true when $f(z)$ is analytic at the origin and zD_α belongs to the star of $f(z)$.

The inequality

$$\log |E_\alpha(z)| = O(|z|^{1/\alpha})$$

is the best estimate of $E_\alpha(z)$ in the sector $|\arg z| \leq \pi\alpha/2$, while the function tends exponentially to zero at infinity outside. The smaller the angle, the bigger the growth. In 1905 Phragmén proved an important qualitative version of this connection, namely: an entire function with the above bound cannot tend to zero outside a small sector with an angle less than $\pi\alpha$ without being identically zero. This theorem was generalized and provided with a natural proof in a paper in 1908 by Phragmén and Lindelöf. It states an extension of the maximum principle, named after the two authors (see Chapter 10).

In connection with the E-function, Mittag-Leffler's student Johannes Malmquist (1905) proved that the entire function

$$\sum z^n/\Gamma(1+n/(\log(2+n))^\alpha) \quad 0 < \alpha < 1,$$

tends to zero in every direction except along the real axis. At the same time Wiman (1905 a,b) made a careful study of the behavior at infinity and the zeros of the function $E_\alpha(z)$ for different values of α. He proved among other things that the zeros far away cluster along the lines $\arg z = \pm \alpha \pi/2$ when $\alpha < 1$. When $\alpha = 1$ there are no zeros.

In the sixth note from 1920, fifteen years have passed since the fifth one. In the meantime, von Koch (1903, 1917a,b) had summed the Taylor series also when the function has poles in the region of summation. The extended star was said to be meromorphic. The method is very simple: replace the function $G(\omega z)$ in (4) by a more general $G(\omega, z)$, permit poles inside the region of summation and reason as before. von Koch employed the function

$$G(a,b,z) = z^a e^{-z^b}$$

where a, b tend to infinity. It then turns out that the summation works only with a selection in the limits which depends on the position of the poles. In addition, some weights are negative.

In his sixth note, Mittag-Leffler introduced a new function G. Let

$$F(z) = e^{e^z} e^{-e^{e^z}}$$

and

$$E(z) = \frac{1}{2\pi i} \int_D F(u) du/(u-z)$$

where D is the negatively oriented boundary of the region $\operatorname{Re} u \geq 0, |\operatorname{Im} u| \leq 3\pi/2$ and z is outside. Inside the residue $F(z)$ must be added. Mittag-Leffler puts

$$G(\omega, z) = E(\omega z + b)/E(b)$$

with $b > 0$ so that $E(b) \neq 0$. This function tends zero at infinity, uniformly outside every sector not containing the real axis, but, in contrast to e.g. Malmquist's function, also along the real axis. The old program (4) can now be carried through also when the point z lies between two poles and on a line through the origin.

In the introduction to his sixth note, Mittag-Leffler thanks the circle of students and collaborators which helped him to get the paper into print: Phragmén, Riesz, Malmquist and the new star Torsten Carleman. We can see traces of the advice in a lengthy note where it is proved that the function $E(z)$ does not vanish.

With this note the circle of methods which Mittag-Leffler had created came to a natural end while, at the same time, the larger problem, the summation of divergent series, had lost its urgency. Only the function family $E_\alpha(z)$ remains as interesting examples of entire functions. As has been remarked above, the properties of this family were at the origin of the paper by Phragmén which led to the Phragmén-Lindelöf principle (see Chapter 10).

Sonya Kovalevski

The first fruit of Sonya Kovalevski's studies in Berlin with Weierstrass was the Cauchy-Kovalevski theorem which is the basic existence proof for analytical solutions of analytical differential equations.

Let $f(x)$ be a function of n variables $x = (x_1, ..., x_n)$. The Cauchy data of f of order m on a surface $S : s(x) = 0$ are defined as the retrictions to the surface of the function and its normal derivatives of order $< m$. These Cauchy data are generically independent of each other and determine the derivatives of the function of order $< m$ restricted to the surface. For a general differential equation

$$F(x, u, \partial u, ..., \partial^m u) = 0$$

of order m in several variables $x = (x_1, ..., x_n)$. Cauchy formulated a boundary value problem which is called Cauchy's problem: find a solution of the equation with given Cauchy data of order $< m$ on a given surface. The problem is meaningful only when the equation gives the highest normal derivative as a function of the lower ones. If we introduce coordinates so that S is locally the plane $x_1 = 0$, this means that the equation can be written locally as

$$\partial^m u/\partial x_1^m = G(x, u, \partial u, ..., \partial^m u)$$

where the left side does not appear among the derivatives $\partial^m u$ on the right side. In an equation of this form it is possible to compute by differentiation all the derivatives of a solution u restricted to S when its Cauchy data are given. Kovalevski shows that the formal solution, computed in this way, is analytic at a point x_0 on $x_1 = 0$ if the Cauchy data are analytic and, in addition, the function G is analytic in x i x_0 and in the other variables for values of the derivatives $u, \partial u, ..., \partial^m u$ corresponding to the Cauchy data at the point x_0. The method, which is borrowed from Cauchy, has the purpose to majorize the coefficients of the coefficients of the Taylor series for u.

The theorem extends to systems of differential equations for a number of unknown functions $u_1, ..., u_N$. The condition is that the system can be solved for the highest normal derivatives of each function and that no derivatives of the corresponding right sides are of higher order than these normal derivatives. If this condition does not hold, for instance in case of the heat equation

$$u_t = u_{xx},$$

the theorem fails. The solutions may be analytic in x without being analytic in t.

After an interval of almost ten years, Sonya Kovalevski started to publish again and this time the problem was Cauchy's problem for the propagation of light in doubly refracting crystals where light has two velocities in every direction. In his studies of elasticity theory, the French physicist and mathematician G. Lamé had found a system of equations that fitted this situation and was assumed to express vibrations of the ether. If $u = u(t, x) = (u_1, u_2, u_3)$ is the displacement vector, the vibrations are determined by the following system of equations

$$u_{tt} + \partial \times a \partial \times u = 0, \quad \operatorname{div} u = 0.$$

Here t is time, $\times$ means vector product, $\partial = (\partial_1, \partial_2, \partial_3)$ is differentiation with respect to x and a is a diagonal matrix with the diagonal elements $a_1 > a_2 > a_3 > 0$. The coordinates are chosen so that the velocities of light along the x_1-axis are $\sqrt{a_2}, \sqrt{a_3}$ etc. The system can also be written as

$$u_{tt} + \operatorname{rot} a \operatorname{rot} u = 0, \quad \operatorname{div} u = 0$$

and fits precisely into present-day electro-magnetic theory, i.e., Maxwell's equations. These equations give the system above for $\operatorname{rot} H$ where H is the magnetic vector in a dielectric medium.

An essential feature of Lamé's system is the wave surface, i.e., the locus at time $t = 1$ of a wave which at time $t = 0$ starts as a pulse at the origin. For uniform progagation of light, the wave surface is a sphere. For Lamé's system the wave surface has two parts that touch each other in four pairwise opposite points on the so-called optical axes. Its equation is $\lambda(x) = 1$, where $\lambda(x)$ is homogeneous of degree one and has two values λ_1, λ_2 which are equal on the optical axes.

Euler discovered that the spherical waves

$$v(t, x) = f(t \pm |x|)/|x|,$$

where f is any function, solves the wave equation so that

$$v_{tt} - \Delta v = 4\pi f(t)\delta(x).$$

The functions $t \pm |x|$ for uniform propagation of light become the functions $t \pm \lambda(x)$ in case of the crystal and it was natural for Lamé to try to find an analogue of the spherical waves. His computations went very well and produced credible formulas. Afterwards it turned out that Lamé's formulas do not give a solution (see Gårding (1989)).

In her paper (1886) Sonya Kovalevski tried to solve Cauchy's problem for Lamé's system with Cauchy data for $t = 0$. She followed a manuscript by Weierstrass where he treats an analogous system where the wave surface has two separate parts which do not touch. Kovalevski's computations lead her to a result which, in spite of seemingly great differences, is identical to that of Lamé. Like Lamé she could not see the paradox inherent in the result although at one point it is clear that her solution has vanishing Cauchy data. After her death the mistake was detected by Volterra (1892). He could only interpret the formulas correctly but not solve the problem. It was solved later by Grünwald (1904) and by Nils Zeilon (1919-21). They showed that it is only by means of the Fourier transform that it is possible to analyze the complicated propagation of light in doubly refracting crystals.

One of the greatest triumphs of Newtonian mechanics outside planetary motion are Euler's equations that describe the movements of a rigid body around a fixed point under the influence of gravitation. One example from reality is the oscillating movements of a spinning top.

The position of the top relative to the fixed point and its velocity is described by six variables q, three for position and three for velocity, which obey a system of equations

$$q_t = f(q).$$

An invariant or integral of the system is a function $g(q)$ such that $dg(q) = 0$ under the movement, i.e. $(g_q, f) = 0$. The energy and the overall moment of rotation give two invariants. When there is a third invariant, independent of the others, it may be shown that the system is integrable, i.e. solvable by quadrature. This situation occurs when the center of gravity is fixed (Euler), when the body is rotation symmetric and when the body is attached to a point on the axis (Lagrange, Poisson). Kovalevski (1889) found a new case with a third invariant in a situation that is not so easy to describe as the earlier ones.

Kovalevski's point of departure is rather simple: she requires all variables to be power series in the time t, eventually multiplied by a negative power of t. After rather complicated computations this gives restrictions on the body and its movement which imply the existence of a third invariant and integration by quadrature. As in the earlier cases, the position variables are elliptic functions of time. Kovalevski's paper is a considerable achievement whose value is enhanced by a later proof that the three cases of quadrature are the only possible ones. In present terminology this means that most movements of a rigid boy around a fixed point are chaotic.

BIBLIOGRAPHY

BOREL E.
1896. *Fondements de la théorie des séries sommables*, . J. de Math. Série 5, tome 2, 1896, 103-122.
1899. *Mémoire sur les séries divergentes*, Annales de l'école normale Sér. 3. t.16, 1899.
1901. *Leçons sur les séries divergentes*, Paris 1901.
GRÜNWALD J.
1904. *Über die Ausbreitung der Wellenbewegung in optisch-zweielastischen Medien*, Festschrift Boltzmann 518-527, Leipzig 1904.
GÅRDING L.
1989. *History of the Mathematics of Double Refraction*, Archive for Hist. Exact Sc. vol. 40, 1989, 355-385.
von KOCH H.
1903. *Sur le prolongement d'une série de Taylor*, Acta Math. 27, 1903, 79-114.
1917. a. *Sur le prolongement d'une série de Taylor*, Arkiv Mat. Fys. Astr. bd 12 no 11.
1917. b. *Contributions à la théorie du prolongement d'une fonction analytique*, Arkiv Mat. Fys. Astr. bd 12 no 23.
KOVALEVSKI S.
1875. *Zur Theorie der partiellen Differentialgleichungen*, Jour. f. Mathematik bd 80, 1875, 1-32.

1885. *Über die Brechung des Lichtes in crystallinschen Medien*, Acta Math. 6, 1885, 249-304.
1899. *Mémoire sur un cas particulier de la rotation d'un corps solide autour d'un point fixe*, Acta Math. 12, 1889, 177-232.

LAMÉ G.
1866. *Leçons sur la théorie de l'élasticité des corps solides*, Deuxième édition. Paris 1866.

MALMQUIST J.
1905. *Etude d'une fonction entière*, Acta Math. bd 29, 1905, 203-215.

MITTAG-LEFFLER G.
1873. *Försök till ett nytt bevis för en sats inom de definita integralernas teori*, KVA Öfversigt 1873. Göttinger Nachrichten, no iii, 1873, Quarterly Journal of Pure and Applied Mathematics Vol. L no 3, 1925.
1884. *Sur la représentation analytique des fonctions monogènes uniformes d'une variable indépendante*, Acta Math. 4, 1884, 1-79.
1900. *Sur la représentation analytique d'une branche uniforme d'une fonction monogène. Première Note.*, Acta Math. 23, 1900, 43-62.
1901. a. *Seconde Note.*, Acta Math. 24, 1901, 183-204.
1901. b. *Troisième Note.*, Acta Math. 24, 1901, 205- 244.
1902. *Quatrième Note.*, Acta Math. 26, 1902, 353-392.
1905. *Cinquième Note.*, Acta Math. 29, 1905, 101-181.
1920. *Sixième Note*, Acta Math. 42, 1920, 285-308.

RUNGE C.
1885. *Zur Theorie der analytischen Funktionen*, Acta Math. 6, 1885, 229-244.

VOLTERRA V.
1892. *Sur les vibrations dans les milieux birefringents.*, Acta Math. 8, 1892.

WIMAN A.
1905. a. *Über den Fundamentalsatz in der Theorie der Funktionen $E_\alpha(z)$*, Acta Math. 29, 1905, 191-201.
1905. b. *Über die Nullstellen der Funktionen $E_\alpha(x)$*, Acta Math. 29, 1905, 191-201.

ZEILON N.
1919. -1921.*Sur les équations aux dérivées partielles et le problème optique des milieux birefringents I,II*, Nova Acta Reg. Soc. Sc. Upsal. IV vol. 5.3, 5.4, 1919,1921, 1-59, 1-131.

CHAPTER 9

Astronomy and Optics

Gyldén Lindstedt Bohlin von Zeipel Gullstrand

At the turn of the century Sweden had some good mathematicians and also other scientists who did successful work in the applications of analysis. They were astronomers and mathematical physicists and one ophthalmologist who was so successful that he received a Nobel prize. Their work will be reviewed in this chapter. The first part deals with papers in celestial mechanics by the astronomers Hugo Gyldén (1841-1896), astronomer at the Academy of Science, his successor Karl Bohlin (1860-1939); Anders Lindstedt (1854-1939), professor at the Royal Technological Institute[1] in Stockholm; and Hugo von Zeipel (1873-1959), professor in Uppsala and known for interesting theoretical results. The first three studied the orbits of planets using perturbation theory, a subject indispensable at the time, and then encountered unsuspected difficulties and interesting problems which are of continued interest. The names of all three appear in the subtitle of the second volume of Poincaré's masterpiece *Méthodes Nouvelles de la Mécanique Céleste* (1892-99) and their papers were treated at length. To understand what they did it is necessary to know something about the history and problems of celestial mechanics. This area is treated in a summary way in the first section. The statements made there are proved in a short appendix.

One of the most famous Swedish scientists in the beginning of the century was the physician Allvar Gullstrand (1862-1930), professor of ophthalmology at Uppsala university. He received the Nobel for medicine in 1911 and three years later he was rewarded with a personal professorship in physiology and physical optics. Gullstrand was an optics expert and wrote extensively about the mathematics of the subject. His contributions are the subject of the short, last section of this chapter.

Celestial mechanics

Newton

Newton's main result in his book *Principia Mathematica* (1686) was the formulation of the laws which connect force, acceleration and motion and the discovery of the law of gravitation: two spherical, homogeneous bodies attract each other with a force directed along the straight line which passes through their centers. The force is proportional to the product of the masses of the bodies and inversely proportional to the square of the distance between these points. Among other things it follows from these laws that the planets move in ellipses with the sun in one focus and that the movement depends only on their positions and velocities at an arbitrary time.

Astronomical observations, old and new, verified this unique breakthrough in science. At the same time the problem arose how to account also for the attractions between planets. This is the so-called n-body problem, which is to determine all possible motions $t \to x_k = x_k(t) \in R^3$ of n points with positive masses $m_1, ..., m_n$

---

[1] In Swedish Kungliga Tekniska Högskolan, abbreviated KTH.

when the acceleration for every point-mass equals the gravitational force coming from the others,

$$d^2 x_k(t)/dt^2 = \sum_{i \neq k} m_i(x_i - x_k)/|x_i - x_k|^3.$$

For $n > 2$ this is a nonlinear system of $3n$ equations which has proved itself curiously unmanageable apart from some immediate properties.

The center of gravity $M(t) = \sum m_k x_k(t)/\sum m_k$ satifies the simple equation $d^2 M(t)/dt^2 = 0$ and moves on a straight line. The kinetic energy T and the potential energy U of the system are

$$T = \sum m_k |\dot{x}_k(t)|^2/2, \quad U = -\sum_{i \neq k} m_k m_i \frac{1}{2}|x_k - x_i|^{-1}$$

where $\dot{x}_k(t)$ is the traditional notation for velocity. It follows from the equations of motion that the total energy $T + U$ as well as the angular momentum $V = \sum m_k x_k(t) \times \dot{x}_k(t)$ are invariants under the equations of motion.

When two bodies attract each other according to the laws of motion, we may assume that their center of gravity is fixed. It then appears as a center of attraction. This is planetary motion, analyzed by Newton. Let $t \to x(t)$ be such a motion with the center of attraction at $x = 0$. Because the angular momentum is constant, the movement is planar so that we may assume that $x(t) \in R^2$. Using polar coordinates r, θ we have

$$T + U = (\dot{r}^2 + r^2 \dot{\theta}^2)/2 - r^{-1} = \text{const}, \quad V = r^2 \dot{\theta} = \text{const}.$$

If the movement is not directed towards the center, r is a function of θ, and after an integration and elimination of t one finds that

$$r = p/(1 + e \cos(\theta - \theta_0)),$$

which is the equation of a conic section with one focus at the origin, perimeter p and eccentricity e. If the eccentricity is negative, zero or positive, the conic is a hyperbola, parabola or ellipse. The movement has three independent invariants, namely p, e, θ_0 which describe the orbits. The position and and velocity of a planet at a fixed time also determine the orbit and constitute four invariants, but they are not independent because they vary under time translations.

Lagrange and Hamilton

Newton's laws hold not only for point masses under gravitation. In great generality they determine the movement of bodies under the influence of outer forces. This general situation requires a corresponding formulation of Newtonian mechanics. Such a formulation was found by Lagrange in an abstract fashion that has survived two centuries. Lagrange and his followers, Laplace, Hamilton and Jacobi developed Newtonian mechanics to an extensive theory of great practical importance. In the nineteenth century it was named rational mechanics and competed with mathematics as a major subject at the universities.

To illustrate Lagrange's ideas, imagine a mechanical system of rigid masses connected to each other so as to restrict their movements, but such that the possible motions can be described by a number of position coordinates $x = (x_1, ..., x_n)$ which are free to move in a certain domain. Under a hypothetical motion $x = x(t)$ and velocity $\dot{x}(t)$ the system has a certain kinetic energy $T(x, \dot{x})$ and potential energy U which, in the simplest case, depends only on the position x of the system. The

9 Astronomy and optics

difference $L = T - U$ is called the Lagrangian. Newton's laws in Lagrange's formulation say that the movement of the system in an interval I of time is such that the integral

$$\int_I L(x, \dot{x}) dt$$

assumes an extreme value among all possible movements. The laws of the calculus of variations give Lagrange's equations of motion, n in number,

(1) $$L_x - (d/dt)L_{\dot{x}} = 0,$$

where the index means partial differentiation with respect to the variable indicated. These equations constitute an unsurpassed formalization of Newton's laws which is applicable to a multitude of situations. They express widely applicable laws of motion with the additional advantage of being invariant under changes of coordinates: if the coordinates x in $L(x, \dot{x})$ are replaced by other coordinates y, we get another Lagrangian $L'(y, \dot{y})$ for which the left sides of (1) are linear combinations of the preceding ones.

If the kinetic energy

$$T = (Q\dot{x}, \dot{x})/2$$

is a positive definite quadratic form in the variables $\dot{x}$ then $L_{\dot{x}} = Q\dot{x}$ and, if $y = Q\dot{x}$, Lagrange's equations are

$$dy/dt = -U_x.$$

Then $dx/dt = \dot{x}$ can be rewritten as

$$dx/dt = Q^{-1}y = \partial(Q^{-1}y, y)/2\partial y,$$

where $T = (Q\dot{x}, \dot{x})/2 = (Q^{-1}y, y)/2$. Hence with $H = T + U$, the total energy, we can rewrite Lagrange's equations in Hamiltonian form

(2) $$dx/dt = H_y, \quad dy/dt = -H_x, \quad H = T + U,$$

where now the total energy H or the Hamiltonian has replaced the Lagrangian L.

In complicated situations with many coordinates Lagrange's equations are sometimes not easy to manage and it may be advantageous to write them in Hamiltonian form, which leads to a simpler formalism for coordinate changes (see the appendix about generating function).

When $H(x, y)$ is twice continuously differentiable, Hamilton's equations (2) have a local unique solution $x = x(t, a)$, $y = y(t, b)$ such that $x = a$, $y = b$ for some fixed $t = t_0$ and it is nonconstant when the point a, b is nonsingular, i.e. when $\text{grad}_{a,b} H(a, b) \neq 0$. In connection with mechanics, the solutions are called orbits. Taken together they form a flow in x, y-space given by (2). Locally, the flow is a manifold in (x, y, t)-space of codimension 1.

A function is said to be invariant under the flow (2) when it is constant on all orbits. If it is differentiable, this means that its derivative along the flow vanishes, i.e.

$$f_t + f_x H_y - f_y H_x = 0, \quad (f_x H_y = \sum_1^n f_{x_i} H_{y_i} \text{ etc.}).$$

With $2n$ independent invariants $f_1, ..., f_{2n}$ the orbits are given by the $2n$ equations $f_k = \text{const}$.

The Hamiltonian form (2) is invariant under certain coordinate changes, said to be canonical and leading to canonical variables. These are bijections $x, y \to x', y'$ such that

(3) $$dy \wedge dx = \sum dy_i \wedge dx_i = \sum dy'_i \wedge dx'_i = dy' \wedge dx',$$

where the sum runs from 1 to n. If we define a new Hamiltonian H' by $H'(t, x', y') = H(t, x, y)$, it can be proved that the new equations of motion have the same form as the old ones, namely

(2') $$dx'/dt = H'_{y'}(t, x', y'), \quad dy'/dt = -H'_{x'}(t, x', y'),$$

(see the appendix for a simple proof). The situation is especially advantageous when H is independent of t and it is possible to find canonical variables x', y' such that $H(x', y') = H(y')$ only depends on half of the variables. Then Hamilton's equations (2') say that the variables y' are constant and that the flow is given by

$$y' = \text{const}, \quad x' = H'_{y'} t + \text{const}.$$

In practice, the variables x' are angular and the flow consists in general of quasi-periodic n-dimensional tori parametrized by y'.

Generating function

If $S(x, y')$ is continuously differentiable and $S_{xy'}$ is invertible, then the equation $S_x = y$ is at least locally solvable and gives $y' = y'(x, y)$ as a function of x, y. Simultaneously, the map

$$(x, S_x) \to (S_{y'}, y')$$

is a local bijection. For the differentials dx, dS_x and $dS_{y'}, dy'$ are both equivalent to dx, dy' and one sees that

$$dx \wedge dS_x = dS_{y'} \wedge dy'$$

by expressing the two sides in terms of dx, dy'. A function S above is said to be a generating function. With $S = y'x = \sum y'_k x_k$ one gets the identical map $(x, y') \to (x, y')$. A canonical map close to the identity can always be expressed in terms of a generating function.

An interesting variant of the above is a theorem by Jacobi: if $S(x, y')$ with y' as a parameter satisfies the Hamilton-Jacobi equation

$$H(S_x, x) - G(y') = 0$$

and $S_{xy'}$ is invertible, then $(x, y = S_x) \to (x' = S_{y'}, y')$ is a time independent canonical transformation with the property that it carries the system (2) into a system where the Hamiltonian function $H' = G(y')$ only depends on y'. For

$$y dx - H dt = S_x dx + S_{y'} dy' - S_{y'} dy' - G(y') dt = dS - S_{y'} dy' - G(y') dt$$

so that

$$dy \wedge dx - dH \wedge dt = -dS_{y'} \wedge dy' - dG(y') \wedge dt = dy' \wedge dx' - dG(y') \wedge dt.$$

In the Hamilton-Jacobi equation the choice of the parameters y' of the function $S(t, x, y')$ is entirely free, which may be very useful in concrete cases.

Remark. The classical full Hamilton-Jacobi equation is

$$S_t + H(t, S_x, x) = 0.$$

If we can find a solution $S(t, x, y')$ depending on n parameters y' such that $S_{xy'}$ is invertible, then S is a time dependent generating function such that

$$dy \wedge dx - dH \wedge dt = dy' \wedge dx'$$

where $(x', y') = (S_{y'}, y')$ and the Hamiltonian function is a constant. This means that the functions $y'(x, y, t)$, $x'(x, y, t)$ are invariants of the original system.

Astronomy in the nineteenth century   Poincaré

In the wake of Newton's discovery of the laws that govern the movements of planets, astronomy could even more than before motivate its position as an important subject of teaching at the universities. There was also the need for a reliable almanac stating the most important celestial phenomena. After its foundation in 1736 the Swedish Academy got its revenues from a royal privilege: to be the sole publisher of an almanac sold to the Swedish public at a reasonable price. This monopoly lasted till 1972. To manage the almanac, the academy needed an astronomer of its own, a post which still exists but without a connection to the almanac.

For more than two bodies there is nothing that corresponds to Newton's explicit solutions. It turned out that the next step, the three-body problem, is exceedingly complicated with no explicit solutions except in very special cases. One of them was found by Lagrange who proved that there is a solution where the three bodies are at the corners of an equilateral triangle that rotates with uniform velocity around its center.

The most important task for astronomers was to use the theory of mechanics to compute future positions of the celestial bodies, in the first place the moon, but also the other planets and comets. Very often, the desired movements could be regarded as a two-body problem, but for instance the earth and the moon move in such complicated orbits under mutual attraction and attraction from the sun that new methods became necessary. Astronomers and mathematicians were then left with two methods, successive approximations (also called perturbation theory), and numerical integration. The first of these methods was risky and the second one involved too much work to be practicable on a large scale. The arrival of computers in our time has made the second method almost universal and the first one obsolete, but in the nineteenth century astronomy was a science involved with computing by hand. Many results could be checked by observations which made the computation especially exciting and meaningful. There were plenty of interesting problems, for instance how the large planets influence each other and the small planets.

The father of perturbation theory is Pierre Simon Laplace (1749- 1827) who summed up the development after Newton in five volumes Traité de mécanique céleste 1799- 1825. Perturbation theory could explain many phenomena, for instance how the flattening of the earth at the poles influence the motions of the moon. It is in Laplace's work that the phenomenon of small denominators is first seen. This means that successive approximations by time dependent amplitudes in a periodic motion can give large values for some of them. By computations of this kind Laplace could analyze irregularities in the movements of Jupiter and Saturn which are caused by mutual attraction. The terms with small denominators have a slow influence with long periods and are therefore said to be secular.

One astronomer after Laplace, the Frenchman Leverrier (1811-77), caused a big sensation in 1845 when he used irregularities in the motion of the planet Uranus to localize a newcomer in the solar system, the eighth planet Neptune. That a scientist had broken the holy number seven and, as it was said, had pierced a new planet with his pen, was one of the great events in nineteenth century popular science. Another of his discoveries was to arouse attention much later. Leverrier observed that the perihelion of the planet Mercury, i.e., the point where the planet is closest to the sun, is not fixed as in Newton's theory, but moves slowly around the sun in a direction opposite to that of the planet. This phenomenon was explaned by Einstein's general theory of relativity.

Poincaré and the three-body problem

One of the reasons why the three-body problem was a prize subject in the royal mathematical competition of Oscar II and Mittag-Leffler was the newborn interest in power series in one or several variables. It was hoped that the solution could be found in such series.

Poincaré's prize essay (1889) contained almost 300 pages and has at the end a proof that the solutions to the three-body problem and other equations of mechanics cannot be represented by uniformly convergent series. One of the reasons that has been made very clear through the work by Arnold, e.g. (1963), is that closed periodic solutions are inextricably mixed with quasiperiodic ones. In his conclusion Poincaré says that there are solutions of the three-body problem which lead to collisions between the bodies and that these collisions are elastic and should be part of a general theory. This problem was solved by the Finnish astronomer Sundman (1907,1913) by replacing time by another variable which makes collisions computable (see Siegel (1956)).

In classical mechanics invariant functions are of great importance. Poincaré begins his essay with a new idea that uses the fact that also differential forms may be differentiated in the direction of the flow and that they can be used for something. The volume form,

$$dx_1 \wedge ... \wedge dx_n \wedge dy_1 \wedge .... \wedge dy_n,$$

is such an invariant. Poincaré used it to show the theorem of return. Using measurability in Lebesgue's sense it says the following. If there are orbits of (2) which, at a given point in time, fill a measurable region Ω in x,y-space with a positive volume and stay within Ω for all time afterwards, then almost every orbit passing through an arbitrarily small subregion U with positive volume will return infinitely many times to the same subregion. Apart from this result, Poincaré's great work is full of pioneering analysis of periodic and asymptotic solutions to mechanical systems.

Lindstedt, Bohlin, Gyldén, von Zeipel

After his prize essay Poincaré continued his work and published three volumes (1892-99) entitled *Nouvelles Méthodes de la Mécanique Céleste* where he shows a unique insight into the nature of the problems and the work of astronomers. One his achievements was to clarify the computations of astronomers by writing the equations in Hamiltonian form where he could show the practical usefulness of the Hamilton-Jacobi theorem.

In the second part of *Nouvelles Méthodes* three Swedish astronomers, Anders Lindstedt (1854-1939), Karl Bohlin (1869-1939) and Hugo Gyldén (1841-96) have their own chapters. All of them worked with the limited three-body problem where two bodies move with the same angular velocity in a plane and a third much smaller body moves in the same plane attracted by the two others. The problem is to compute the motions of the third body and to find possible periodic motions.

The work of the three astronomers cannot be reviewed completely here. It will only be briefly touched upon in so far as it is mentioned by Poincaré. The most successful one of the three was Anders Lindstedt. He got his doctorate in Lund in 1877 but moved to Dorpat in Estonia,[2] the present Tartu. After some time with astronomical observations he was nominated professor of applied mathematics. In 1886 he moved to the Royal Technological Institute in Stockholm. A paper (1883) in the annals of the St. Petersburg Academy was the beginning of his reputation. At the time Gyldén was the leading Nordic theoretical astronomer and had found

---

[2] At the time a Russian province.

an application of the theory of elliptic functions to the three-body problem. One of his reductions led to a model equation,

(1) $$x_{tt} + n^2 x = f(x),$$

where $n > 0$ and $f(x)$ is a power series with small coefficients. This equation describes a harmonic oscillation with the frequency n which is perturbed by an exterior position dependent force $f(x)$. The homogeneous equation has the solution $x = r \cos nt$ and a solution of (1) is wanted of the form

$$x = c_0(r) + c_1(r) \cos mt + c_2(r) \cos 2mt + ...$$

with an unknown frequency m.

Let us first note that the equation (1) with $f(x) = \cos \omega t$ has the periodic solution $A \cos \omega t$, $A = 1/(n^2 - \omega^2)$ when $n^2 \neq \omega^2$ but the nonperiodic, nonbounded solution $(t/2n) \sin nt$ in the resonance case $n^2 = \omega^2$.

As Lindstedt (1883) himself writes, already the special case

$$x_{tt} + n^2 x = a + bx$$

shows what to do: rewrite the equation as $x_{tt} + (n^2 - b)x = a$ which, when b is small, gives oscillating solutions with the new frequency.

According to Lindstedt this special case also indicates a method for a perturbation method in the general case where the classical scheme is combined with a simultaneous change of frequency. With ν as an unknown quantity and small, put $\omega^2 = n^2 - \nu$, rewrite (1) as

(2) $$x_{tt} + \omega^2 x = -\nu x + g(x),$$

and insert $x = \cos \omega t$ into the right side. The result is a sum

$$b_0 + (b_1 - \nu) \cos \omega t + b_2 \cos 2\omega t + ...$$

where we get rid of the dangerous second term if we put $\nu = b_1$ which determines a frequency $\omega = \omega_1 = \sqrt{n^2 - \nu}$ and a first approximation

$$x_1 = \sum_{k \neq 1} \frac{b_k \cos k\omega t}{\omega^2 (1 - k^2)}$$

to the solution. In the second approximation the same procedure is repeated with n replaced by ω_1 and with a new $\nu = \nu_1$ and $g(x)$ in (2) replaced by $g(x_1)$, now written as cosine series in integral multiples of $\omega_1 t$.

It is clear that the computational work increases with every step and a proof that the process converges seems out of reach. The reason why Poincaré takes Lindstedt's papers into consideration is that his principle can be generalized to give at least a formal solution of a general perturbation problem.

To give an idea of this, we can imagine that Poincaré rewrote (1) as a Hamiltonian system with the Hamiltonian

$$H(\dot{x}, x) = \frac{\dot{x}^2 + n^2 x^2}{2} - \mu F(x),$$

where $F'(x) = f(x)$ and the factor μ on the right side is small. Here it is possible to introduce polar coordinates R, θ, where θ is an angular variable with the period 2π,

$$\dot{x} = \sqrt{2nR} \sin \theta, \quad x = \sqrt{2R/n} \cos \theta.$$

This gives $d\dot{x} \wedge dx = dR \wedge d\theta$ so that $\dot{x}, x \to R, \theta$ is a canonical map. We get a new Hamiltonian of the type

$$H(R,\theta) = H_0(R) + \mu H_1(R,\theta)$$

($H_0(R) = nR$ in our special case) where the function

$$H_1(R,\theta) = \sum h_k(R) e^{ik\theta}$$

is periodic in θ with the period 2π.

The Hamiltonian determines the flow and therefore it was natural for Poincaré to refine Lindstedt's idea by finding in a first step a canonical map where the new Hamiltonian $H'(R',\theta')$ has the form $F(R') + \mu^2 G(R',\theta')$. This provides a new circular motion which approximates the wanted periodical motion one step better. To this end, we may further imagine that Poincaré looked for a generating function

$$R'\theta + \mu S(R',\theta),$$

where the term

$$S(R',\theta) = \sum_{k \neq 0} S_k(R') e^{ik\theta}$$

is periodic in θ with the period 2π and the coefficients on the right side are to be determined. This gives the conditions

$$R = R' + \mu S_\theta(R',\theta), \quad \theta = \theta' - \mu S_{R'}(R',\theta)$$

where an index denotes differentiation. An insertion gives a new Hamiltonian

$$H'(R',\theta') = H_0(R') + \mu \omega S_\theta + \mu H_1(R',\theta) + O(\mu^2).$$

where $\omega = \partial H_0(R')/\partial R'$. The sum of the two middle terms of the right side is a series

$$\mu(h_0(R') + \sum_{k \neq 0} (\omega S_k(R') + ik h_k(R')))$$

and we can make a new main term $F(R') = H_0(R') + \mu h_0(R')$ and with it a new frequency $\partial F(R')/\partial R'$ and choose the $S_k(R')$ so that the sum above vanishes. This means that

$$S_k(R') = -\frac{i}{\omega k} h_k(R'), \quad (k \neq 0)$$

and we can now write

$$H_0(R) + \mu H_1(R,\theta) = F(R') + \mu^2 F_2(R',\theta') + O(\mu^2).$$

The entire procedure may now be repeated so that the error term becomes $O(\mu^3)$ etc. In addition, the formulas are the same for more variables $R_1, ..., R_n$ and $\theta_1, ..., \theta_n$ provided that we replace $k\theta$ and ωk by sums

$$k\theta = \sum_1^n k_j \theta_j, \quad \omega k = \sum_1^n \omega_j k_j.$$

This is the large scale version of Lindstedt's idea that perturbation schemes ought to imply changes of frequencies. That Poincaré named his procedure after Delaunay and Lindstedt shows that his generosity could equal his insight.

Using Jacobi's theorem and expansion in powers of a small parameter, Poincaré could present the end result of the perturbation scheme as an asymptotic series which at the same time is a a formal solution. This is the famous problem of small denominators which appeared already in Laplace's work.

When the numbers $\omega_1, ..., \omega_n$ above are sufficiently irrational, no linear combination of them with integral coefficients vanishes and the formal solution exists. By combining this situation with a restriction to frequencies for which $|\omega_k||k|^\nu$ has a fixed positive lower bound for some $\nu > 0$, Arnold (1963) could show that there are frequencies for which a modified perturbation scheme converges towards a Hamiltonian which produces quasiperiodic orbits determined by a parameter corresponding to R. On the other hand it turns out that these stable situations always are infinitely close to chaotic movements. This basic result shows the innate complexity of celestial mechanics.

In practice there are important cases, e.g. the restricted three-body problem, where the initial Hamiltonian does not depend on all variables R with the result that certain frequencies ω_k vanish. One gets the same difficulty when the quotient of certain frequencies is rational, but a change of variables tranforms this situation into the previous one. In both cases a modified perturbation scheme is necessary. One of those who in competition with Poincaré introduced new methods for such very essential cases was Karl Bohlin. We cannot here go into details, but a sketch can be given.

Suppose that $R = (R_0, R_1)$ separates into two groups such that $H_0 = H_0(R_0)$ and let $\theta = (\theta_0, \theta_1)$ be the corresponding division of the angular variables. In an unperturbed motion, the variables θ_1 are constant. In a perturbed Hamiltonian $H_0 + \mu H_1$ with a small constant μ the motion is approximately separated into a slow motion represented by the variables R_1, θ_1 and a fast motion described by the remaining variables. A strict proof of this result can be found in Arnold (1963).

Besides Lindstedt and Bohlin Poincaré devotes a chapter also to Hugo Gyldén but the treatment is different. The text is filled with cautious criticism and reasoning with the object of putting things right. The main reason for Poincaré's mistrust is of course Gyldéns mathematics, but it is probable that the origin of some occasional acerbity is Gyldén's opinion that he had been the obvious candidate for the royal prize.

Gyldén was Sweden's first astronomer with mathematical ambitions and he was the founder of a tradition. He was first to use elliptic functions for planetary motion which is a natural thing to do when time is retained as an independent variable. In retrospect Gyldén appears as an expert of fancy approximations. After Gyldén's death Poincaré published a long review (1905) of Gyldén's paper (1892) where the author claims to have replaced the small denominators by harmless terms and to have constructed a perturbation scheme which leads to convergent series. Poincaré's paper is written with some of the holy wrath which can seize a scientist who has to use the scientific method in order to analyze statements which are clearly wrong. In the end nothing remains of Gyldén's paper.

After a solid education in Paris with Poincaré, Hugo von Zeipel wrote a string of papers on the n-body problem. Later he left mechanics for the thermodynamic equilibrium of stars.

In the n-body problem a system of n bodies move according to Newton's law of gravitation. If the orbits cease to be analytic at a finite time t_0, the least distance between the bodies must tend to zero. The state at $t = t_0$ is called a pseudocollision since, for $n > 3$, it has not been shown that a proper collision takes place. In a long, so-far neglected paper (1908, see McGehee 1987) von Zeipel proved that if all bodies do not have unique, finite positions at a pseudocollision, then the largest distance between them tends to infinity. In other papers he made it probable that the long term stability of the planetary system depends on the circumstance that the planets move in planes which are only slightly inclined to each other.

Gullstrand

Allvar Gullstrand (1862-1930) studied medicine in Uppsala, got his degree in 1890 with a thesis about the optics of the eye and was nominated professor of ophthalmology in Uppsala in 1894. Simultaneously with his heavy duties as a doctor he published long papers (1904, 1905, 1906, 1915) in geometrical optics that led to a collaboration with the Carl Zeiss foundation in Jena and construction of pioneering ophthalmological instruments and improved glasses.

It seems very probable that Gullstrand had studied mathematics on his own. He was the first medical doctor who understood the mathematics of geometrical optics and could use its results.

Briefly, geometrical optics is a theory of geodesics in a metric $n(x)|dx|$ where $n(x)$ is the index of refraction and $|dx|$ is the euclidean metric element in three-dimensional space. In this connection the geodesic distance is also called the optical distance and the geodesics are rays of light. In a system of lenses, $n(x)$ is locally constant outside a system B of refracting lens surfaces. In such systems the theory is usually restricted to two-dimensional ray bundles, e.g. the bundle of rays from a point a. If the points a, y, x lie on a ray with $y \in B$ and x outside, the optical distance $S(a,x)$ between a and x is a stationary function of $y \in B$ which gives the familiar law of refraction. Outside of B, the rays are orthogonal to the wave surfaces where $S(a,x)$ is constant. The point a is mapped to points where some rays meet and the wave surface is singular. In general this happens in two surfaces formed by the two curvature centers of the wave surfaces. The circle points of the wave surface where the curvature centers coincide are of special interest. It is only rarely that the point a is mapped at least approximately to a single point, a focus, but in elementary optics which is restricted to spherical lenses and rays which are close to a central, unrefracted ray, these points are of great practical interest.

The situation just described is analyzed in detail in Gullstrand's massive and almost unreadable papers. Among the problems he discusses are the focussing of the ray bundle and how the form of the lenses influences it. Gullstrand has a theory of the first, second, and third order according as the ray bundle is differentiated once, twice, or three times. In the long paper (1904) the normal bundle is investigated near a surface given up to order four. In a paper (1905), published in *Acta Mathematica*, Gullstrand computes the curvature lines on a surface close a circle point where the surface has a contact of order four with the osculating sphere. In general there are two curvature directions in every point on a surface where the curvature lines constitute a regular, rectangular net. Close to a singular point the net deteriorates in various ways and at a circular point all directions are curvature directions. Gullstrand finds that a great number of configurations are possible at a circular point, e.g. several incoming curvature lines and curvature lines that encircle the point.

Gullstrand's aim from the beginning was to improve the methods and instruments of ophthalmology. But it is clear that he found pleasure and satisfaction in his mathematical work. They bear witness to a remarkable capacity for work and a strong feeling for the subject. Gullstrand worked in a closed theory, and that may be the reason why his papers have left few traces in the optical literature and when this did happen it was often possible to simplify them.

Gullstrand received the Nobel prize for medicine in 1911. It is probable that the very rare combination of pioneering ophthalmologist, mathematician and consultant to a prestigious industry was decisive for the outcome. The choice did not escape criticism. The astronomer Bohlin, himself no stranger to successful computations, gave his comment on the event with a very unofficial: one should not get the Nobel prize just for computing.

Gullstrand's prestige and his uncontested ability to master new areas of knowledge made him member of the committee for the Nobel prize in physics in 1911 and he stayed on till 1929. In 1921 the Academy of Science gave the physics prize to Albert Einstein for his fundamental work in Brownian motion and quantum theory from the beginning of the century. Relativity theory, on the contrary, both the special and general met with scepticism. The philosopher Hägerström wrote about logical gaps in Einstein's papers and in 1921 Gullstrand wrote a critical paper, perhaps supposed to explain the Nobel committee's hesitation. Gullstrand examines Einstein's equations for the movements of bodies and his explanation of the movement of the perihelion of the planet Mercury. His criticism was that Einstein's equations for this phenomenon permit several solutions. As remarked by Oseen in a subsequent paper, Gullstrand did not observe that the choice of coordinates must be adjusted to the observer and that this gives the correct result.

The thoroughness and strict discipline which one can see in Gullstrand's mathematical work was also evident in other areas. Oseen's obituary of Gullstrand (1935) contains the following passage.

> He was a man who made himself felt wherever he was. His position in the faculty and senate in Uppsala became a legend. ... One peculiarity which soon became obvious was that the propositions from the medical faculty were always unanimous. There was never the least inconsistency which could give rise to criticism or discussion. And these propositions were always granted by the senate. The medical members of the senate could testify that this unbroken unity was the work of Gullstrand. Still more obvious was Gullstrand's position in the senate. At every session there were a great number of cases to be decided on. If every case had initiated a discussion, the session would be unbearably long. The modus vivendi that developed under these circumstances was characterized by the condition that only one person spoke and that person was Gullstrand. In this situation Gullstrand could mark the end of a session by thanking the other members of the senate that they had not taken up the senate's time by speaking ... .

Appendix

The section on celestial mechanics would be incomplete without some precise statements and simple proofs. In connection with Hamiltonian flows, consider time dependent flows

$$(1) \qquad dx/dt = F(t,x,y), \quad dy/dt = G(t,x,y)$$

and let $a = (a_1(t,x,y), ..., a_{2n}(t,x,y))$ be a complete set of invariants, for instance $x(0), y(0)$ as functions of t, y, x.

LEMMA. *Let $H(x,y,t)$ be a Hamiltonian. The differential form*

$$dy \wedge dx - dH \wedge dt$$

is invariant under (1), *i.e., a form depending only on the invariants a and their differentials, if and only if*

$$(2) \qquad F = H_y, \quad G = -H_x.$$

PROOF: Under the flow (1) expressed in terms of t, a, we have $dx = X + Fdt$, $dy = Y + Gdt$ where $X = x_a da$, $Y = y_a da$. An insertion into the differential form gives

$$Y \wedge X + Gdt \wedge X + Y \wedge Fdt - (H_x X + H_y Y) \wedge dt.$$

For this form to have the shape

$$Z = \sum A_{jk}(a,t) da_j \wedge da_k,$$

it is necessary and sufficient that (2) holds, for X and Y are independent linear combinations of the differentials of the invariants. Now the form is closed, $dZ = 0$, which means that its coefficients do not depend on t. This ends the proof.

A canonical map $(y, x) \to (y', x')$ with $H'(y', x') = H(y, x)$ gives the equality

$$dy' \wedge dx' - dH' \wedge dt = dy \wedge dx - dH \wedge dt.$$

Here both sides are invariant under the Hamiltonian flow

$$dx'/dt = H'_{y'}, \quad dy'/dt = -H'_{x'}$$

which, if written in the variables x, y, t, has the general form (1). According to the Lemma, this form is canonical with the Hamiltonian $H(t, x, y)$. What has been said above about Jacobi's theorem also follows from the Lemma.

BIBLIOGRAPHY

ARNOLD V.I.
1963. *Small denominators and problems of stability of motion in classical and celestial mechanics.*, Russian Mathematical Surveys 18.6, 1963, 85-191.

GULLSTRAND A.
1904. *Allgemeine Theorie der monochromatischen Aberrationen und ihre nächste Ergebnisse für die Ophtalmologie*, Nova Acta Soc. Sc. Ups. Ser. 3 vol 20, 1904, 1-204.
1905. *Zur Kenntniss der Kreispunkte*, Acta Math. 29, 1905, 59-100.
1906. *Die reelle optische Abbildung*, KVA Handl. bd 41, 1906, no 3.
1915. *Das allgemeine optische Abbildungssystem*, KVA Handl. Bd 55, 1915, no 1.
1921. *Allgemeine Lösung des statischen Einkörperproblems in det Einsteinschen Gravitationstheorie*, Arkiv 16, 1921, no 8.

GYLDÉN H.
1892. *Nouvelles recherches sur les séries employées dans les théories des planètes*, Stockholm 1892.

LINDSTEDT A.
1883. *Beitrag zur Integration der Differentialgleichungen der Störungstheorie*, Mém. Ac. Impér. St. Petersbourg, série VII. tome XXXI, 1883, no 4.

McGEHEE R.
1986. *Von Zeipel's theorem on singularities of celestial mechanics*, Exp. Math. 4, 1986, 335-0345.

OSEEN W.
1921. *Über das allgemeine statische kugelsymmetrische Gravitationsfeld in der Einsteinschen Theorie*, Arkiv 16, 1921, no 9.
1935. *Allvar Gullstrand*, KVA levnadsteckningar bd 6 1935.

POINCARÉ H.
1889. *Sur le problème des trois corps et les équations de la dynamiqe*, Acta Math. 13, 1889, 5-270.
1905. *Sur la méthode horistique de Gyldén*, Acta Math. 29, 1905, 235-272.
1892. -99. *Méthodes nouvelles de la mécanique céleste* I,II,III, Paris 1892-99.

SIEGEL C.L.
1956. *Vorlesungen über Himmelsmechanik*, Springer 1956.

SUNDMAN K.F.
1907. *Recherches sur le problème des trois corps*, Acta Soc. Sci. Fenn. 34, 1907, no 6.
1913. *Mémoire sur le problème des trois corps*, Acta Math. 36, 1913, 105-179.

von ZEIPEL H.
1908. *Sur les singularités du probléme des n corps*, Arkiv 4, 1908, no 32.
1959. *Hugo von Zeipel*, Pop. Astr. Tidskr 40, 1959.

CHAPTER 10

Stockholm University 1880-1920 I

Bendixson   Phragmén   von Koch   Fredholm

Mittag-Leffler's three decades in Stockholm were a golden age for mathematics in Sweden. The small and recently founded Stockholm University became a mathematical center where the incitements came from the exciting papers of *Acta Mathematica* and Mittag-Leffler's lectures and international contacts. In this environment his students received impulses which could carry them directly to the front of research. This happened to his earliest and very successful students, Ivar Bendixson, Edvard Phragmén and Helge von Koch. All of them were established mathematicians at the age of twenty-five. Phragmén ended up in insurance, the others stayed with mathematics.

All three became known through distinguished contributions, but none of them achieved the fame of Ivar Fredholm. His contribution was a general theory of linear integral equations. It made the author famous and the integral equations a subject of intense interest for two decades. Fredholm was more interested in mathematical physics than in mathematics and he was more independent of Mittag-Leffler than the others in the same circle.

Apart from these stars, many mathematicians worked at Stockholm University. They are the object of a review in the next chapter of the subjects and problems which grew there in the fertile soil of analytic functions.

Bendixson

Ivar Otto Bendixson (1861-1935) came from a middle class family in Stockholm, he was professor at KTH[1] from 1900 to 1905 and at Stockholm University from 1905 to 1927 where he served as rector from 1911 to 1927.

Bendixson wrote his first mathematical paper at the age of twenty-two. The origin was the translations into French of the papers by Georg Cantor which Mittag-Leffler had inserted in the second volume of *Acta Mathematica* in order to make them known in France.

A point inside or outside of a point set M is said to be a limit point of M if every neighborhood of the point contains infinitely many points of M. The set of limit points was called by Cantor the derived point set of M, denoted by M'. When $M' = M$, M is said to perfect, in modern terminology closed without isolated points.

Iterations of the map $M \to M'$ were denoted by $M^{(2)}, M^{(3)}, ..., M^{(\gamma)}, ...$ where γ may be transfinite. A set R such that $R^{(\gamma)}$ is empty for some γ was called reducible by Cantor. He stated the theorem that M' is the union of a perfect and a reducible set without common points.

In a letter to Cantor, printed in *Acta Mathematica* Volume 2, Bendixson (1883) gave a counterexample. In each of the infinitely many open intervals of the comple-

---
[1] The Royal Technological Institute in Stockholm.

ment of a Cantor set C he constructed point sequences M, chosen so that the limit points of set M' are the end points of the intervals. In this way every point of C (but no other point) belongs to M'' which has to be identical to C. Since $C' = C$, this is a counterexample to Cantor's theorem.

Bendixson also found the correct result, the Cantor-Bendixson theorem, which, in modern terminology says that every closed set is the disjoint union of a perfect set and a discrete (and hence countable) set. In this theorem the transfinite part has disappeared. Twenty years after Bendixson the Finnish mathematician Lindelöf found the simplest proof by introducing the concept of condensation point of M. Such a point has the property that each of its neighborhoods contains more than countably many elements of M. It is clear that the condensation points constitute a perfect subset C which may be empty and that every point of $M\setminus C$ has a neighborhood with at most countably many points in M. From this it follows easily that M is countable.

Bendixson returned a couple of times to point set theory, but his main work was in analysis. The only exception is a paper in the theory of equations which he published three times, in 1891, 1893 and, ten years later, in a paper of *Acta Mathematica* that Mittag-Leffler devoted to the centenary of Abel's birth.

A fundamental fact of Galois theory which, except for the terminology, was known to Abel and Galois is the following: if only the identity of the Galois group leaves certain zeros invariant, then the others are rational functions of these. Example: if the group is commutative all zeros are rational functions of one of them. For if an element a in the group leaves a zero x invariant, then $ax = x$ whence $abx = bx$ for all b in the group so that a is the identity.

In the paper (1891) Bendixson put himself to the task of determining all irreducible polynomials $f(x)$ with rational coefficients with the property that all zeros are rational functions of one of them. He wanted to avoid group theory, referred to as 'Galois's substitutions', and only use Abel's methods. Nevertheless, his result, which deals with arrangements of the zeros when all are rational functions of one of them, is best presented using group theory in the following way.

Let $A = (a, b, ...)$ be a finite set whose elements we may call zeros, and let the group $G = (x, y, ...)$ with a unit e permute the zeros transitively and have the property that only the unit leaves a zero invariant. Since no two different elements of G can map a zero into another one, it follows that G and A have the same number of elements. The basic observation now runs as follows: if H is a subgroup of G and $x \in G$ is outside H, then x permutes the orbits of H in A. Proof: if $y, z \in H$ and $ya = xza$ for some $a \in A$ then $xz = y$ so that $x \in H$. It also follows that $xHx^{-1} = H$ and H and x generate the group HX where X is the group generated by x. Given a subgroup H_0 generated by an element x_0 we may successively adjoin elements $x_1, ..., x_m$ until the group G is filled out and get the following structure: the elements of the group can be written uniquely as products

$$x(k_0, ..., k_m) = x_0^{k_0}......x_m^{k_m}$$

where k_j is determined modulo the order m_j of x_j. If the zeros are ordered accordingly, $a(k_0, ..., k_m) = x(k_0, ..., k_m)a$ for some fixed $a \in A$, we have

$$x_j a(k_0, ..., k_m) = a(k'_0, ..., k_j + 1, k_{j+1}, ..., k_m).$$

with $x_j x_0^{k_0} x_j^{-1} = x_0^{k'_0}$ and so on. This is Bendixson's result. Conversely, under this law, only the identity can leave $a(k_0, ..., k_m)$ invariant. When the group G is commutative, every x_j operates on $a(k_0, ..., k_m)$ so that k_j is increased by 1.

Bendixson's paper illustrates the insecurity that many mathematicians far into our century felt when confronted with the notion of a group. On the other hand, the same volume of *Acta Mathematica* has a paper by Anders Wiman (1863-1957), then a specialist on the groups which appear in algebraic geometry. Wiman uses Galois theory to deduce Abel's results about the form of the zeros of solvable equations of prime degree (which are not in Bendixson's class).

Like his competitor Phragmén, Bendixson was a frequent contributor to *Öfversigt*, where he wrote about various subjects in analysis. In the end he devoted himself to the study of the behavior of solutions (orbits) of systems of two ordinary real differential equations

(1) $$dx/dt = f(x,y), \quad dy/dt = g(x,y)$$

near a singular point, i.e. a point where both f and g vanish. The problem had been initiated by Poincaré (1881). When f, g are polynomials he had classified the singular points according to the behavior of the orbits close to the point in question which could be the point at infinity.

Poincaré often used complex variables x and y, but with Bendixson we shall restrict ourselves to the real case. In order to orient the reader we shall now analyze the singular points for the elementary system

$$f(x,y) = ax + by, \quad g(x,y) = cx + dy$$

where f, g are linearly independent. What happens depends entirely on the eigenvalues λ_1, λ_2 of the matrix

$$\begin{matrix} a & b \\ c & d \end{matrix}.$$

The main cases are as follows. When the eigenvalues are real and separate there are independent linear forms ξ, η of x, y such that the orbits are

$$\xi = c_1 e^{\lambda_1 t}, \quad \eta = c_2 e^{\lambda_2 t}, \quad c_1^2 + c_2^2 > 0.$$

If $\lambda_1 \lambda_2 > 0$ all orbits tend to the origin as $t \to +\infty$ or $t \to -\infty$, but if $\lambda_1, \lambda_2 < 0$ all orbits except the axes $\xi = 0, \eta = 0$ pass by the origin. Finally, if an eigenvalue is complex, $= \lambda + i\mu$ there are ξ, η such that the orbits are

$$\xi + i\eta = ce^{(\lambda + i\mu)t}.$$

When $\lambda \neq 0$, they spiral for increasing t towards the origin or infinity, otherwise they are circles around the origin.

Bendixson's contribution was a qualitative, topological theory for the behavior of orbits close to a singular point. It requires only that f and g are continuously differentiable. He could also extend the results of Poincaré when f, g are analytical and their linear parts are degenerate. He summarized his work in a long paper (1901) in *Acta Mathematica*. This is Bendixson's main work and his contribution to what was later to be called the Poincaré-Bendixson's theory. The greater part of this theory has later entered into the textbooks, and in this way it has shared the fate of large part of mathematics, namely to melt into a growing mathematical general knowledge where the originators are not mentioned.

With a contemporary term, the function $(x,y) \to (f(x,y), g(x,y))$ is a vector field and the solutions $z(t) = (x(t), y(t))$ of (1) are, as above, the corresponding orbits. The basic properties of orbits is that they have continuous tangent, that there is precisely one orbit through every nonsingular point, that the orbits are oriented by the parameter t and that an orbit never comes arbitrarily close to a singular point when when t is bounded.

Bendixson begins by showing that orbits with the above properties exist and depend continuously on an initial point when f, g are continuously differentiable. The global behavior of orbits requires some definitions. An orbit is closed when it is compact and hence a topological circle and an orbit is a loop when it tends to the same critical point for $t =\to \pm\infty$.

Bendixson's main result is a list of theorems with a strong visual content, for instance

1. An orbit in a bounded region that does not come close to any singular point is either closed or a spiral tending to a closed orbit.

2. An orbit in an open bounded set with just one singular point P either tends to P or else it is closed or it is a spiral tending to a closed orbit or to a loop attached to P.

3. Inside a closed orbit there is at least one singular point.

4. An orbit that meets a singularity free region bounded by a loop is also a loop.

5. If an isolated singular point is not circumscribed by arbitrarily small close orbits, there is at least one orbit which ends or starts at the point.

6. If there is an orbit that spirals towards a singular point, then every orbit sufficiently close to the point is also a spiral.

A sector at an isolated point P is defined as a region bounded close to P by two orbits which begin or end in P with determined tangents. A sector is said to be ingoing if it only meets orbits which end in P. An outgoing sector is defined analogously. A sector is said to be open if it only meets orbits which do not meet P and it is said to be closed if it only meets orbits which begin and end in P, i.e. loops contained in each other.

According to Poincaré, the index of a vector field (f, g) at a singular point is the number of turns in the positive direction that it makes when moved in the positive direction around the point. This index can be computed when a small disk around the singular point can be divided into finitely many sectors which are ingoing, outgoing, open or closed. Bendixson proved that

$$s - g = 2(i - 1)$$

where s is the number of closed sectors and g the number of open sectors and i is the index of the vector field.

In the second part of his paper Bendixson tries to construct a complete theory for the system (1) close to an isolated singular point and succeeds completely when f and g are analytic and all orbits have tangents at the point. A series of changes of variables leads to a system of the type

(2) $$dx/dt = x^m, \quad dy/dt = ax + by + h(x, y)$$

where $m > 0$ is an integer, $a^2 + b^2 > 0$ and $h = O(x^2 + y^2)$. Such a system has no loops and hence the number of branches is $2(1 - i)$.

In Bendixson's papers many orbits are computed explicitly in many cases and especially for the system (2) and its extensions.

Phragmén

Lars Edvard Phragmén was born in 1863 in Örebro where his father taught at the gymnasium. He studied first in Uppsala and then in Stockholm. He became a docent in 1890 and in 1892 he succeeded Sonya Kovalevski as professor. In 1903 he left the professorship to be chief inspector of insurance. He died in 1937.

After a short time in Uppsala, Phragmén left for Stockholm and its new professor Mittag-Leffler. The appreciation became mutual. At the age of twenty-one

Phragmén had papers printed in the proceedings of the Academy, and one year later *Acta Mathematica* published his new proof of the Cantor-Bendixson theorem.

Soon this precocious student became Mittag-Leffler's indispensable helper and proof-reader in the work with Acta Mathematica. It was in this capacity that Phragmén had to examine Poincaré's prize essay (see the previous chapter) and discovered that it contained a serious error.

After the death of Sonya Kovalevski, Mittag-Leffler had the satisfaction that two of his students were capable of succeeding her. Phragmén and Bendixson did perhaps not have any weighty papers to their credit, but they wrote diligently in *Öfversigt*, and both had shown unusual critical gifts which for Bendixson had resulted in a very early paper of his own. After careful deliberation Mittag-Leffler found that Phragmén was the best candidate. Nobody at Stockholm University could contradict him.

Already in the beginning of his time as a professor Phragmén wrote some papers about the practical application of the proportional election method and later he was employed as an expert by the committee that wrote the final proposal to the government. He was also interested in numerical computation and was very successful in insurance.

In mathematics Phragmén is remembered for a result with prospects of surviving for a very long time in its original formulation, namely the Phragmén-Lindelöf principle.

According to Liouville every bounded entire function must be a constant. In (1904) Phragmén extended this result to a theorem about entire functions which have certain highest permitted growth in a given sector and are bounded outside. The inspiration came from Mittag-Leffler's E_α-function (see p. 87), in the simplest case the exponential function e^z.

Phragmén's theorem says that an entire function $F(z)$ with the bound

$$(1) \qquad |F(z)| \leq C_1 e^{C_2 |z|^c}, \quad 0 < c < 1, \quad C_1, C_2 \quad \text{constants},$$

in a proper sector $S : -\pi/2 < a \leq \arg z \leq b < \pi/2$ of the right half-plane is a constant if it is bounded in the complementary sector. In his proof Phragmén considers the integral

$$H(z) = \int_0^\infty F(zw) e^{-w} dw$$

which is an entire function of z since the integral converges uniformly. It is clear that $H(z)$ is bounded outside the sector S. If $z \in S$ and $\operatorname{Re} w > 0$ then

$$|F(zw)e^{-w}| \leq e^{C|zw|^c - \operatorname{Re} w}$$

for some constant C. This means that one can use Cauchy's theorem to change the path of integration from the real, positive axis to any ray L from the origin into the open right half-plane. Hence we may move the ray zL into the complementary sector which shows that that $H(z)$ is bounded also when $z \in S$. Hence $H(z)$ is bounded and must be a constant. Since the Taylor coefficients of $H(z)$ are multiples of the corresponding coefficients of $F(z)$ it follows that also $F(z)$ is constant.

Phragmén felt that it was not necessary to require that the function $F(z)$ is bounded in the entire complementary sector. With the bounds of $|F(z)|$ in the sector S he could show that $F(z) = O((\log |z|^N)$ in S for every $N > 1$ if $F(z)$ is bounded in small sectors adjoining S.

The sector S can be increased to the entire right half-plane H and the proof can be improved if one uses the function e^z in a more efficient way than in Phragmén's integral. It suffices to consider the functions

$$F_\varepsilon(z) = e^{-\varepsilon z^a} F(z), \quad \varepsilon > 0, \quad 1 > a > c.$$

If $F(z)$ satisfies (1) in H and $|F(z)| \leq C$ on the imaginary axis, the same is true of the function $F_\varepsilon(z)$, which, in addition, tends to zero at infinity in H. Hence the parts of H where $F_\varepsilon(z) > C$ are bounded and the classical maximum principle shows that they are empty. Hence $|F_\varepsilon(z)| \leq C$ for every $z \in H$ so that $|F(z)| \leq C$ in H since ε can be chosen arbitrarily small.

Remark. More refined methods (see p. 193) prove the following theorem: if $F(z)$ is analytic in the right half-plane, if $\log|F(z)| = o(|z|)$ and if $|F(z)| \leq C$ on the imaginary axis, then the function has this bound everywhere. The case $F(z) = e^z$ shows that $o(|z|)$ cannot be replaced by $O(|z|)$.

Phragmén's result and the extensions above are special cases of the Phragmén-Lindelöf principle, which extends the maximum principle for the absolute values of analytic functions and is stated in the well-known paper (1908) which Phragmén wrote together with the Finnish mathematician Ernst Lindelöf. One might ask who got the idea or had the insight to find the right formulation and the right proof of Phragmén's problem. Since Phragmén is not the sole author of his most important paper it is easy to guess that the decisive idea came from Lindelöf, an opinion that his compatriots could express in private.

The reasoning above with $F_\varepsilon(z)$ extended to an arbitrary open set opens the paper by Phragmén and Lindelöf. Suppose that a function $F(z)$ is analytic and single-valued in an open set Ω and that $\limsup |F(z)| \leq C$ at the boundary except at a single point P. Suppose also that there is an analytic function $\omega(z) \neq 0$ in Ω of absolute value ≤ 1 which is very small at P and such that $F_\varepsilon(z) = \omega(z)^\varepsilon F(z)$ is $\leq C$ at the entire boundary for every $\varepsilon > 0$. Then $|F(z)| \leq C$ in Ω. One can draw the same conclusion if $F_\varepsilon \leq C$ at the boundary of an open neighborhood $\Omega(\varepsilon)$ of P which tends to P when ε tends to zero.

When $f(z)$ is analytic it is well known that $\log|f(z)|$ is subharmonic, i.e., it satisfies the maximum principle relative to harmonic functions. The part played by the auxiliary function ω becomes clearer if we pass to the function $V(z) = \log|F(z)|/C$ which is ≤ 0 at the boundary of Ω except at P and to the auxiliary function $U(z) = \log 1/\omega(z)$ which is harmonic and ≥ 0 in Ω and large at P. The Phragmén-Lindelöf principle now says for instance that $V \leq 0$ in Ω if $V - \varepsilon U \leq 0$ in a neighborhood of P for every $\varepsilon > 0$. As above it suffices that $V - \varepsilon U \leq 0$ at the boundary of a neighborhood of P which tends to P when ε tends to zero. An extreme but not always optimal auxiliary function vanishes at the boundary of Ω outside of P and tends to infinity at P. When Ω is the right half-plane and $P = \infty$, all such functions are positive multiples of x.

In the Phragmén-Lindelöf principle the function $V = \log|F(z)|$ is provided with harmonic majorants in various regions. This way of estimating the size of analytic functions has been called the principle of the harmonic majorant. As a method it appeared for the first time explicitly in a paper by Carleman (see Chapter 12) and was employed with success by mathematicians in Sweden and Finland. A general theory of subharmonic functions was first formulated by Fredric Riesz (1926).

Remark. In the paper by Phragmén and Lindelöf there is also a kind of interpolation between growth in various directions which facilitates the application of their principle. We know for instance from the main theorem that if $f(z)$ is analytic in the first quadrant, $\leq C$ at the boundary and $\log|f(z)| = O(|z|^a)$, $a < 2$, then $|f(z)| \leq C$ in the entire quadrant. If

$$|f(x)| \leq e^{ax}, \quad |f(iy)| \leq e^{-by}$$

when $x > 0$, $y > 0$, this theorem, applied to $e^{-(a+ib)z}f(z)$, gives an interpolation

$$|f(z)| \leq C e^{-by+ax}$$

between the two bounds.

With this interpolation we can now prove a well-known theorem by F. Carlson (1914): if $|f(z)| \leq e^{a|z|}$ in the right half-plane and $|f(iy)| \leq e^{-b|y|}$ for some $b > 0$ then $f = 0$. For then we have the inequality above with $|y|$, and, if we exchange $f(z)$ for $e^{cz}f(z)$, $c > 0$, the result is that

$$|e^{cz}f(z)| \leq Ce^{-b|y|+(a+c)x}.$$

This function is $\leq C$ when $b|y| = (a+c)x$ and hence also, by the Phragmén-Lindelöf principle, when $b|y| \leq (a+c)x$. Since c may be arbitrarily large, this means that $|f(z)| \leq C$ in the right half-plane. Hence $e^{cz}f(z)$ with Re $z > 0$ has a fixed bound independent of c so that $f = 0$.

von Koch

The third of Mittag-Leffler's successful students, Helge von Koch (1870-1924), was a professor at KTH from 1905 and succeeded Mittag-Leffler at Stockholm University in 1911. He made a brilliant debut at the age of twnety-two with a thesis which combined two areas of current interest, the theory of differential equations with analytic coefficients and the theory of infinite determinants.

Infinite determinants

A paper by the American astronomer Hill gave Poincaré (1885) the impulse to investigate determinants of infinite matrices of the form $E + A$, where E is the unit matrix and $A = (a_{jk})$, $j, k = 1, 2, \ldots$ has the property that

$$\sum_{j,k} |a_{jk}| < \infty.$$

Poincaré had all diagonal elements equal to zero, but, as pointed out afterwards by von Koch, Poincaré's proofs work without this restriction.

By assumption, the product

$$P = \prod(1 + |a_{jk}|)$$

converges. Let $C = E+A$ be the matrix with elements $c_{jk} = \delta_{jk}+a_{jk}$. If $C_n = (c_{jk})$ when $j, k \leq n$ and P_n is the corresponding partial product of P, then $|\det C_n| \leq P_n$ because all terms on the left side occur as absolute values on the right side. Some reflection also shows that

$$|\det C_m - \det C_n| \leq P_m - P_n$$

when $m > n$, because all terms of the left side occur in P_m, but none in P_n. Hence it follows from (1) that $\det C = \det(E + A)$ is well defined.

When the determinant exists as above, it is a linear function of any row or column and hence subdeterminants exist. If C_{jk} is the algebraic complement of c_{kj} in $\det C$, we have

$$\sum_j c_{ij} C_{jk} = \delta_{ik} \det C$$

from which it follows that the equation $Cx = 0$, where $x = (x_j)^t$ is a column matrix with bounded elements, only has the solution $x = 0$ when $\det C \neq 0$. A little more work shows that the basic theorems about linear systems of equations extend to the equations $Cx = y$ with bounded x and y. All these theorems are consequences of present general knowledge, namely the theory of compact operators, first sytematized by F. Riesz (1918). For the matrix $A = (a_{jk})$ induces a compact linear operator on the space of bounded sequences when $\sum |a_{jk}|$ converges.

Differential equations at singularities

The basic paper about singularities of solutions of analytical differential equations was published in 1866 by L. Fuchs, later professor in Berlin. It deals with ordinary, homogeneous differential equations of order n, $L(D)f = 0$, where $D = d/dz$ and

$$L(D) = D^n + c_1(z)D^{n-1} + ... + c_n(z)$$

with analytical coeffcients. Fuchs proved the theorem, later to become classical, that the solutions form a complex linear space of dimension n, that the equation has a unique local solution with given derivatives of order $< n$ at a point and that every solution can be analytically continued in every region where the coefficients remain analytic. This theorem is of great interest when the coefficients are single-valued and analytic in a ring-shaped region Ω. Let S denote analytical continuation one revolution in the positive direction. If $f_1, ..., f_n$ form a basis for the solutions at a point in the region, the functions $g_k(z) = Sf_k(z)$ are linear combinations of the basis elements,

(1) $$Sf_j = \sum S_{jk} f_k.$$

Remark. If the coefficients are analytic and single-valued outside finitely many points B in the extended complex plane C^*, there are as many linear maps S which, under composition, give a representation of the homotopy group of $C^* \backslash B$ into the group of complex, invertible matrices of the type $n \times n$. The problem to reconstruct the differential equations from this object was put by Riemann. Fuchs had a simple special case and the general problem was to be the object of many papers.

Let us restrict ourselves to the local theory with analytic continuation S in a ring-shaped region surrounding $z = 0$. A change of basis in the space of solutions leaves invariant the characteristic polynomial

$$\det(S - tE) = \det(S_{jk} - t\delta_{jk}).$$

For this reason its coefficients are called *invariants* of the operator $L(D)$ at the point $z = 0$. It was once considered an important problem to compute them from the coefficients of the operator and solutions were given by, among others, Poincaré and Mittag-Leffler.

If the characteristic polynomial of S has a zero t, it also has an eigenvector $u(z)$ such that

$$Su(z) = tu(z).$$

Since the function

$$z^\lambda, \quad \lambda = \log t / 2\pi i$$

behaves in the same way under S, i.e. $Sz^\lambda = tz^\lambda$, the function

(2) $$v = z^{-\lambda} u(z)$$

is single-valued in the region. Note that the number λ is determined by t modulo the integers.[2]

---

[2] With the aid of Jordan's normal form (which was first constructed in this context) it is possible to show that every solution w of $L(D)w = 0$ is a sum of single-valued functions multiplied by fractional powers of z and entire powers of $\log z$.

When the coefficients of $L(D)$ are analytic outside $z = 0$ Fuchs posed the following problem: When are all solutions meromorphic at the origin? The answer is simple: All $z^k c_k(z)$ must be analytic, i.e. $c_k(z)$ has at most a pole of order k. We say that the operator $L(D)$ and the equation $Lu = 0$ are of Fuchsian type at $z = 0$. If $z^n L(D)$ is written in the form

$$L^*(D) = \sum_0^n c_m(z) z^m D^m, \quad c_m(z) = \sum c_{mj} z^j, \quad c_n(z) = 1,$$

the operator is of Fuchsian type exactly when $c_{mj} = 0$ for $j < 0$. It is clear that $L(D)$ and $L^*(D)$ have the same solutions. We observe that

$$z^m D^m z^\lambda = \varphi_m(\lambda) z^\lambda, \quad \varphi_m(\lambda) = \lambda(\lambda - 1)...(\lambda - m + 1),$$

i.e. $z^m D^m$ reproduces every power of z modulo a constant factor. Hence

$$L^* z^\lambda v(z) = z^\lambda w(z)$$

where v, w are single-valued functions. If $v(z) = \sum v_j z^j$ and $w(z) = \sum w_k z^k$, a small computation shows that

(3) $$w_k = \sum a_{ki}(\lambda) v_i$$

where

$$a_{ki}(\lambda) = \sum_{m,\, j+i=k} a_{mj} \varphi_m(i + \lambda),$$

with diagonal elements

$$a_{kk}(\lambda) = \sum_m a_{m0} \varphi_m(\lambda + k) = \varphi(k + \lambda).$$

In the Fuchsian case, the matrix $(a_{ik}(\lambda))$ is relevant only when $i, k \geq 0$ and, in addition, triangular with vanishing elements for $i > k$. In this case, and if $\varphi(\lambda + k) = 0$ only when $k = 0$, we can solve the equation $L^* z^\lambda v(z) = 0$ by successively computing the coefficients $v_0 = 1, v_1, ...$. This turns out to give an analytic function $v(z)$.

von Koch's papers

von Koch's thesis (1891,1892), published when the author was twnety-two years old, is a fantastically mature work and was treated at length in *Jahrbuch*. The author posed himself the problem of solving the system (3) in the general case by using Poincaré's theory of infinite determinants.

The first task was to extend Poincaré's paper to matrices $A = (\delta_{jk} + a_{jk})$ where $-\infty < j, k < \infty$ and $\sum |a_{jk}| < \infty$ and this did not cause any trouble.

Otherwise the computations required that D^{n-1} appears in $L(D)$ with the coefficient zero. This could be achieved by a transformation $D \to e^{-g} D e^g$ with a suitable function g. Under this assumption it is easy to show that

$$\sum_{i \neq k} |a_{ik}(\lambda)|/(1 + |k|^n) < \infty,$$

which opens the way to Poincaré's theory of determinants. With $\varphi(\lambda)$ as above, let $\varphi(\lambda) = \prod(\lambda - \lambda_m)$ with n factors. With $h_0(\lambda) = 1$ and the non-zero factors

$$h_k(\lambda) = k^{-n} \exp(-\sum(\lambda - \lambda_m)/k),$$

which approximate $1/\varphi(\lambda + k)$, it turns out that the matrix with elements

$$h_k(\lambda)a_{ki}(\lambda) = \delta_{ik} + b_{ki}(\lambda)$$

has the property that $\sum |b_{ik}(\lambda)| < \infty$ for all λ.

Poincaré's theory can now be applied to the system (3), written in the form

$$v_j + \sum b_{jk}(\lambda)v_k = 0$$

where $D(\lambda) = \det(E + B(\lambda))$ is an entire function. If $D(\mu) = 0$, then $L^*(D)u = 0$ has a solution $z^\mu v(z)$ and conversely. But $z^{\mu+k}z^{-k}v(z)$ is the same solution for all integers k. Hence $D(\mu + k) = 0$ for all integers k, which proves that $D(\lambda)$ is divisible by $\sin \pi(\lambda - \mu)$.

According to the theory of Fuchs, $D(\lambda)$ has at most n zeros

$$\lambda_1, ..., \lambda_n$$

which are incongruent modulo the integers. In this case von Koch shows that

$$D(\lambda) = \prod \pi^{-1} \sin \pi(\lambda - \lambda_j)$$

and that

$$\det(S - e^{2\pi i\lambda}E) = (-1)^{n(n-1)/2}(2\pi i)^n e^{n\pi i\lambda}D(\lambda)$$

where $S = (S_{jk})$. This formula gives an elegant solution to the problem of expressing the invariants of the system in terms of its coefficients.

After the thesis von Koch wrote many papers, among others some on infinite determinants, for instance (1901), but the subject did not have many possibilities for extension and growth and present interest is nil. What von Koch wrote about the distribution of primes will be touched upon in the next chapter. He was the first to extend Mittag-Leffler's summation to points in the shadow of the poles of the function in question. All this is now history, but in the 1880's, the rising interest in chaos made von Koch's snowflake a well-known object. It is the simplest and most beautiful of the many constructions of nowhere differentiable curves which were in vogue at the turn of the century.

Weierstrass's analytical construction of a continuous function which is nowhere differentiable challenges geometrical intuition which is not capable to visualize such an object. In a paper (1906) von Koch constructed a continuous plane curve without tangents by repeated application of the following construction: an interval is divided into three equal parts, on the middle part an isosceles triangle is constructed whose base is then erased. If we start with the three sides of an isosceles triangle and see to it that all small triangles point outward, we get in the limit von Koch's snowflake or star. The figure on the next page, which is taken from von Koch's paper, illustrates parts of the construction.

The star is a continuous curve which does not have a tangent in any point. The proof of this statement occupies the main part of von Koch's paper. The author takes great pains to show that there are two nowhere differentiable functions $f(t), g(t)$ with $-1 \leq t \leq 1$ such that the curve is represented by the formula

$$x = f(t), \quad y = g(t).$$

The proof cannot be given here. It depends on the following fact: the broken curve to the right in the picture below is seen from anyone of its points under an angle with a positive lower bound which is not diminished in the next step of the construction.

At the end of his paper von Koch follows Weierstrass in constructing a continuous function which is nowhere differentiable. The construction is geometric and analogous to the preceding ones. It can also be expressed analytically. Let $f(x) = f_0(x) = 0$ when $x < 0$, $x = 0, 1/3, 2/3, 1$, $x > 1$ and $f(x) = 1$ when $x = 1/2$ with linear interpolation between the points thus specified and put

$$f_n(x) = \sum_{k=-\infty}^{\infty} 6^{-n} f(6^n(x-k)).$$

Then $f(x) = \sum_0^\infty f_n(x)$ is a function of period 1 which has the desired property. We see here the analogy with Weierstrass's example.

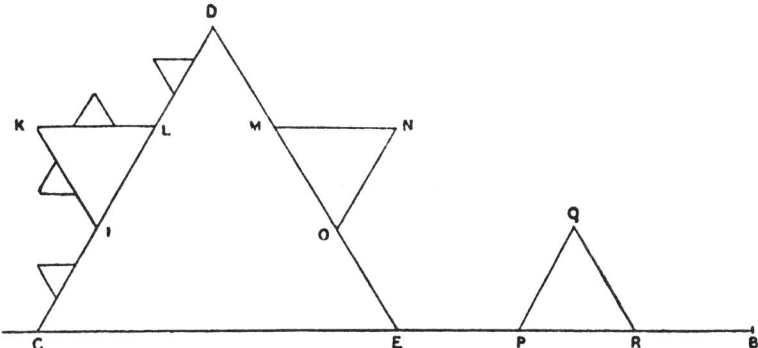

Fredholm

Erik Ivar Fredholm (1866 - 1927) was born in Stockholm. After matriculation he studied for one year at the Royal Technological Institute but continued in Uppsala where he received the degree of doctor of philosophy in 1898. The same year he was nominated docent of mathematical physics at Stockholm University. In contrast to Bendixson and von Koch, Fredholm matured late. The subject of his carefully written papers is mathematical physics but his main influence lies in mathematics.

Fredholm's début is a short note from 1890 in Swedish with the misleading title *On a special class of singular lines*. The word singular refers to curves which are natural boundaries of analytic functions, but the main interest of the note is the construction of an analytic function in the unit disk which is infinitely differentiable in the closed disk but cannot be continued analytically outside. Mittag-Leffler had constructed functions which cannot be continued outside the unit disk by concentrating poles at the boundary and Weierstrass gave the example

$$g(z) = \sum_0^\infty a^k z^{b^k}$$

where $|a| < 1$ and $b > 1$ is an integer. We see that the disk of convergence is $|z| < 1$ and that the function is continuous when $|z| \leq 1$. But it cannot be continued outside the circle. For there must be a singular point z_0 at the boundary and, if

$$\theta^{b^k} = 1,$$

then $g(z)-g(\theta z)$ is a polynomial of degree $< b^k$. Hence θz_0 must be a singular point and these points are dense in the unit circle. In this example by Weierstrass and later extensions by Hadamard, the power series has exponentially large gaps and the function it represents is continuous in the closed disk, but the formal derivatives diverge. Fredholm's function

$$f(a,z) = \sum a^n z^{n^2}, \quad |a| < 1,$$

has less extensive gaps and is infinitely differentiable on the unit circle, but still cannot be continued across the boundary. This property is said to depend on the fact that the function

$$\varphi(v,t) = f(e^{-v}, e^t)$$

solves the heat equation $\varphi_t = \varphi_{vv}$ when $\operatorname{Re} v > 0$ and $\operatorname{Re} t \leq 0$. If the function $t \to \varphi(v_0, t)$ were analytic at a point v_0, t_0 with $\operatorname{Re} t_0 = 0$, it is clear that the heat equation would give the estimate

$$\partial_v^{2k}\varphi(v,t) = \sum n^{2k} e^{-vn} e^{tn^2} = O(C^k k!)$$

when $v = v_0$, $t = t_0$. By quoting at this point a paper by Sonya Kovalevski, Fredholm draws the conclusion that this means that $v \to \varphi(v,t)$ is an entire analytic function of v when t is close to t_0. This result contradicts the fact that every function $a \to f(a,z)$ with $|z| = 1$ has at least one singular point when $|a| = 1$. We can remark that this conclusion is too hasty since the inequality above only concerns the even derivatives, but since all terms of this series are positive and decrease when v increases, a summation by parts shows that the inequality holds uniformly as $v \geq v_0$ whence the corresponding inequalities for the odd derivatives.

Fredholm's example falls under a theorem by Fabry (see Landau (1916)) which says that a power series with gaps

$$\sum_k a_{k_n} z^{k_n}$$

and the radius of convergence 1 has the unit circle as a natural boundary when $n = o(k_n)$, a condition largely satisfied in Fredholm's case where $k_n = n^2$.

Fredholm's little note was a sensation at least in Sweden and so amazed Mittag-Leffler that he communicated the result to Poincaré.

The reason why the problem arose was the fixation on power series. A simple way of constructing similar examples is to use Poisson's formula to construct a harmonic function $u(z)$ in the unit disk with a given boundary function $u(e^{i\theta})$ which is infinitely differentiable without being analytic. If v is a conjugate function, $u+iv$ is analytic in the unit disk and infinitely differentiable in the closure but without an analytic continuation across the boundary.

Fredholm's construction of fundamental solutions

Fredholm did not use the heat equation by chance. Almost everything he did is inspired by problems in mathematical physics. In order to solve an equilibrium problem in elasticity theory, Fredholm needed to construct fundamental solutions of his differential operators.

Let $P(\xi)$ be a polynomial in $\xi = (\xi_1, ..., \xi_n)$ and $P(\partial)$ with $\partial = \partial/\partial x$ the corresponding partial differential operator. In modern terminology a fundamental solution of $P(\partial)$ is a distribution $E(x)$ such that

(4) $$P(\partial)E(x) = \delta(x).$$

If $P(\partial)U(x) = 0$ then $E(x) + U(x)$ is also a fundamental solution.

This definition also covers cases where P and E are quadratic matrices. When $P = \Delta$ is Laplace's operator in three variables, $E = -1/4\pi|x|$ is a fundamental solution and this is one of the few isolated examples before Fredholm's paper (1900 a). In the first part Fredholm constructs fundamental solutions $F(x)$ of the differential operators $P(\partial)$ in three variables, which are homogeneous and elliptic in the sense that the polynomial $P(\xi) = P(\xi_1, \xi_2, \xi_3)$ is not zero when $\xi \neq 0$ is real.

When the polynomial P has degree m, the operator $P(\partial)$ is homogeneous of degree $-m$ and the right side of (4) homogeneous of degree $= -n$. Hence it is natural to look for fundamental solutions $E(x)$ which are homogeneous of degree $m - n$.

As a start Fredholm considered solutions u of $Pu = 0$ outside $x = 0$, homogeneous of degree -1, which are linear combinations of solutions of the type

$$1/(x,\xi), \quad P(\xi) = 0$$

where $(x,\xi) = \sum x_k \xi_k$. He constructed fundamental solutions as Abelian integrals

$$(5) \qquad F(x) = \int_\gamma Q(\xi) d\xi_1 / P_2(\xi)(x,\xi)$$

where $\xi_3 = 1$, $Q(\xi)$ is a polynomial of degree $m - 2$, $P_2 = \partial P(\xi)/\partial \xi_2$ and γ is a loop in the algebraic, complex curve $P(\xi) = 0$ which encloses points where $(x,\xi) = 0$. It is immediately clear that the integral is an analytic function of x when x is real and $\neq 0$. Fredholm proved that if γ is properly oriented and encloses the points common to $P(\xi) = 0$ and $(x,\xi) = 0$ where $\text{Im}(x,\xi) < 0$, then $F(x) = Q(\partial)G(x)$, where $G(x) \neq 0$ is a constant times a fundamental solution of $P(\partial)$. Later Fredholm (1908) gave an improved form of the somewhat awkward and unsymmetrical formula (5).

The great goal, however, was to construct a fundamental solution of the elasticity operator, a matrix $L(\partial) = (L_{jk}(\partial))$ with $j, k = 1, 2, 3$ whose elements are homogeneous differential operators of degree two with constant coefficients and whose determinant $P(\partial) = \det(L_{jk}(\partial))$ is an elliptic operator. Outside the action of an exterior force, the displacement vector $u(x)$ of an elastic medium in equilibrium satisfies the equation $L(\partial)u(x) = 0$.

Since the elements of $L(\xi)^{-1}$ have the form $(M_{jk}(\xi)/P(\xi))$ where P has degree six and every M_{jk} has degree four, the desired fundamental solution is obtained in the form of a matrix $E(x)$ of order three whose elements are given by (5) with P as above and Q/P replaced by M_{jk}/P. When u is a vector, the physical significance of $E(x)u$ is the equilibrium state of an elastic medium which covers all of space and is subject to a force $u\delta(x)$ supported at $x = 0$.

Fredholm, whose main interest was theoretical physics, never extended his results to several variables. A sketch of the present general theory of fundamental solutions of constant coefficient homogeneous differential operators is given in an appendix at the end of this chapter. It also covers hyperbolic equations.

Integral equations

Dirichlet's problem is the problem of finding a harmonic function in a given open set with given values at the boundary. Before the turn of the century it had been treated by Riemann, H.A. Schwartz, Carl Neumann, and Poincaré using different methods. Fredholm's contribution was to see the problem as an integral equation and at the same time give a general theory of such equations.

When $f(x)$ is continuous and $f(x)/(1 + x^2)$ is absolutely integrable, it is well known that the formula

$$u(x,y) = \frac{1}{\pi} \int_{-\infty}^\infty \frac{y}{(x-x')^2 + y^2} f(x') dx'$$

represents a harmonic function in the upper half-plane which assumes the value $f(x)$ when $y > 0$ tends to zero. Note that

$$\frac{y}{(x-x')^2 + y^2} = \frac{d}{dx'} \arctan \frac{y}{x-x'} = -\frac{d}{dy} \log \frac{1}{\sqrt{(x-x')^2 + y^2}}.$$

Now let Ω be a plane, open and bounded set with a smooth, connected boundary D, represented by a smooth function $t \to h(t)$ where t is arc length in a positive direction. The explicit formula in the half-plane makes it reasonable to try to represent the solution $u(x) = u(x_1, x_2)$ of Dirichlet's problem in Ω as an integral

(9) $$u(x) = \int_0^L \kappa(x, h(t))f(t)ds$$

where L is the length of D and

$$\kappa(x, h(t)) = \frac{1}{\pi} D_t \log \frac{1}{|x - h(t)|} = \frac{1}{\pi}(x - h(t), n(t))/|x - h(t)|^2.$$

In this formula, D_t is differentiation along the interior unit normal $n(t)$ of C. When C has a continuous curvature, then $k(s,t) = \kappa(h(s), h(t))$ is a continuous function, and it is clear from the above that the right side of (9) tends to

(10) $$f(s) + \int_0^L k(s,t)f(t)dt, \quad k(s,t) = \kappa(h(s), h(t)).$$

when x tends to a point $h(s)$ at the boundary. When the boundary is a straight line, $k(s,t)$ vanishes. Hence Dirichlet's problem with boundary values $g(s)$ is reduced to the problem of solving the integral equation

(11) $$f(s) + \int_0^L k(s,t)f(t)dt = g(s)$$

where f is an unknown continuous function.

Fredholm's starting point (9), where the right side is the double layer of potential theory, was not new. Before Fredholm, Carl Neumann had reduced Dirichlet's problem in the plane to the integral equation (11). With the notation

$$(Kf)(s) = \int_0^L k(s,t)f(t)dt$$

it can be written as $f + Kf = g$. Neumann had succeeded in proving that the geometric series

$$(1+K)^{-1}g = g - Kg + K^2g - K^3g + ...$$

converges and solves the equation when the region Ω is strictly convex. Here K^2 means an operator with the iterated kernel

$$\int_0^L k(s,r)k(r,t)dr$$

and so on.

Fredholm's lasting contribution was a general theory of integral equations (11) with continuous functions f, g and a kernel $k(s,t)$ which is continuous in the square $0 \leq s, t \leq L$.

The theory is analogous to the theory of a finite system of linear equations and has the following main result, sometimes called the Fredholm alternative: The equation is uniquely solvable for any right side if and only if zero is the only solution of the homogeneous equation. But the theory contains more and deals with the family of integral equations

$$(12) \qquad f(s) + \lambda \int_0^1 k(s,t)f(t)dt = g(s)$$

where λ is a parameter and the upper bound of the integral is normalized to 1.

Fredholm's idea does not appear in the final paper. It was to approximate the integral equation by large linear systems of equations. At about the same time Jacques Hadamard (see (1949)) had the same idea, applied it to equations of the type (9) without $f(s)$ on the left side and got nowhere. The explanation is that the first term on the left side represents the identical map. The second term, where the integration has a smoothing effect on the function f, plays the part of a mild perturbation of the identical map. In the passage to the limit from a finite system of linear equations, the first term has a stabilizing effect.

It is easy to approximate (12) by a linear system of equations. If we let s,t pass through the values k/n where $k = 0, 1, ..., n-1$ and put $dt = 1/n$, we get a linear system of equations with the determinant.

$$D_n(\lambda) = \det(\delta(s,t) + \lambda k(s,t)n^{-1}).$$

By determinant theory we may develop $D_n(\lambda)$ in a series

$$1 + \lambda \sum_t n^{-1} k(t,t) + \frac{\lambda^2}{2!} \sum_{t_1,t_2} n^{-2} \det k(t_1;t_2) + ...,$$

where $k(s;t)$ is the matrix

$$\begin{matrix} k(s,s) & k(s,t) \\ k(t,s) & k(t,t) \end{matrix}$$

and analogously for several variables. The corresponding linear system is uniquely solvable if and only if $D_n(\lambda) \neq 0$. Fredholm guessed that D_n has as a limit an entire function $D(\lambda)$ with the same property relative to the integral equation. This turned out to be correct and, in addition, there is an explicit formula

$$D(\lambda) = 1 + \lambda \int k(t,t)dt + \frac{\lambda^2}{2} \int \int \det k(t_1;t_2)dt_1 dt_2 + ...,$$

which is an obvious limit of the preceding one as $n \to \infty$. The convergence follows from an inequality for determinants which by Fredholm first proved himself and then found in a paper by Hadamard, namely

$$|\det(A_1,...,A_n)| \leq |A_1|...|A_n|$$

where $A_1,...,A_n$ are vectors with n components. This means that the coefficient of λ^n is majorized by

$$M^n n^{n/2}/n!, \quad M = \max|k(s,t)|,$$

which in turn means that $D(\lambda)$ is an entire function of λ.

Cramer's rule gives the solution of a linear system of equations approximating (11) in the form of a quotient of two determinants. Fredholm's theory has a corresponding formula, mediated by the kernel

$$K(s,t,\lambda) = k(s,t) + \lambda \int \det k(s,r;t,r)dr + \frac{\lambda^2}{2}\int \det K(s,r_1,r_2;t,r_1,r_2)dr_1 dr_2 + ...$$

where $K(s_1,...,s_n;t_1,...,t_n)$ is a matrix with elements $k(s_j,t_k)$. This kernel has the property that (12) has the unique solution

$$f(s) = g(s) - \int_0^1 K(s,t,\lambda)g(t)dt/D(\lambda)$$

when $D(\lambda) \neq 0$. Here a conjecture by Poincaré is proved: the Neumann series

$$1 - \lambda K + \lambda^2 K^2 + ...,$$

which converges for small λ, is a meromorphic function of λ.

When $D(\lambda)$ has a zero λ_0 of order n it turns out the that the number of linearly independent solutions of $f + \lambda Kf = 0$ equals the number of linearly independent conditions on the right side which guarantee that the inhomogeneous equation has a solution. This theorem is Fredholm's alternative in a general form. With the means of expression of linear algebra the alternative can be written as

(13) $$\dim \mathrm{nil}\, A - \dim L/AL = 0$$

where $A = 1-\lambda K$ with arbitrary λ, L is the linear space of continuous functions from the interval $(0,1)$ and $\mathrm{nil}\,A$ is the null space of A, i.e. the linear space of solutions f of $Af = 0$. The left side of (13) is called the index of A, $\mathrm{ind}\,A$. The essential contant of Fredholm's theory is expressed by the formula $\mathrm{ind}(1 + K) = \mathrm{ind}\,1 = 0$ for the operators K which Fredholm considers.

A weakness of Fredholm's theory is the requirement that the kernel is required to be continuous. In later papers Fredholm and others proved that it is sufficient that some iteration of the kernel is continuous. The reason for this is very simple: the equation $f + Kf = g$ means that $f - K^2f = g - Kg$ and so on.

After his great work from 1900 to 1910 Fredholm wrote little and almost entirely within the domains he had conquered.

Fredholm's work on integral equations was met with great interest and boosted the morale and self-respect of Swedish matematicians who so far had been working under the shadow of the continental cultural empires Germany and France. Integral equations had now become a new mathematical tool not confined to symmetrical kernels. It was developed during several decades and was seen as a universal tool with which it was possible to solve the majority of the boundary value problems of physics. But the qualitative insight that the theory gave could also be achieved in a simpler way. The significance of Fredholm's work was more the qualitative insight than the explicit formulas.

After Fredholm

Fredholm's paper about integral equations received much attention in Göttingen, Germany, the mathematical center at the time. There the theory was reworked in a way which made the determinants disappear. This was done so that the kernel $k(s,t)$ was approximated by a kernel of finite rank, i.e. of the form

$$k_1(s,t) = \sum f_i(s)g_j(t), \quad i,j=1,..,m,$$

with the difference $k_2(s,t) = k(s,t) - k_1(s,t)$ so small that the Neumann series of the corresponding operator K_2 converges. This reduces the problem with $\lambda = 1$ to finite linear system of equations. It suffices for instance that $|k_2(s,t)| \leq 1/2$.

In the work of Hilbert the theory of integral equations took another direction. His main problem was to extend the diagonalization of quadratic forms to infinitely many variables, but his famous book (1912) contains many other things, for instance a new proof of Fredholm's alternative and many applications to boundary value problems for ordinary and partial differential equations.

Hilbert's diagonalization of quadratic forms is now seen as a classical spectral theorem for bounded self-adjoint operators. It is sketched in an appendix of Chapter 14 in this book and is the point of departure of Carleman's work in the beginning of the twenties. There Hilbert's spectral theory is extended to linear integral operators

$$Af(s) = \int_a^b K(s,t)f(t)dt$$

with real and symmetric kernels satisfying certain conditions. The final stage in this development is von Neumann's paper (1928) commented on in the same appendix. In this paper abstract Hilbert space is introduced for the first time and Hilbert's theorem is extended to unbounded selfadjoint operators.

The Swede David Enskog (1884-1947), professor from 1930 at KTH, worked as a mathematical physicist. He discovered thermal diffusion and worked with the kinetical theory of gases. In his thesis (1917) Enskog introduced a method of solving integral equations which was later developed in several papers (1926 a,b, 1930 a,b). His method is best explained in the framework of Hilbert space with an interior product (f,g). Suppose that T is a bounded linear operator and that its adjoint T^* is surjective and has a bounded inverse. If $v_1, v_2, ...$ is a complete system in H, the system $T^*v_1, T^*v_2, ...$ can be orthonormalized to a complete orthonormal system $T^*u_1, T^*u_2, ...$ such that the system $u_1, u_2, ...$ is also complete. Under these circumstances the equation $Tf = g$ has the unique solution

$$f = \sum_1^\infty (g, u_j) T^* u_j.$$

In fact, $(f, T^*u_j) = (g, u_j) = (Tf, u_j)$ for all j so that $Tf = g$. When $T = E + A$ where A is compact, this method can be used to prove Fredholm's alternative. When T is symmetric and $\geq E$, Enskog (1917) orthnormalized in the interior product (f, Tf) which, as above, gives a direct formula for the solution.

Fredholm's theory has been fitted into functional analysis through a classical paper (1918) by F. Riesz. Fredholm's kernel $K(s,t)$ gives a linear operator $f \to Kf$ in the space of continuous functions in an interval. When the kernel is continuous, it is easy to see that every bounded sequence $f_1, f_2, ...$ is mapped to a sequence $Kf_1, Kf_2, ...$ with the property that it contains a convergent subsequence. A bounded linear operator on a linear normed space with this property is said to be compact. In his paper (1918) F. Riesz extended Fredholm's theory to such operators by proving that $E + K$ has index zero.

A linear operator $A : L_1 \to L_2$ from one linear space to another has come to be called a Fredholm operator when $\operatorname{nil} F$ has finite dimension and if AL_1 is a closed part of L_2 of finite codimension. Every such operator has a well-defined index

$$\operatorname{ind} A = \dim \operatorname{nil} A - \dim L_2/AL_1.$$

It can be shown that $\operatorname{ind} A = \operatorname{ind}(A + K)$ when K is compact (see Hörmander (1985)). There are operators whose index does not vanish, for instance the shift operator $S : e_k \to e_{k+1}$ where $e_0, e_1, \ldots$ is an orthonormal basis of a Hilbert space H. It is clear that $\operatorname{nil} S = 0$ and that SH is spanned by av $e_1, e_2, \ldots$ which means that $\operatorname{ind} S = -1$.

Sometimes the index of an operator has a topological significance. If $\nu \ne 0$ is a smooth vector field on the unit circle, then d_ν, considered as an operator on smooth harmonic functions in the unit disk, has index $2(1-p)$ where p is the index of ν (see Hörmander (1963)). This example is just one of many index theorems with varying content (see Hörmander l.c.).

Fredholm's obituary

Ivar Fredholm's student Nils Zeilon wrote a long and tender obituary (1930) of his admired teacher. I quote some passages:

This is not the place to enter into the importance of Fredholm's discovery or its impact on the later developments or the part played by integral equations in mathematical physics. All this is already known and has found its place in encyclopedias. Very soon Fredholm became famous. In the spring of 1901 in Göttingen Erik Holmgren gave the first seminar outside of Sweden about Fredholm's equations. It is well known how much Hilbert and his school have been inspired by the deep questions posed by integral equations .....

We may ask what in Fredholm's eyes was the essential basis of his work. The answer is immediate: potential theory. .... Already in 1895 after a seminar lecture in 1895 he had talked about Dirichlet's problem as one of elimination. .... Two years later in Stockholm a lecture about the 'principal solutions' of Roux and their connections with Volterra's equation led to a vivid discussion. Finally, after a long silence Fredholm spoke and remarked in his usual slow drawl: in potential theory there is also such an equation....

In his Stockholm lectures Fredholm loved to talk at length about the great problems and methods of classical mathematical physics which had been the main theme of his scientific work. But this did not prevent him from talking about all parts of modern physics. Fredholm was not what is usually called a brilliant speaker. He talked slowly in a monotone voice and it could happen that he got embroiled in computational mistakes at the blackboard. But this had little importance. In fact, his lectures revealed an unusual mastery of his subject and he had the ability of communicating to his students a feeling for the unity and logic of physical theory which is so apparent in his own written work.

Appendix: Fundamental solutions by distribution theory

Generalities about fundamental solutions

In order to put Fredholm's result about elasticity theory into a wider context we shall sketch how fundamental solutions are constructed in distribution theory. This will also give a background to the papers about fundamental solutions which were written by Fredholm's student Zeilon and are the subject of part of the next chapter.

A differential operator with constant coefficients has the form $P(\partial_x)$ where P is a polynomial and ∂_x the gradient in terms of n real variables $x = (x_1, \ldots, x_n)$. A fundamental solution of $P(\partial_x)$ is a distribution $E(P, x)$ such that

(1) $$P(\partial_x) E(P, x) = \delta(x).$$

This means that the convolution $u(x) = \int E(P, x - y) f(y) dy$ solves the equation

$P(\partial_x)u(x) = f(x)$, whence the prefix fundamental. A fundamental solution is determined modulo solutions $V(x)$ of the homogeneous equation $P(\partial_x)V(x) = 0$. In the sequel we shall use the imaginary gradient $D = \partial_x/i$ and write $P(D)$.

We observe that $G(x) = Q(\partial_x)E(PQ, x)$ is a fundamental solution of P for $PG(x) = PQE(PQ, x) = \delta(x)$. This remark permits us to reduce to the scalar case the construction of fundamental solutions of operators $P(D)$ which are quadratic matrices whose elements are polynomials in D. In fact, to such a matrix $P(D)$ there is a matrix $P'(D)$ such that

$$P(D)P'(D) = P'(D)P(D) = I \det P(D)$$

where I is the unit matrix. The distribution $F(P, x) = P'(D)E(\det P(D), x)$ is a fundamental solution of $P(D)$ since $P(D)F(P, x) = I\delta(x)$.

Already Cauchy observed that the Fourier transform gives at least a formal fundamental solution of $P(D)$ via the formula

(2) $$E(P, x) = \frac{1}{(2\pi)^n} \int e^{i(x,\xi)} d\xi / P(\xi).$$

In fact, by a formal calculation,

$$P(D)E(P, x) = \frac{1}{(2\pi)^n} \int e^{i(x,\xi)} d\xi = \delta(x).$$

In distribution theory, (2) is taken in integrated form as $E(P, f) = \int E(P, x)f(x)dx$ where f is infinitely differentiable with compact support. The idea of this theory is that $E(P, x)$ is known as a generalized function or distribution when $E(P, f)$ is known as a linear function of f and has certain continuity properties. Formally we get

(2') $$E(P, f) = \frac{1}{(2\pi)^n} \int F(\xi)d\xi / P(\xi)$$

where

$$F(\xi) = \int e^{(x,\xi)} f(x)dx$$

is an entire analytic function.

It was not until the fifties that it was shown that every differential operator $P(D) \neq 0$ has a fundamental solution. We shall sketch a proof which defines $E(P, f)$ as a fundamental solution even when the denominator can have zeros. By a linear transformation we may assume that $P(\xi) = c\xi_1^m + ...$ where $c \neq 0$ and m is the degree of P. Let $\xi' = (\xi_2, ..., \xi_n)$. In the integral (2') we shall now integrate first with respect to ξ_1 over a path $C(\xi')$ in the complex plane homologous to the real axis which avoids the zeros of $\xi_1 \to P(\xi_1, \xi')$ and lies in the strip $|\operatorname{Im} \xi_1| \leq a$ where $a > 0$ is fixed. When ξ' varies it may happen that the path $C(\xi')$ and the boundary of the strip enclose a zero which is on its way out of the strip or that the path is squeezed between two zeros which are coming together. Then we decide to change to a new path $C(\xi')$ before this happens. In this way the right side of (2') is well defined (but depends on many choices) and we get

$$E(P, P(-D)f) = \frac{1}{(2\pi)^n} \int d\xi' \int_{C(\xi')} F(\xi)d\xi_1$$

where we now are free to replace every $C(\xi')$ by the real axis and the properties of the Fourier transform show that

(2'') $$E(P, P(-D)f) = f(0)$$

which is (2) in distribution form.

Homogeneous equations, the Gelfand-Shilov formula, the singularity surface

In order to give explicit constructions of fundamental solutions of homogeneous differential operators it is convenient to employ the Gelfand-Shilov formula (1958) which is a decomposition of the $\delta(x)$ into plane waves

$$(3) \qquad \delta(x) = (2\pi)^{-n} \int_{r=1} (n-1)!(+0+i(x,\xi))^{-n} \omega(\xi).$$

Here $r(\xi)$ is a smooth positive function which is homogeneous of degree one when $\xi \neq 0$. The differential form

$$\omega(\xi) = \sum (-1)^{k-1} d\xi_1 \wedge ... d\xi_{k-1} \wedge d\xi_{k+1} \wedge ... d\xi_n$$

is assumed to be positive on $r = 1$. The formula (3) and the ensuing ones are to be interpreted as the ones we get by multiplying both sides by $f(x)dx$ and integrating where $f(x)$ is a smooth function with compact support. The formula (3) is proved by an introduction of polar coordinates $\xi = r(\xi)\eta$, $r(\eta) = 1$ in the formula

$$\delta(x) = \lim_{\varepsilon \to 0} (2\pi)^{-n} \int e^{-\varepsilon r(\xi) + i(x,\xi)} d\xi$$

followed by an integration with respect to r. This proof shows that the right side of (3) does not depend on the choice of the function r.

In (3) there appears the distribution $(n-1)!(+0+it)^{-n}$ which is the boundary value of the analytic function $(n-1)!s^{-n}$ where $\operatorname{Re} s > 0$. It will be suitable to introduce successive derivatives and integrals of this functions. Put

$$\varphi_k(s) = s^k (\log s - c_k)/k!$$

wher $\operatorname{Re} s > 0$, $k \geq 0$ and $c_0 = 0$ and, otherwise, $c_k = 1 + \frac{1}{2} + ... + \frac{1}{k}$. This means that

$$(4) \qquad \varphi'_k(s) = \varphi_{k-1}(s).$$

If we use the integration to define φ_k when $k < 0$, the result is

$$\varphi_{-k}(s) = (-1)^{k-1}(k-1)! s^{-k}.$$

The boundary values for $\operatorname{Re} t = +0$ of these analytic functions are distributions which will be denoted by $\varphi_k(+0+it)$.

If $P(\xi)$ is a homogeneous polynomial of degree m we can now at once write down a fundamental solution of the operator $P(D)$, namely

$$(5) \qquad E(x) = (-1)^{n-1} (2\pi)^{-n} \int_{r(\xi)=1} \varphi_{m-n}(+0+i(x,\xi)) \omega(\xi)/P(\xi)$$

where $1/P(\xi)$ in some way or other defines a distribution when $r(\xi) = 1$. On the right side, $\varphi_{m-n}(+0+i(x,\xi))$ and $1/P(\xi)$ are two distributions on the manifold $r(\xi) = 1$ whose product meets an infinitely differentiable function on the manifold. The first distribution is infinitely differentiable when $(x,\xi) \neq 0$ and the other one when $P(\xi) \neq 0$. At places where $P(\xi)$ and (x,ξ) can be enlarged to a coordinate system, the integrand of the right of (5) is well defined as a product of two distributions in two variables, $P(\xi)$ and (x,ξ). Hence the simplest properties of distributions give us the following result:

(A) If the manifold $P(\xi) = 0$ and the plane $(x, \xi) = 0$ meet transversally everywhere, the right side of (5) is a well defined and, in addition, infinitely differentiable function of x and even real analytic.

It follows from this that the singularities of the fundamental solution $E(P, x)$ given by (5) form a subset of the *singularity surface* $S(P)$ of P, defined as the complement of the set of points where the condition (A) is satified. When $x \in S(P)$, the real plane $(x, \xi) = 0$ and the real manifold $P(\xi) = 0$ are tangent to each other outside the origin. When $P(D)$ is elliptic, i.e. $P(\xi) \neq 0$ when $\xi \neq 0$, the singularity surface reduces to the point $x = 0$.

Variants and reductions of the Gelfand-Shilov formula

Suitable manipulations of the Gelfand-Shilov formula give Abelian integrals generalizing those of Fredholm.

In the integral (5) it is possible to let the manifold $r(\xi) = 1$ be the two planes $\xi_n = \pm 1$, oriented by $\omega(\xi) > 0$. Since $\omega(-\xi) = (-1)^n \omega(\xi)$, (5) can be split according to the parity of n. At the same time we shall split $f_k(t) = \varphi_k(+0 + it)$ into two parts,

$$f_k(t) = i^k(g_k(t) + ih_k(t))$$

where

$$h_k(t) = t^k(\log|t| - c_k)/k!, \quad h_{-k-1} = (-1)^k k! \operatorname{Pv} t^{-k}$$

(Pv means principal value) and

$$g_k(t) = (\pi/2)t^k \operatorname{sgn} t/k!, \quad g_{-k-1} = -\pi \delta^{(k)}(t).$$

Hence, when n is odd we can rewrite (5) as

$$E(x) = (-1)^{n-1} 2(2\pi)^{-n} i^{m-n+1} \int g_{m-n}((x,\xi)) \omega(\xi)/P(\xi)$$

with integration over $\xi_n = 1$. When n is even we get a corresponding formula with $h_{m-n}((x,\xi))$.

We can now get Fredholm's formulas back when $n = 3$ and $m - n \geq 0$. If $Q(\xi)$ is homogeneous of degree $m - 2$ the result is

$$Q(D)E(x) = \operatorname{const} \int \delta((x,\xi)) d\xi_1 d\xi_2 Q(\xi)/P(\xi).$$

Here it is easy to reduce the right side to a sum of residues

$$\sum \operatorname{sgn}(\operatorname{Im} \xi_2) Q(\xi)/(x_2 P_1 - x_1 P_2)(\xi)$$

where ξ runs through all points common to the plane $(x, \xi) = 0$ and the algebraic curve $P(\xi) = 0$. This formula which, however, requires that the plane and the curve meet in m separate points, is the point of departure of Fredholm's last construction (1908) of a fundamental solution in the form of an Abelian integral.

Hyperbolic operators

In the early twentieth century fundamental solutions were discussed only for elliptic equations, but fundamental solutions for other equations were known at least implicitly. Euler discovered that the spherical waves

$$v(t, x) = f(t \pm |x|)/|x|,$$

where f is an arbitrary function, solve the inhomogeneous wave equation, in modern notation

$$v_{tt} - \Delta v = 4\pi f(t)\delta(x).$$

When $f(t) = \delta(t)$ a multiplication by Heaviside's function $H(t) = (1 + \operatorname{sgn} t)/2$ produces a fundamental solution

$$E(t,x) = H(t)\delta(t^2 - |x|^2)/2\pi$$

which describes spherical wave propagation with velocity one from the origin.

The wave operator belongs to a class of homogeneous differential operators $P(D)$ which are hyperbolic with respect to some time variable $t = (x, \theta) \neq 0$ in the sense that $P(\xi + i\theta) \neq 0$ for every real ξ, a condition simultaneously satisfied by $\pm\theta$. The condition means that the equation $P(s\theta + \xi) = 0$ has $\operatorname{grad} P$ real solutions s for every real ξ, i.e. that the real surface $P(\xi) = 0$ is maximal in a certain sense. The corresponding hyperbolic operators $P(D)$ have the characteristic property of having a fundamental solution which vanishes outside a proper cone with its vertex at the origin, a propagation cone on which $t \geq 0$. This fundamental solution describes wave propagation with finite propagation velocity from a momentaneous source at the origin. All the singularities lie in the intersection of the singularity surface with this cone. Hyperbolic operators and analogous quadratic systems with a hyperbolic determinant occur in physics in connection with wave propagation with finite speed and therefore the singularity surface is in these cases also called the wave surface. The mathematical content of all these statements is the following one (see Hörmander (1985)II, Ch. XII): The real surface $P(\xi) = 0$ divides R^n into connected open sets. The set $\Gamma = \Gamma(P,\theta)$ which contains θ, is a convex open cone with the property that $P(\xi + i\eta) \neq 0$ for all real ξ when $\eta \in \Gamma$. We can now write down a fundamental solution

$$E(P,x) = \frac{1}{(2\pi)^n} \int e^{ix\cdot(\xi-i\eta)} d\xi/P(\xi - i\eta)$$

where the denominator does not vanish. Cauchy's theorem shows that the right side is independent of η when $\eta \in \Gamma$. If $(x, \eta) < 0$ the integrand tends to zero when we exchange η for $r\eta$ and let $r \to +\infty$. The same follows for $E(P, x)$ as a distribution. This means that $E(P, x)$ vanishes outside a propagation cone $C = C(P, \theta)$, defined by the property that $x \cdot \Gamma \geq 0$ when $x \in C$. Since Γ is open, C is a proper cone which, apart from its point, is entirely on one side of every plane $x \cdot \eta = 0$ with $\eta \in \Gamma$.

We get precisely this fundamental solution if in (5) we interpret the distribution $1/P(\xi)$ as $1/P(\xi - i0\theta)$.

Bibliography

BENDIXSON I.O.
1883. *Quelques théorèmes de la théorie des ensembles*, Acta Math. 2, 1883, 415-429.
1891. *Bestämning af de algebraiskt upplösbara likheter i hvilka hvarje rot kan uttryckas som en rationel funktion af en af rötterna*, KVA Öfversigt 1891, 131-147.
1901. *Sur les courbes définies par des équations différentielles*, Acta Math. 24, 1901, 1-88.
1903. *Détermination des équations résolubles algébriquement*, Acta Math. 27, 1903, 317328.
ENSKOG D.
1917. *Kinetische Theorie der Vorgänge in mässig verdünnten Gasen*, Ak. avh. Uppsala 1917.
1926. a.*Eine allgemeine Methode zur Auflösung linearer Integralgleichungen*, Math. Zeitschr. 24, 1926, 670-673.
1926. b. *Zur Begründung der Theorie der Fredholmschen Integralgleichungen*, Mat. Zeitschr. 25, 1926, 299-304.

1930. a. *Über die Hauptfunktionen unsymmetrischer Kerne*, Mat. Zeitschr. 31,1930, 601-610.
1930. b. *Über die Auflösung einer singulären Integralgleichung*, Acta Math. 54, 1930, 177-184.
FREDHOLM I.
1955. *Oeuvres Complètes de Ivar Fredholm*, Litos Reprotryck Malmö 1955.
1890. *Om en speciell klass af singulära linier*, KVA Öfversigt 1890, no 3.
1899. *Sur une équation aux dérivées partielles du quatrième ordre*, KVA Öfversigt 1899, no 9..
1900. a. *Sur les équations de l'équilibre d'un corps solide élastique*, Acta Math. 23, 1900, 1-42.
1900. b. *Sur une nouvelle méthode pour la résolution du problème de Dirichlet*, KVA Öfversigt 1900, no 1.
1903. *Sur une classe d'équations fonctionelles*, Acta Math. 27, 1903, 365-390.
1906. *Solution d'un problème fondamental de la théorie de l'élasticité*, Arkiv Mat. Astr. Fysik 2, 1906, no 28.
1908. *Sur l'intégrale fondamentale d'une équation différentielle elliptique à coefficients constants*, Rend. Circ. Mat. Palermo t. 25, 1908, 346-351.
GELFAND I.G., SHILOV G.E.
1958. *Generalized functions and operations on them* (Russian). vol. 1, Moskva 1958.
HADAMARD J.
1949. *The psychology of invention in the mathematical field*, Princeton 1949.
HILBERT D.
1912. *Grundzüge einer allgemeinen Theorie der linearen Differentialgleichungen*, Berlin Leipzig 1912.
HÖRMANDER L.
1963. *Linear Partial Differential Operators*, Springer 1963.
1985. *The Analysis of Linear Partial Differential Operators III*, Springer 1985.
von KOCH H.
1891. *Sur une application des déterminants infinis à la théorie des équations différentielles*, Acta Math. 15, 1891, 53-63.
1892. *Sur les déterminants infinis et les équations différentielles linéaires*, Acta Math. 16, 1892-93, 217-295.
1901. *Sur quelques points de la théorie des déterminants infinis*, Acta Math. 24, 1901, 89-122.
1906. *Une méthode géométrique élémentaire pour l'étude de certaines questions de la théorie des courbes planes*, Acta M. 30, 1906, 145-174.
LANDAU E.
1916. *Neuere Ergebnisse der Funktionentheorie*, Chelsey reprint 1946.
PHRAGMÉN E.
1884. *Beweis eines Satzes aus der Mannigfaltigkeitslehre*, Acta M. 5, 1884, 47-48.
1904. *Sur une extension d'un théorème classique de la théorie des fonctions*, Acta Math. 28, 1904, 351-368.
1908. Phragmén and E. Lindelöf, *Sur une extension d'un principe classique de l'analyse et sur quelques propriétés de fonctions monogènes dans le voisinage d'un point singulier*, Acta Math. 31, 1908, 381-406.
POINCARÉ H.
1886. *Sur les déterminants d'ordre infini*, Bull. Soc. Math. France t. 14, 1885-86, 77-90.
1881. *Sur les courbes définies par des équations différentielles*, Journal de Mathématiques pures et appliqués, série 3, t. 7, 1881, 375-482; t. 8, 1882, 251-298.
RIESZ F.
1918. *Über lineare Funktionalgleichungen*, Acta Math. 41, 1918, 71-98.
1926. *Sur les fonctions subharmoniques et leur rapport à la théorie du potentiel*, Acta Math. 48, 1926, 329-343.
ZEILON N.
1930. *Ivar Fredholm*, Acta Math. 54, 1930, I-XII.

CHAPTER 11

Stockholm University 1880-1920 II

Mathematics and mathematicians

The success of Mittag-Leffler and his students attracted many students to Stockholm University and to the study of mathematics. The theory of analytic functions was a new and exciting field, and there were many examples to show that the road to the frontier of research could be short. At the other universities it could happen that a majority of theses were written in the field of one dominating teacher, but Mittag-Leffler never played this role. To a great extent the mathematicians at Stockholm University were left to their own devices and this accounted for a creative atmosphere with varied thesis subjects All students were not as successful as for instance Phragmén or Fredholm, but many did good work. In the beginning of the period, classical mechanics and complex analysis were the main subjects. Later there was analytical number theory. With the arrival of Marcel Riesz in 1911 Fourier series and summation were added. The history of the period has beeen sketched earlier by Domar (1978).

Analysis

Kobb, Cassel, Petrini, Grönwall, Malmquist

In addition to the quartet of Bendixson, Phragmén, von Koch, and Fredholm there were two more people in the same generation who worked with mathematics at Stockholm University, Gustaf Kobb (1863-1934) and Karl Gustav Cassel (1866-1945).

Kobb

Kobb defended his thesis in Uppsala in 1889 before Stockholm University got the right to create Ph.D.'s. He was a docent in Stockholm from 1889 to 1898, and got the position as professor of rational mechanics at KTH in 1912, where he stayed until his retirement in 1928. Kobb was also known as a politician and he wrote on political subjects.

Kobb made his debut in rational mechanics, a subject popular among mathematicians and mathematical physicists alike. Sonya Kovalevski had been a model through her prize-winning paper on the spinning top. Kobb's theme (1887) was the motion of a small body on a surface of rotation in a rotationally invariant potential field. In 1842 Jacobi had proved that the motion can be solved by quadrature. If the surface and the potential are algebraic, the motion can be expressed by abelian integrals. Kobb gave a necessary and sufficient condition for the integrals to be elliptic. When the potential is gravitational, the possible surfaces are listed. There are five cases of which two were new. This paper required good insight into the theory of algebraic curves. A similar subject was also treated (1893, 1908) by Frans de Brun who later wrote papers about special functions.

Kobb wrote other papers (1894 a,b) about rational mechanics but also a long paper (1892), inspired by Schwarz's investigations of minimal surfaces and Picard's papers about elliptic equations of order two. Both have to do with the problem of minimizing an integral

$$\int_D F(x, x_u, x_v) du dv.$$

The integrand F is given, D is a plane region and $x = x(u, v)$ represents a surface in R^3 with a given boundary. It is required that the integral is independent of the parametrization which gives a number of conditions. The author computes the second variation and uses Weierstrass's strict conditions for a minimum. He also tries to use his variational problem to solve Dirichlet's problem for quasilinear elliptic equations of order two. This paper, which was an ambitious effort for its time, is now only of historical interest.

Cassel

Karl Gustav Cassel wrote some mathematical papers and a thesis (1894). The theme was the theory of discrete groups of Möbius transformations of the complex plane, a subject opened up by Poincaré in several papers. But this ambitious opening turned out to be too heavy for the young mathematician. His thesis has very few new results and can be seen as a farewell to mathematics. Later Cassel switched to economics and the change turned out well. He was nominated docent in 1902 and was professor of economics from 1904, both positions at Stockholm university. Cassel became an internationally famous economist. He wrote mainly about general subjects in economics but could use mathematics in price theory. Cassel is one of the founders of the so-called Stockholm School of Economics.

Petrini

There is a third noteworthy mathematician in the same generation as Kobb and Cassel, Henrik Petrini (1863-1957). He may not have studied at Stockholm University, but he wrote good papers in *Acta Mathematica* and *Arkiv*. He was also known to the general public as a writer about general subjects in a radical vein. Petrini was a lektor at the gymnasium in Växjö from 1901 and later in Stockholm from 1914 to 1918. His sister Märta Petrini was a prominent soprano and his wife Gulli had a doctorate in physics and taught at a well-known school in Stockholm. Like her husband she was a left liberal and worked for the liberation of women.

Petrini's main interest was potential theory, in particular the regularity properties of the potential. The point of departure of his main work (1908) is as follows. If the function $f(x)$ is continuously differentiable or, more generally, Hölder continuous, then the Newtonian potential in three variables,

$$U(x) = \int |x - y|^{-1} f(y) dy$$

satisfies Poisson's equation $\Delta U(x) = -4\pi f(x)$ (Δ is Laplace's operator). But if the density $f(x)$ is only continuous, the second derivatives of $U(x)$ need not exist. Petrini (1908) proved that Poisson's equation still holds in a certain generalized sense. This long paper got a three-page review in *Jahrbuch*. He also studied the regularity of potentials of mass distributions on surfaces and double layer potentials. His papers are written with an analytical rigor unusual at the time. Distribution theory has made Petrini's problem somewhat old-fashioned. In a later paper (1917) which may still be of interest, Petrini studied the singularities of the potential of an electrostatic charge in equilibrium on a surface with singularities, for instance at sharp points.

Grönwall

In the 1890's Hakon Grönwall (1877-1932) studied at Stockholm University. As a mathematician he was as precocious as Phragmén and von Koch and published around ten papers in a very short time. His main theme from (1895) to the thesis (1898), inspired by a paper in *Acta* (1889) by Horn, was an extension of Fuchs's theory to so-called total differential equations

$$dz_j = \sum_1^n a_{jk} z_k, \quad j = 1, ..., m$$

where the coefficients

$$a_{jk} = \sum_1^n a_{jki}(x) dx_i, \quad x = (x_1, ..., x_n),$$

are Pfaffian forms in n variables whose coefficients a_{jki} are analytic functions. The condition that

$$da_{jk} + \sum_i a_{ji} \wedge a_{ik} = 0$$

makes the system integrable and the general situation is the same as in Fuchs's case. With a given $z = z(x)$ the system has a unique analytic solution which can be continued outside a singular manifold $S \subset C^n$, consisting of points x where all a_{jk} vanish. As in Fuchs's theory the problem is to investigate the branching of the solution space around the manifold S.

In a neighborhood of a point a where S has the form $h(x) = 0$, $h(a) = 0$ with $h(x)$ analytic and $\deg h(a) \neq 0$ the theory is the same is in the case $n = 1$. The solution space has a basis whose elements behave as rational powers or logarithms of $h(x)$ in suitable combinations under one turn around $h(x) = 0$. In Grönwall's paper this situation is analyzed also when $\deg h(a) = 0$. The author uses Weierstrass's preparation theorem and shows good knowledge of the theory of functions of several complex variables. Another theme is criteria for the system to be of Fuchs's type in a suitable sense.

Grönwall's second subject was transcendental numbers and functions. Liouville had constructed transcendental numbers as decimal expansions with increasing gaps. The reason for this success is that a real and not rational algebraic number cannot be approximated too well by rational numbers. This follows from a theorem by Liouville: if $f(x) = a_n x^n + ...$ is an irreducible polynomial of degree $n > 1$ with integral coefficients and α is a zero of f, there is a number $c = c(f) > 0$ such that

$$|\alpha - p/q| \geq 1/c(f)q^n$$

for every rational number p/q with $q > 0$. Liouville's proof is very simple. In the equality

$$|f(p/q)| = |a_n||\alpha - p/q| \prod_{k>1}(|p/q| + |\alpha_k|)$$

where $\alpha, \alpha_2, ..., \alpha_n$ are the zeros of the polynomial, the left side is a rational number $\neq 0$ with the denominator q^n whose absolute value must exceed or equal $1/q^n$. When $|\alpha - p/q| < 1$, all factors on the righthand side except $|\alpha - p/q|$ are not greater than a constant which only depends on f and the inequality follows. In the opposite case, i.e., when $|\alpha - p/q| > 1$, the inequality is also true.

The inequality just proved shows that if a number α is not rational and there are integers $p = p_n$, $q = q_n$ such that $q^n|\alpha - p/q| \to 0$ for a sequence of integers $n \to \infty$ then α is a transcendental number. In this form the theorem appears in one of Grönwall's papers at a time when the author was ignorant of Liouville's theorem.

In May of 1899 Grönwall and friend had a drunken party which caused some damage at the university. For this his friend was banned from the university for half a year and Grönwall, then a Ph. D. and an extraordinary student at KTH, was banned for the same period. His crime was attendance at the university without being properly matriculated. Naturally enough, Grönwall thought this to be too much for too little and left Stockholm to study at the Charlottenberg Technical Institute in Berlin.

After some time Grönwall came to the United States where he had various positions, first as a mathematician at Princeton University and later as physicist and engineer. Einar Hille, who also had studied at Stockholm University and left for the United States, describes Grönwall's vicissitudes in later life in an obituary (1932). When very young Grönwall had a unique ability to master difficult material in a very short time. Maybe this is the reason why he came to work in many fields. At present Grönwall is often quoted for the following simple and useful result which is called Gronwall's lemma and does not do him justice: If $u'(t) + au(t) \leq f(t)$ when $t \geq 0$, then

$$u(s) \leq e^{-as}u(0) + \int_0^s e^{a(t-s)}f(t)dt, \quad s \geq 0.$$

Malmquist

Mittag-Leffler's last student was Johannes Malmquist (1882-1952), a professor at KTH 1913-1947 and for a long time secretary of *Acta Mathematica*. In (1905) he made a contribution to the circle of problems originating from Mittag-Leffler's E_α-function. He constructed an entire function which tends to zero at infinity on all rays from the origin except the positive real axis. Afterwards he devoted himself to the global theory of differential equations of the first order, in the simplest case

(1) $$dy/dx = P(x,y)/Q(x,y),$$

where P, Q are polynomials without a common factor. The variables x, y are complex and may take the value ∞. The system has the same form in terms of any of the variables $(z, x), (y, t), (z, t)$ when $z = 1/y$, $t = 1/x$. Note that the variable y has a special position as a dependent variable.

Malmquist's impulse came from the lectures in the subject given in Stockholm in 1903 by the French mathematician Paul Painlevé, invited by Mittag-Leffler. The problem was to investigate analytical continuation and ramifications over the complex x-plane of solutions $y = y(x)$ of (1).

The local theory of (1) is obvious: if a, b is a point where $Q(a, b) \neq 0$, there is a unique, locally analytic solution $y = y(x)$ which takes the value b at the point a. To these ordinary points we must also add points where the situation just described arises after an inversion $x \to t = 1/x$ or $y \to z = 1/y$. Through every ordinary point there passes a unique solution such that one of the variables (y, z) is an analytic function of one of the variables (x, t) or conversely. If, after possible inversions, $Q(a, b) = 0$ but $P(a, b) \neq 0$ and $Q(a, y)$ is not identically zero, the system has a solution

$$x = a + c_m(y - b)^m + c_{m+1}(y - b)^{m+1} + ..., \quad c_m \neq 0$$

where $m > 1$. This means that

$$y = b + ((x-a)/c_m)^{1/m} + \dots$$

is finitely ramified at (a,b). Such a singularity is said to be moveable since it depends on the value at a of the solution $y(x)$.

The Riccati equation

$$dy/dx = a_0(x) + a_1(x)y + a_2(x)y^2$$

plays the part of an exception in the theory. For if the system (1) does not have ramified solutions $\neq \infty$ at point $a \neq \infty$, this means that Q does not depend on y, and, if y is not ramified at ∞, it follows from the system for $z = 1/y$,

$$dz/dx = -z^2 P(x, 1/z),$$

that P has at most degree two in y, i.e., the equation (1) is a Riccati equation.

In addition to the ordinary points and the moveable singularities, the equation (1) also has fixed singularities a, b where P and Q vanish simultaneously or $Q(a,y)$ is identically zero. At these points there might be solutions which are infinitely ramified or have exponential growth. Examples: the equation

$$dy/dx = ay/x$$

has the solution $y = Cx^a$ which is infinitely ramified at the origin when a is irrational and ramified of order q when $a = p/q$ is rational and $(p,q) = 1$. The equation

$$dy/dx = (a+y)/x^2$$

has a solution with exponential growth at the origin.

If we require that there are at most a bounded number of solutions of (1) over any point x, it is possible to say something definite about the structure of solutions. This is Malmquist's subject in his best known paper (1914). The problem had been treated earlier by Painlevé (1903) and Boutroux (1908) with essentially the same results, but Malmquist thought that Painlevé's proofs were not complete in their treatment of the singular points.

Malmquist's main results are as follows

1. If (1) is not a Riccati equation, every single-valued solution is rational.

2. If (1) cannot be transformed into a Riccati equation by a change of variables $y \to z$ of the form

$$z = A(x,y)/B(x,y)$$

where A, B are polynomials in y whose coefficients are rational functions of x and $\deg A - \deg B < 0$, every solution of (1) with a finite number of branches is algebraic.

3. When the previous condition is satisfied locally[1], then every solution $y(x)$ with a finite ramification is algebroid, i.e., $y(x)$ is a zero of a polynomial in y with analytical coefficients.

The proofs use a theorem by Boutroux which says that finite ramification means at most polynomial growth at the fixed singular points.

In the sequel Malmquist devoted himself exclusively to analytical differential equations. He returned to the results above extended to equations of higher order with algebraic coefficients and to an analysis (1921) of solutions of (1) at the singular points. This situation falls under Bendixson's general theory without being covered

---

[1] In the sense that the coefficients of A, B are analytical functions of x.

completely. Malmquist writes in detail about normal forms of the equation (1) when (0,0) is a singular point and there are solutions $y(x)$ which tend to zero with x. This paper has a sequel (1941) which is an extension to systems.

In the 1940's Malmquist and his collaborators Sture Danielson and Valdemar Stenström published an extensive textbook of analysis, a true *Cours d'Analyse* modelled after the famous French textbooks with this title. This gave engineers with little or no knowledge of French an opening to a large part of the international mathematical cultural heritage.

For over thirty years Malmquist was the sole professor of mathematics at KTH, the Royal Technological Institute in Stockholm. The engineering students knew him as MalmQ. Only a few of them had any idea of the mathematics that their kind teacher and examiner had worked with and written about.

Hille

Carl Einar Hille (1894-1980) started his mathematical studies at Stockholm University and wrote two review papers about a problem in analysis in *Arkiv*. Later he emigrated to the United States. He was Benjamin Peirce Instructor at Harvard 1921 - 1922 and this was the beginning of a successful career in USA. He published many papers, some in collaboration with others, and he is best known for a book about semigroups whose infinitesimal generators are characterized in a theorem named after Hille and Yosida.

Analytical number theory

Stridsberg, Wigert, Cramér

The part of mathematics where problems in number theory are treated with the tools of analysis is called analytical number theory. The first success in the field was Dirichlet's proof in 1837 that that there are infinitely many primes in every integral arithmetic series. After this, the outstanding problem was the prime number theorem, conjectured by Gauss. It is the statement that the number $\pi(x)$ of prime numbers $\leq x$ is asymptotically equal to $x/\log x$. In 1851 Chebyshev gave a rather elementary proof that the quotient $\pi(x)\log x/x$ belongs to a vary small interval around 1 for large x (see Chapter 6 under Berger). Riemann (1859) gave the problem an analytical turn by introducing the ζ-function

$$\zeta(s) = \sum_1^\infty n^{-s} = 1/\prod(1-p^{-s})$$

where $\operatorname{Re} s > 1$ and p runs through the primes. Riemann proved that the ζ-function has an analytical continuation to the entire complex plane except for a pole at $s=1$ where $\zeta(s) \sim 1/(1-s)$ and that the function

$$\frac{s(s-1)}{2}\Gamma(\frac{s}{2})\pi^{-s/2}\zeta(s)$$

is invariant when $s \to 1-s$. The Gamma factor provides the ζ-function with uninteresting zeros $-2, -4, \ldots$ while the interesting ones lie in the critical strip $0 < \operatorname{Re} s < 1$.

By expanding $\log(1-p^{-s})$ in powers of p^{-s} we get Riemann's formula

$$\log \zeta(s) = \int_1^\infty x^{-s} df(x)$$

for the ζ-function and the function

$$f(x) = \sum_1^\infty \pi(x^{1/n})/n$$

with the main term $\pi(x)$. Riemann sketched a proof of an inversion formula which makes it probable that, for large x, $\pi(x)$ is asymptotically equal to the integral

$$\text{Li}(x) = \int_0^x dt/\log t,$$

with a principal value at $t = 1$. That this approximation has some credibility is seen from the fact that

$$\int_A^\infty x^{-s} d\,\text{Li}(x) \sim \log\frac{1}{s-1}$$

when s decreases towards 1, independently of the choice of lower bound $A > 1$ in the integral. Riemann also remarked that $\text{Li}(x)$ is not a very precise approximation. He argued that the difference $f(x) - \text{Li}(x)$ probably cannot be less than $O(\sqrt{x})$.

When Hadamard and de la Vallée Poussin proved the prime number theorem in the middle 1890's they used variants of Riemann's formula and the behavior of the ζ-function in the critical strip. It was shown later that the prime number theorem is a consequence of the fact that $\zeta(s) \neq 0$ when $\text{Re}\,s = 1$, but that the size of the difference $\pi(x) - \text{Li}(x)$ has to do with the position of the zeros of the ζ-function in the strip. Riemann's as yet unproved hypothesis that all zeros lie on the line $\text{Re}\,s = 1/2$ gives the best estimate.

Other functions than $f(x)$ are used in the theory, for instance

$$\theta(x) = \sum_{p<x} \log p, \quad \psi(x) = \sum_n \theta(x^{\frac{1}{n}})$$

and the function $N(T)$ which is the number of zeros of the ζ-function in the part of the critical strip where $0 < \text{Re}\,s < T$. By the prime number theorem, $\theta(x) \sim \psi(x) \sim x$.

The interest in analytical number theory at Stockholm University began when Phragmén wrote two papers (1891 a,b), inspired by Riemann. The point of departure of the first paper is the connection between $\zeta(x)$ and $f(x)$, written in the form

$$\log\frac{\zeta(s)}{s} = \int_1^\infty x^{-s-1} f(x) dx.$$

Some simple arguments are used to show that the function

$$f(x) - (\text{Li}(x) - \log 2)$$

must change its sign infinitely many times for large x. This modest result just says that $\text{Li}(x)$ is a good candidate for the position as principal term of $f(x)$. It was shown later that the deviations can be rather large. According to Erhard Schmidt (1903) there is a number $K > 0$ such that

$$|f(x) - \text{Li}(x)| > K\sqrt{x}/\log x, \quad \text{Li}(x) = \int_2^x dx/\log x$$

has infinitely many solutions for arbitrarily large x. Here and in the sequel the modified $\text{Li}(x)$ above is used. In the same paper there is a similar result for $f(x)$ and the inequality

$$|\pi(x) - \text{Li}(x) + \text{Li}(\sqrt{x})/2| > x^{\nu-\varepsilon}$$

where $\varepsilon > 0$ is arbitrary and ν is the upper bound of the real parts of the zeros of the ζ-function. We see the importance of Riemann's hypothesis for the size of $\pi(x) - \text{Li}(x)$. These results were later improved.

Helge von Koch wrote about the prime number theorem in several papers. In (1901) he proved that
$$\pi(x) - \text{Li}(x) = O(x^{\frac{1}{2}+\varepsilon})$$
for $\varepsilon > 0$ when Riemann's hypothesis holds. Using a result by von Mangoldt he could also replace x^ε by a logarithmic term. On the other hand, Phragmén proved at the same time Riemann's statement that the inquality does not hold with a smaller power of x. In a later paper (1910) von Koch shows that the difference $\psi(x) - x$ with an error close to $O(x^{1/2})$ depends only on the zeros ϱ of $\zeta(s)$ with $|\text{Im}\,\varrho| < \sqrt{x}$.

Stridsberg

The recent theory of irrational and transcendental numbers had one representative at Stockholm University, Erik Stridsberg (1871-1950), later employed in insurance.

In 1873 Hermite proved that the number e is transcendental and some years later Lindemann showed that π is transcendental. Lindemann is famous for a more general result which says that if $\xi_1, ..., \xi_n$ are algebraic numbers and linearly independent over the rationals, then the numbers
$$e^{\xi_1}, ..., e^{\xi_n}$$
are algebraically independent, i.e., the these numbers do not constitute a zero of a rational polynomial in n variables.

According to a theorem by Weierstrass at most one of the numbers α and e^α can be algebraic. This theorem contains the transcendency of e and π for if $\alpha = 1$ it follows that e is transcendent and the case $\alpha = 2\pi i$ shows that π is transcendental. Using Weierstrass's method of proof, Hurwitz (1883) proved among other things that the Bessel function
$$f_a(x) = \sum_0^\infty \frac{x^n}{n!a(a+1)...(a+(n-1))}, \quad a \neq 0, -1, ...$$
has transcendental zeros when a is an integer. In connection with this paper Stridsberg (1910) proved that $f_a(x)$, $f_a'(x)$ and 1 are linearly independent over the rational numbers. This result was extended by Siegel (1949) to algebraic independence with certain exceptions ($2a$ not an integer).

At the end of the first world war Stridsberg wrote in *Arkiv* about coefficient sequences, i.e., the coefficients of various generating functions, and operations on them. These papers would not have passed a latter referee system, but they convey the image of an interested, inquisitive and enthusiastic writer. The subject brought Stridsberg in contact with the moment problem, an interest which he shared with his good friend Marcel Riesz.

Wigert

Still another representative of analytic number theory was Carl Severin Wigert (1871 - 1941), a scholar of independent means, attached to the university as a docent. The theme of his most important papers is majorants and asymptotics for number-theoretic functions connected with $d(n)$, the number of divisors of an integer n including 1. When n is a prime, then $d(n) = 2$ but $d(n)$ can also be very large. It is obvious that $d(n) \leq n$ and that this estimate can be improved. Runge had shown

that $d(n) = O(n^\varepsilon)$ for every $\varepsilon > 0$ when Wigert published the definitive result in (1906),
$$d(n) \leq (1 + o(1))2^{\log n/\log\log n}.$$
In a later paper (1914) there are asymptotic formulas for $\sigma(n)$, which is the sum of the divisors of n divided by n, and for a number of similar function, for instance
$$f(x) = \sum_{n \leq x} \sigma(n).$$
Among other things it is shown that $\sigma(n)$ has the order of magnitude at most $\log\log n$ and that $f(x)$ has the main term $\pi^2 x/6$.

Wigert found an equivalent formulation of Riemann's hypothesis (1909) and in (1918) he returned to asymptotic estimates of the numbers $d(n)$. Some reflection shows that they appear as coefficients in the Lambert series
$$\sum_1^\infty x^n/(1-x^n) = \sum_1^\infty d(n)x^n$$
which Wigert rewrites as
$$L(z) = \sum_1^\infty 1/(e^{nz}-1) = \sum d(n)e^{-nz}$$
where $\operatorname{Re} z > 0$. Via an approximative functional equation for $L(z)$ Wigert gets an asymptotic formula
$$\sum \frac{1}{k!} \sum_{n<x} d(n)(x-n)^k = x^{k+1}\log x/(k+1)! + \ldots$$
where $k \geq 1$ and ... stands for certain explicit remainder terms.

Cramér

Harald Cramér (1893-1985) was the pioneer of modern probability theory in Sweden and also its leading representative for several decades. In the early 1930's Kolmogorov's axiomatizing of probability theory made it a branch of mathematics. Afterwards, Cramér's influential textbooks, *Random Variables and Probability Distributions* (1937) and *Methods of Mathematical Statistics* (1946) gave to the earlier results a strict mathematical form in the new frame of reference. This contribution could hardly have been possible without the mathematical training the author received when he studied analytical number theory at Stockholm university from 1917 to 1923. At the end of this period Cramér and Harald Bohr wrote about recent results of the subject in the mathematical encyclopedia (1923).

After his thesis in 1917 Cramér made a strong effort in analytical number theory with several papers in a short time. He summed up his most important in a contribution to the Scandinavian congress of mathematicians in 1922. At that time Littlewood had shown that the asymptotic laws of the theory had such large deviations that essential improvements were not possible. It is for instance not true that
$$\psi(x) = x + O(\sqrt{x}).$$
where the left side is defined above (p. 134). Cramér (1922 b) proved that the inequality is true in the mean, i.e., that
$$x^{-1}\int_2^x |\psi(t) - t|dt = O(\sqrt{x}),$$

and he proved a similar formula

$$\frac{1}{\log x} \int_2^t \left(\frac{\psi(t)-t}{t}\right)^2 dt \to \sum |1/\varrho|^2$$

where ϱ runs through all the zeros of the ζ-function in the critical strip. In a short note (1920) Cramér wrote about prime number solutions x, y of the equation $ax + by = c$ where a, b, c are integers. When $x + y = c$ this gives Goldbach's problem (every integer is a sum of two primes) and when $x = y + 2$ the problem of the existence of infinitely many twin primes. Using the prime number theorem, the author proves that there are many integral triples a, b, c such that infinitely many pairs of primes have an arbitrarily small distance to the line $ax + by + c = 0$.

Fourier analysis

Riesz, Zeilon

Mittag-Leffler had made analytical functions the main theme of mathematics at Stockholm University, but after 1910 also other parts of analysis were included. Two newcomers, Marcel Riesz and Nils Zeilon, represented Fourier analysis, each in his own way.

Riesz

Marcel Riesz (1886-1969) was the youngest in a successful generation of Hungarian mathematicians, Lepold Fejér, Riesz's elder brother Fredrik Riesz and Alfred Haar. From them Riesz got his early interest in the convergence and summation of series. Riesz's first paper (see (1976)), written when he was twenty, deals with ways of continuing a power series analytically outside the circle of convergence. At the time this theme was also close to Mittag-Leffler. He and Riesz met for the first time in 1908 or 1909. This contact led to a job for the young mathematician. Mittag-Leffler paid Riesz a year's salary to edit a portrait volume of *Acta Mathematica* where all contributors of the first thirty-three volumes are presented with portraits and personal biographies.[2] Later Riesz became a docent at the university and he stayed there till 1927 when he moved to Lund University to become a professor there.

Between 1908 and 1916 Riesz worked successfully with summation theory of power series, trigonometric series and Dirichlet series.

A series $a_0 + a_1 + a_2 + ...$ is convergent when the sequence $S = s_0, s_1, ...$ of partial sums

$$s_n = a_0 + ... + a_n$$

converges. A method of summation M is a process which averages the partial sums by a family of weights $M(n, k) \geq 0$ with sums over k equal to 1 and centers of gravity far away (see p. 84-85). The sequence S is then replaced by another sequence

$$s_n^{(M)} = \sum_{k=0}^{\infty} M(n, k) s_k.$$

When this sequence has a limit, this limit sums the series. The point of the averaging is that oscillating sequences S can be summed in a consistent way. A simple example is summation by arithmetic means,

$$s_n^{(1)} = (s_0 + ... + s_n)/(n+1), \quad n = 0, 1, 2, ...$$

---

[2] The greater part of photographs of the present book come from this source.

where $n \to \infty$. This method sums for instance the sequence

$$1, 1, -1, 1, 1, -1, 1, 1, -1, \ldots$$

to $1/3$. The sequence of arithmetic means can then be summed in the same way and so on.

Means of fractional order $\alpha > 0$ of the sequence $s_0, s_1, \ldots$ are defined in such a way that $s_n^{(\alpha)}$ is the quotient between the corresponding coefficients of the two formal power series

$$(1-x)^{-\alpha} \sum_0^\infty s_n x^n, \quad (1-x)^{-\alpha-1}.$$

For $\alpha = 0$ this gives the sequence S, for $\alpha = 1$ the sequence $s_n^{(1)}$ and so on. A sequence $s_0, s_1, \ldots$ is said to be Cesàro summable of order α or (C, α) summable, if the sequence $s_n^{(\alpha)}$ converges.

The interest in summation was first concentrated on Fourier series of functions $f(x)$ which are periodic with the period 2π,

$$a_0/2 + \sum (a_n \cos nx + b_n \sin nx),$$

where

$$a_n = \int_{-\pi}^{\pi} f(x) \cos nx \, dx / \pi, \quad b_n = \int_{-\pi}^{\pi} f(x) \sin nx \, dx / \pi.$$

When f is only continuous, the series need not converge for all x, but summation by arithmetic means gives convergence. When the function is integrable in the Lebesgue sense, the means converge at every Lebesgue point.

In a note (1909) with proofs (1923 a) Riesz extended these theorems by Fejér and Lebesgue to (C, α) summability when $\alpha > 0$. The theorem is proved in Zygmund's standard work on Fourier series (1959), part I, page 94. But Riesz's best known contribution is the introduction of what is now known as Riesz summation. He discovered that Cesàro summation of order α is equivalent to another summation which can be used to sum limits at infinity of functions $s(x)$ of bounded variation for which $s(0) = s(+0) = 0$, namely as the limit of

$$(1) \qquad s^{(\alpha)}(x) = x^{-\alpha} \int_0^x (x-t)^\alpha ds(t)$$

when $x \to \infty$. For a sequence $s_0, s_1, \ldots$ we put $s(x) = s_n$ when $n \le x < n+1$. This summation method, introduced by Riesz (1909) under the name of typical means, can also be applied to Dirichlet series

$$(2) \qquad \sum_0^\infty a_n e^{-\lambda_n s}$$

where the a_n are complex numbers, (λ_n) a sequence of real numbers tending to infinity and $s = \sigma + it$ a complex variable. Dirichlet series first appeared in number theory. The prime example is Riemann's ζ function

$$\zeta(s) = \sum_1^\infty n^{-s}.$$

Dirichlet series are more complex than power series. The two maximal half-planes $\operatorname{Re} s > \text{const.}$, where the series converges or converges absolutely, may be different, and convergence or non-convergence at the corresponding boundaries may lead to very difficult problems.

The Dirichlet series (2) can also be written in integral form. If the function $a(t)$ is constant between the numbers λ_n and is increased by a_n at λ_n, (2) can be written as

$$f(z) = \int_0^\infty e^{-zt} da(t)$$

where $\operatorname{Re} z > 0$. The typical means are

$$x^{-\alpha} \int_0^x (x-t)^\alpha e^{-zt} da(t)$$

where the right side is similar to the Riemann-Liouville integral

$$I^\alpha f(x) = \frac{1}{\Gamma(\alpha)} \int_0^x (x-t)^{\alpha-1} f(t) dt.$$

The property of this integral to constitute a semigroup,

$$I^{\alpha+\beta} = I^\alpha I^\beta,$$

where I^0 is the identity, gives important properties of the typical means, now called Riesz summation.

After 1911 Riesz became known as an expert on Dirichlet series and their summation and in 1915 he wrote a book on the subject together with G.H. Hardy.

Another of Riesz's problems, both for power series and Dirichlet series, was the convergence and summability at the points of the boundary where the function can be continued analytically. He wrote twice, in (1911) and (1916 a), about a theorem by Fatou which says that a power series $f(z) = \sum a_n z^n$ with the property that $a_n = o(1)$ converges locally uniformly on every arc of the circle of convergence where the funcion is analytic. If, instead, $a_n = o(1) n^\gamma$ with $\gamma > 0$ the same theorem holds for summation of order γ. In the second and simplest proof, the functions

$$g_n(z) = \frac{(f(z) - (a_0 + \ldots + a_n z^n))(z - z_1)(z - z_2)}{z^{n+1}}$$

are introduced where z_1, z_2 are on an arc of analyticity. If the sequence is bounded, it is not difficult to see that there is a number $R > 1$ such that $g_n(z)$ is majorized by a constant c_n on the boundary of a sector with center at 0 and endpoints Rz_1, Rz_2. If $a_n = o(1)$, it is easy to see that $c_n = o(1)$ as $n \to \infty$.

In (1912) Serge Bernstein proved two interesting inequalities. The first one says that $|f'(\theta)| \leq nM$, where M is the absolute maximum of some trigonometric polynomial f of degree at most n,

$$f(\theta) = a_0 + \sum_1^n (a_k \cos k\theta + b_k \sin k\theta),$$

and the second one that

$$|g(z)| \leq M(A+B)^n$$

where $g(z)$ is a polynomial of degree $\leq n$ such that $|g(x)| \leq M$ when $|x| < 1$ and A, B are the principal axes of an ellipse through z with foci at ± 1. Riesz (1914, 1916b) gave new proofs in both cases, the first one with the aid of an interpolation formula

$$f'(\theta) = \frac{1}{2n} \sum c_k f(\theta + \theta_k), \quad \theta_k = \frac{(2k-1)\pi}{2n}, \quad c_k = \frac{(-1)^{k+1}}{2n} \sin^2 \frac{\theta_k}{2}.$$

Bernstein's first inequality follows since $\sum c_k = 2n^2$ (1914,1916 b).

At the Scandinavian mathematical congress in 1916 Riesz and his brother Fredric contributed a paper frequently quoted in harmonic analysis. Their theorem says that a measure $d\mu(\theta)$ on the unit circle $E : 0 \leq \theta \leq 2\pi$ whose Fourier coefficients $\int_E e^{im\theta} d\mu(\theta)$ vanish for all sufficiently large m, positive or negative, is absolutely continuous, i.e., there is an integrable function $f(\theta)$ such that $d\mu(\theta) = f(\theta)d\theta$.

Riesz's work in Fourier series gave him a prestigious task: to write together with E. Hilb about recent results in the field for the new edition of the mathematical encyclopedia (1924).

At the end of the first world war Riesz had a temporary position at Stockholm University and he was known for short and carefully written papers in classical analysis. It was perhaps in order to improve his position in future competitions for more permanent posts that he wrote an ambitious paper (1921 - 1923) about the moment problem, divided into three notes. His friend Stridsberg had acquired an intense interest in the subject.

The moment problem had been introduced by Stieltjes (1894) in a paper which also introduced Stieltjes integrals, i.e., integrals

$$\int_a^b f(t) d\mu(t)$$

of continuous functions with respect to a measure $d\mu(t) \geq 0$. The measure is supposed to have moments

$$c_n = \int_a^b t^n d\mu(t).$$

of all orders $n = 0, 1, 2, \ldots$. The problem was to construct the measure from the moments. It was solved by Stieltjes when $a = 0$, $b = \infty$ and Hamburger (1920) was first with the more difficult case $a = -\infty$, $b = \infty$. In both cases it is clear that the set of solutions is convex for if $d\mu_1$ and $d\mu_2$ have the same moments, every convex linear combination $\alpha d\mu_1 + \beta d\mu_2$ also has them.

Riesz's paper is essentially a clarification of the moment problem with few new results. But there is one exception. An argument given below, which was to appear later in Banach's use of transfinite induction to extend a linear functional from one linear space to a bigger one, gave Riesz an insight into the nature of the moment problem and an answer to a question raised by Stridsberg.

For a polynomial $f(t) = \sum a_k t^k$ the integral

(1) $$\int f(t) d\mu(t) = \sum a_k c_k$$

is known through the moments. When the measure is non-negative, the integral is ≥ 0 when $f(t) \geq 0$. In the general case the measure is known when the integral

(2) $$\int g(t) d\mu(t)$$

is known on the space C of all continuous functions with compact supports which in turn has a countable, everywhere dense subset D.

It seems natural to try to define the integral (2) as the supremum S of (1) for all polynomials $\leq g$ or as the infimum I of (1) for all polynomials $\geq g$. If these two numbers are the same for all g, the measure can be reconstructed and the problem is solved. But it may happen that $I > S$ in which case the moment problem is indeterminate. Riesz's way out was to give to (2) a value between I and S and then repeat the procedure, now with the space P of polynomials replaced by the linear hull $L(P,g)$ of P and g. This process goes on indefinitely and stops only when the integral is defined for all functions in D and then a measure is constructed which solves the moment problem.

In this connection the integral

$$\int \frac{d\mu(t)}{t-z}, \quad \text{Im } z \neq 0,$$

is interesting. Rolf Nevanlinna (1922) proved that its values for a fixed non-real z and all $d\mu(t)$ with given moments fill out a compact disk in the lower or upper half-plane which contains z. For points on the boundary of the disk, the corresponding measures, said to be extremal, are unique.

A possible computational approach to the moment problem is as follows. Through its moments, the unknown measure gives an interior product

$$(f,g) = \int f(t)\overline{g(t)}d\mu(t)$$

and a norm $\|f\| = \sqrt{(f,f)}$ for polynomials. If the norm vanishes for one polynomial $\neq 0$, the measure is supported by a finite number of points which is uninteresting. Otherwise we may orthonormalize a set of the polynomials of degrees $0,1,2,\ldots$ to a sequence $h_0(t), h_1(t), \ldots$. It then turns out that the moment problem has a unique solution if $|h_n(z)| \to \infty$ for Im $z \neq 0$, a property independent of z. A condition for an extremal solution is that every element $f(t)$ which is square integrable with respect to the measure satisfies Parseval's formula, i.e., $\|f\|^2$ is the sum of the squares of its coefficients in an expansion in terms of $h_0(t), h_1(t), \ldots$. (Riesz (1923 b)).

With time the moment problem has acquired a large literature (Shohat-Tamarkin (1943), Landau (1985)). One way of treating the problem is to investigate the symmetric map $f(t) \to tf(t)$ in the associated Hilbert space.

Very soon after his arrival in Stockholm University, Marcel Riesz began to play an important part as a conversation partner and guide for some of the younger mathematicians, for instance Harald Cramér and Einar Hille. In 1927 he was nominated professor at Lund University and there his research took a new turn.

Zeilon

When Nils Zeilon (1886-1957) began his studies at Stockholm University he had two interests outside of mathematics, mathematical physics and painting, and he was successful in both.

The years after the turn of the century Fredholm's reputation was at its peak. Zeilon became his student and admirer. In his first papers Zeilon extended Fredholm's construction of fundamental solutions to more general partial differential operators with constant coefficients. Later he used the insight gotten from this work to study the propagation of light in doubly refracting media, a problem which had thwarted earlier efforts. An appendix at the end of this chapter, quoted below as (A), serves as a background to the text below.

When $P(\xi)$, $\xi \in R^n$, is a polynomial, the integral

(1) $$E(x) = (2\pi)^{-n} \int e^{(x,\xi)} d\xi/P(\xi)$$

is a formal solution of the equation $P(D)E(x) = \delta(x)$ when $D = \partial_x/i$. In other words, $E(x)$ is a formal fundamental solution of the differential operator $P(D)$. The object of the greater part of Zeilon's papers is to give a sense to the integral. How this can be done with the aid of distribution theory is shown in (A). It is certain that Zeilon would have recognized most of what is said there about fundamental solutions, but it is not certain that this would have interested him. General theories and attention to detail was not his line. For him the end result was everything and the road taken uninteresting. His future colleague in Lund between 1927 and 1952, Marcel Riesz, took almost the opposite view.

Zeilon on fundamental solutions

Zeilon's object in his papers (1913 a,b) was not only to give a sense to the integral but also to investigate its properties as a function of x for complex values of this variable. Verifications as in Fredholm's papers that the functions $E(x)$ which are constructed really are fundamental solutions are missing in Zeilon's work. His perspective was that of the mathematical physicist.

A direct attack on (1) meets with difficulties even when $P(\xi)$ is homogeneous of degree m and elliptic so that $P(\xi) \neq 0$ when $\xi \neq 0$. In fact, the integral is absolutely convergent at the origin when $n - m > 0$ and at infinity when $n - m < 0$. If the polynomial P has real zeros, the difficulties multiply. Zeilon treats homogeneous polynomials in three and four variables, to begin with elliptic ones. Suppose for instance that $n = m = 3$ as in (1913 a). The author reasons roughly as follows.

If we introduce new variables $\xi = t\eta$ in (1) where $\eta_1 = t$, we get an integrand

$$e^{it(x,\eta)} t^{-1} dt\, d\eta_2\, d\eta_3 / P(\eta).$$

Here we may integrate with respect to t which gives

$$E(P, x) = \text{const} \int \text{sgn}(x, \eta) d\eta_2 d\eta_3 / P(\eta).$$

Exchanging $\text{sgn}(x, \eta)$ for $\text{sgn}(x, \eta) + 1$, modifies the right side only by a constant and we get a fundamental solution

$$E(P, x) = \text{const} \int_{\text{sgn}(x,\eta) > 0} d\eta_2 d\eta_3 / P(\eta).$$

Outside the zeros of the denominator we can now modify the two paths of integration $(x, \eta) > 0$ and $-\infty < \eta_3 < \infty$ in the corresponding complex planes. Then η_2 will begin at $-\infty$ and end in a point determined by η_3 through the condition $x \cdot \eta = 0$. At the same time we can let η_3 take a loop around all the zeros of the denominator where $\text{Im}\, \eta_3 > 0$. If we carry out the residues (the zeros are assumed to be separate) the result is

(2) $$E(P, x) = \text{const} \int_C d\eta_2 / P_3$$

where $P(\eta) = 0$ and $P_3 = \partial P(\eta)/\partial \eta_3$. Under all these manipulations with the path of η_3 the path of η_2 is changed to three paths C from a fixed point on the complex curve $P(\eta) = 0$ to the points on the same curve where the line $x \cdot \eta = 0$ meets the curve and $\text{Im}\, \eta_3 > 0$. If, for a moment, we exchange η_2, η_3 for x, y, the integrand of (2) equals $dx/P_y = -dy/P_x$, the classical invariant differential form on the curve $P = 0$. The right side of (2) is a sum of Abelian integrals.

When $n = 3$, $m > 3$ the result is similar. The integrand of (2) gets the factor $(x, \eta)^{m-3}$:

(3) $$E(P, x) = \text{const} \int_C (x, \eta)^{m-3} d\eta_2 / P_3$$

and this is the same formula that Fredholm obtained by combining Abelian integrals. The point of departure (1) had produced a known formula and this gave Zeilon a good reason to continue with operators $P(D)$ which are not elliptic.

A difficulty arises when $P(\xi) = 0$ for real $\xi \neq 0$. But it is clear that this may be avoided by a slight modification of the original path of integration. Zeilon chooses to take the principal value and gets the formula (3) where the paths which lead to the intersection of $P(\eta) = 0$ and $(x, \eta) = 0$ have the weight $1/2$.

As a pendant to Fredholm's contruction of a fundamental solution of the elliptic operator $D_1^4 + D_2^4 + D_3^4$ Zeilon used the formula (1) for a careful construction of a fundamental solution of the operator $P(D) = D_1^3 + D_2^3 + D_3^3$. According to (A) the fundamental solution ought in this case to be regular except at points x such that the real plane $(x, \xi) = 0$ touches the surface $P(\xi) = 0$.

In this case the tangency means that $x_k = \lambda \xi_k^2$ with λ real and $P(\xi) = 0$. This turns out to define a conical surface containing three lines through the origin and the points $(1, 1, 0), (1, 0, 1), (0, 1, 1)$. Its intersection with the plane $x_1 + x_2 + x_3 = 1$ is a triangle with its three sides curved inwards.

By computations like the one above Zeilon could see that the fundamental solutions he constructed were without singularities as long as the real plane $(x, \xi) = 0$ and the surface $P(\xi) = 0$ do not touch, a condition which defines the singularity surface in x-space. He also did other computations where he let the point x go around the singularity surface in complex space, a procedure which in certain cases gave him an idea of the behavior of the fundamental solution at the singularity surface.

Zeilon's analysis of the divergent integral (1) gave him all the tools he needed to solve the famous problem of the refraction of light in doubly refracting crystals. This problem was touched upon in connection with Sonya Kovalevski in Chapter 9. The more complete account here is motivated both by Zeilon's work and the historical interest of the problem (see Gårding (1990)).

Light in doubly refracting crystals, Lamé, Kovalevski

The French mathematical physicist Fresnel took the first step in the mathematical analysis of the problem of double refraction. In 1820 he discovered that the two light velocities v_1, v_2 in the crystal in the direction ξ obey the velocity equation

$$\sum_1^3 \xi_j^2/(v^2 - a_j) = 0$$

where $a_1 > a_2 > a_3 > 0$ are constants depending on the crystal. The numbers a_2, a_3 are the squares of the velocities when $\xi = (1, 0, 0)$ and analogously for the two other pairs. One of the square velocities is in the interval a_1, a_2, the other one in a_2, a_3. When they coincide, which happens for two directions in the plane $\xi_2 = 0$, the value is a_2. If we mark out the two velocities along the vector ξ, we get a velocity surface which is symmetric under reflections in the coordinate planes and has two sheets which come together in four points where the velocities coincide.

According to Huygens's theory of light, a point-source which lasts just one moment emits plane waves with velocities v_1, v_2 at right angles to the plane. For a fixed time t, for instance $t = 1$ these plane waves envelope a wave surface which describes the position of the wave front after one unit of time. The equation of the wave surface, found a short time after Fresnel's discovery, is

$$\sum_1^3 a_k x_k^2/(|x|^2 - a_k) = 0.$$

It has the same form as the velocity surface with two sheets which come together in four conical points. The inlets to these conical points, which will be important later, may be covered by circular lids which touch the wave surface in circles. The lids together with parts of the outer sheet of the wave front surface bound its convex hull.

The analysis of double refraction given by the velocity and wave surfaces is part of geometrical optics whose object is the study of wave fronts rather than diffuse light. The complete theory is now available in the form of Maxwell's equations for the electromagnetic field. Almost at the same time as Maxwell, the French physicist Lamé deduced the laws of propagation of light by an analogy with elasticity theory. In this theory the components u, v, w of the deviation vector satisfy a system of three linear differential equations which result from various hypotheses about the elastic energy. The equations express the accelerations u_{tt}, v_{tt}, w_{tt} as linear combinations of the space derivatives.

In his book (1866), where he published his lectures on elasticity theory, Lamé established in the last part a system of equations supposed to describe the propagation of light in doubly refracting crystals. He saw light as vibrations of weightless matter (the aether) and wrote down the following system of equations for the deviation vector $u = u(t, x) = (u_1, u_2, u_3)$:

$$u_{tt} + \partial \times a \partial \times u = 0, \quad \operatorname{div} u = 0.$$

Here $\times$ means vector product, $\partial = (\partial_1, \partial_2, \partial_3)$ is differentiation with respect to x and a is a diagonal matrix with the diagonal elements a_1, a_2, a_3. The equation can also be written as

$$u_{tt} + \operatorname{rot}(a \operatorname{rot} u) = 0, \quad \operatorname{div} u = 0$$

and is precisely what one gets from Maxwell's equations when the electric vector is eliminated.

That Lamé's system has something to do with double refraction is clear from the fact that it has solutions $0 \neq u = Ae^{it\tau + i(x,\xi)}$ such that τ equals $v(\xi)|\xi|$ where $v(\xi)$ solves the velocity equation. His system is also hyperbolic. It can be written in the form $P(D_t, D)u = 0$ where $P(\tau, \xi)$ is a square matrix whose elements are second degree polynomials in τ, ξ and the polynomial $\det P(\tau, \xi)$ is hyperbolic with respect to τ.

Lamé tried to construct solutions of his system analogous to Euler's spherical waves. With $\lambda(x)$ homogeneous of order 1 and equal to 1 on one sheet of the wave surface, Lamé's solution is $u(t, x) = f(t - \lambda(x))U(x)$ where the time independent function $U(x)$ has the form

$$U(x) = \frac{C\lambda(x)(x \times \lambda_x)}{\sqrt{|\prod N_k(x)|}}$$

with $N_k(x) = |x|^2 - a_k\lambda^2$ and C a constant. These functions, which will be called Lamé waves, are entirely analogous to the spherical waves, but they do not solve Lamé's system in all of space (which is not so easy to see without distribution theory (Gårding (1990))). The proper solutions $u(t, x)$ of Lamé's system which describes propagation of light form a source at $x = 0$, $t = 0$ are such that $u(0, x) = 0$, $u_t(0, x) = g(x)$ where $g(x)$ is a distribution with support at $x = 0$ satisfying $\operatorname{div} g(x) = 0$.

The Lamé waves were found in another form by Sonya Kovalevski (1886) and by Volterra (1892), who observed that although these waves are local solutions of the system they cannot describe wave propagation from a center. In 1905, when Volterra gave a series of lectures in Stockholm, he pointed out that wave propagation according to Lamé's system was a mystery and a challenge. In a later paper (1914) he proved that the side condition div $u = 0$ could be eliminated by replacing Lamé's system by a corresponding system of differential equations for a vector potential $U(t,x)$ such that $u(t,x) = \partial \times U$. This new system has a fundamental solution form which one gets a solution of Lamé's system with the correct properties.

Zeilon's solution

In his magnum opus (1919) Zeilon transposed his calculations to four variables and could reduce his Fourier integrals to Abelian integrals. For if $P(\xi)$ and $Q(\xi)$ have the degrees m and $m - 3$ and the number of variables is 4, we get

$$Q(D)E(P,x) = \frac{1}{(2\pi)^4} \int e^{i(x,\xi)} Q(\xi) d\xi / P(\xi)$$

and, if we put $\xi = t\eta$, $t = \xi_1$ as above, then

$$Q(D)E(P,x) = \frac{1}{(2\pi)^3} \int \delta((x,\eta)) Q(\eta) d\eta / P(\eta)$$

where the right side is an integral over the plane $x \cdot \eta = 0$. One of the variables can then be integrated out by residue calculus which avoids possible real zeros of the denominator. This leaves an abelian integral over a cycle $\gamma(x)$ in the complex curve $C(x) : (x,\eta) = 0$, $P(\eta) = 0$, i.e., the complex projective intersection of the plane $(x,\xi) = 0$ and $P(\xi) = 0$. If this procedure is correctly done, it will yield solutions of Volterra's modification of Lamé's system which vanish when $t < 0$, have support at $x = 0$ for $t = 0$ and, for $t > 0$, vanish outside the cone which is determined by the largest velocity in the crystal. Zeilon's results are the following ones.

(1) The vector potential $U(t,x)$ vanishes for $t < 0$ and for $t > 0$ when x/t is outside of the convex hull of the wave surface or inside the inner sheet.

(2) The time integral of $U(t,x)$ has explicitly given jumps when x/t passes a convex or concave part of the wave surface. It does not vanish inside the lids which bound the outer inlets to the conical points and has explicity given jumps at the lids.

We see here how the proper wave surface, defined as the locus of the singularities of the fundamental solution, contains the lids of the outer inlets to the conical points of the geometric wave surface. The Lamé waves, which are singular only at the geometrical wave surface, do not contain these singularities and this is the reason why they do not solve the propagation problem.

Zeilon's two-part paper (1919-1921) is a remarkable feat and solves a problem which had thwarted Lamé, Kovalevski, and Volterra. The weakness is that it is practically unreadable for anyone who does not already know the result. We see here an extreme example of Zeilon's philosophy: the result is everything. This first part, which deals with Abelian integrals in general, got a short mention in *Jahrbuch über die Fortschritte der Mathematik*. The second part, which contains the main results, was never mentioned. This is the reason why Zeilon is not quoted in the papers by Herglotz (1926) and Petrovski (1945) where fundamental solutions of general strongly hyperbolic operators are constructed.

A short and unnoticed paper by Grünwald (1904), which was unknown to Zeilon, bereaved him of his priority. As Zeilon did, Grünwald used the Fourier transform to master the complicated propagation of light in doubly refracting crystals and his paper contains formulas that can be used to prove Zeilon's result.

Appendix: *On Huygens's principle and lacunas*

Lamé was misled in the first place by the fact that his calculations succeeded in a miraculous way and in the second place by his belief that light is always confined to the geometrical wave surface. That this holds for the wave equation in space is sometimes called Huygens's (minor) principle and follows from the form of the fundamental solution,

$$H(t)\delta(t^2 - |x|^2)/4\pi,$$

The first one to observe the opposite phenomenon was Volterra. He found the fundamental solution $H(t)/2\pi\sqrt{(t^2 - |x|^2)_+}$ of the wave equation in two space variables which does not vanish inside the light cone. But Huygens's principle is partly true for the Lamé system since its fundamental solution vanishes inside the inner sheet of the wave surface.

The first one to take an interest in the problem to characterize the regions inside the propagation cone for hyperbolic operators where the fundamental solution vanishes was Petrovski (1945). He considered homogeneous hyperbolic operators $P(D)$ in n variables $x_1, ..., x_n$, such that the complex characteristic surface $P(\xi) = 0$ is singular only at the origin and fundamental solutions $E(P, x)$ with support in a propagation cone U. It turns out that $E(P, x)$ is analytic inside U and outside the singularity surface S which consists of points x where the real plane $(x, \xi) = 0$ touches the real surface $P(\xi) = 0$. The singularity surface S bounds the propagation cone U but can also have pieces inside. A component of $U \backslash S$ where the fundamental solution $E(P, x)$ is a polynomial (which vanishes when $E(P, x)$ is homogeneous of degree < 0) was called by Petrovski a *lacuna*.

When P is the wave operator in three space variables, the interior of the propagation cone is a lacuna and the same phenomenon appears partly for the propagation of light in doubly refracting crystals. In the general case, lacunas are an exception and diffuse light the rule. In his paper Petrovski could characterize lacunas by a topological condition: x belongs to a lacuna when a certain cycle in the complex intersection of the characteristic surface and the plane $(x, \xi) = 0$ is homologous to zero in the same intersection. When n is even, $C(x)$ is simply the real intersection of $P(\xi) = 0$ and $(x, \xi) = 0$. Example: for the wave equation with the variables $t, x = (x_1, x_2, x_3)$ and the dual ones τ, ξ the cycle $C(t, x)$ is empty when $t > 0$, $x = 0$ (or if $t > |x|$), for the real plane $\tau = 0$ does not meet the real sheet $\tau^2 - \xi_1^2 - \xi_2^2 - \xi_3^2 = 0$ outside the origin. Petrovski's criterium gives Huygens's principle in this case.

Petrovski's paper was extended to arbitrary hyperbolic operators with constant coefficients by Atiyah, Bott, and Gårding (1970).

BIBLIOGRAPHY

ATIYAH M., BOTT R., GÅRDING L.
1970. *Lacunas for Hyperbolic Differential Operators with Constant Coefficients I,II*, Acta Mathematica 124, 1970, 110-189; 131, 1973, 145-206.

BOUTROUX P.
1908. *Leçons sur les fonctions définies par les équations différentielles du premier ordre*, Coll. Borel 1908.

de BRUN F.
1893. *Rotation kring en fix punkt*, KVA Öfversigt 1893.
1908. *Rotation kring en fix punkt II*, Arkiv Mat. Astr. Fys. bd 4, 1908, no 5.
1909. *Sur une généralisation de la fonction gamma*, Arkiv bd 5, 1909, no 10.

CASSEL K.G.
1891. *Sur un problème de représentation conforme*, Acta Math. 15, 1891, 33-44.
1894. *Kritiska studier öfver teorien för de auotomorfa funktionerna jämte deras användning för integration af linjära differentialekvationer*, Ak. Afhandl. Uppsala 1894.

CRAMÉR H.
1917. *Sur une classe de séries de Dirichlet*, Diss. Stockholm 1917.

1920. *Nombres premiers et équations indéterminées*, Arkiv 14, 1920, no 12.
1922. *Ein Mittelwertsatz in der Primzahltheorie*, Math. Zeitschr. 12, 1922, 147-153.
1923. a. *Contributions to the analytic theory of numbers*, Femte Skand. Mat.- Kongr. Helsingfors 1922.
1923. b. *Die neuere Entwicklung der analytischen Zahlenteorie*, Enz. Math. Wiss 2 C8, 723-849 (med Harald Bohr).
1937. *Random variables and probability distributions*, Cambridge Tracts in Mathematics 36, 1937, omtryckt 1970.
1945. *Mathematical Methods of Statistics*, Stockholm 1945, Princeton Univ. Press 1946.
DOMAR Y.
1978. *Matematisk forskning under Stockholms Högskolas första decennier*, manuskript.
GRÜNWALD J.
1904. *Über die Ausbreitung der Wellenbewegungen in optisch-zweiachsigen elastischen Medien*, Festschrift Boltzmann, 518-527, Leipzig 1904.
GRÖNWALL H.
1895. *Om system af lineära totala differentialekvationer*, KVA Öfversigt 1895.
1897. a. *Notes sur les fonctions et nombres algébriques*, KVA Öfversigt 1897.
1897. b. *Deux théorèmes sur les nombres transcendants*, KVA Öfversigt 1897.
1898. a. *Sur les nombres transcendants II*, KVA Öfversigt 1898.
1898. b. *Om system af linjära differentialekvationer*, Akad. Afh. Uppsala 1898.
GUSTAFSON T.
1958. *Nils Zeilon*, Fysiogr. Sällsk. Förhandl. 1958 Lund.
GÅRDING L.
1990. *History of the mathematics of double refraction*, Archives Hist. Math. 40, 1990, 355-385.
HAMBURGER H.
1920. *Ueber eine Erweiterung des Stieltjes'schen Momentproblems*, Math. Ann. 81, 1920, 235-319; 82, 1921, 120-164, 168-187.
HERGLOTZ G.
1926. *Über die Integration linearer partieller Differentialgleichungen mit konstanten Koeffizienten I-III.*, Ber. der Sächsischen Akad. der Wiss. 78, 1926, 93-126,287-318; 80, 1928, 69-114.
HILLE E.
1917. *Über die Variation der Bogenlänge bei konformer Abbildung von Kreisbereichen*, Arkiv 11, 1916-17, no 27.
1918. *Some problems concerning spherical harmonics*, Arkiv 13, 1918, no 17.
1922. *On the Zeros of Sturm-Liouville Functions*, Arkiv 16, 1922, no 17.
1932. *H.T. Gronwall in memoriam*, Bull. Am. Math. Soc. 38, 1932, 775-786.
HORN I.
1889. *Über ein System linearer partieller Differentialgleichungen*, Acta Math. 12, 1889, 113-175.
HURWITZ A.
1883. *Über aritmetische Eigenschaften gewisser transcendenten Funktionen*, Math. Ann. 22, 1883, 211-229.
KOBB G.
1887. *Sur le mouvement d'un point matériel sur une surface de révolution*, Acta Math. 10, 1887, 89-108.
1892. *Sur les maxima et minima des intégrales doubles*, Acta Math. 16, 1892-93, 65-140.
1893. *Sur les maxima et minima des intégrales doubles, Seconde mémoire*, Acta Math. 17, 1893, 321-343.
1894. a. *Rotations d'un corps autour d'un point fixe*, Soc. M. France Bull. 23, 1994, 210-215.
1894. b. *Solutions périodiques dans le problème des trois corps*, KVA Öfversigt 52, 1894, 215-22.
von KOCH H.
1901. *Sur la distribution des nombres premiers*, Acta Math. 34, 1901, 159-182.
1910. *Contribution à la théorie des nombres premiers*, Acta Math. 33, 1910, 293-320.
KOVALEVSKI S.
1886. *Über die Brechung des Lichtes in crystallinschen Mitteln*, Acta Mat. 6, 1886, 249-304.
LAMÉ G.
1866. *Leçons sur la théorie de l'élasticité des corps solides.*, Deuxième édition, Paris 1866.
LANDAU E.
1909. *Handbuch der Lehre von der Verteilung der Primzahlen I,II*, Leipzig 1909.
LANDAU H.
1987. *Moments in Mathematics. H. Landau Ed.*, Proc. Symp. Appl. Math 37, 1987, Amer. Math. Soc.
LIOUVILLE J.
1844. *Nouvelle démonstration d'un théorème...*, C.R. Paris vol 18, 1844, 910-911.
MALMQUIST J.

1905. *Étude d'une fonction entière*, Acta Math. 29, 1905, 203-215.
1914. *Sur les fonctions à un nombre fini de branches définies par les équations différentielles du premier ordre*, Acta Math. 36, 1914, 297-343.
1921. *Sur les points singuliers des équations différentielles* I,II, Arkiv 15, 1921, nos 3, 27, 28.
1941. *Sur l'étude analytique des solutions d'un système d'équations différentielles dans le voisinage d'un point singulier* I,II,III, Acta Math. 73, 1941, 87-129; 74, 1941, 1-64, 109-128.
    NEVANLINNA R.
1922. *Asymptotische Entwicklung beschränkter Funktionen und das Stieltjessche Momentproblem*, Ann. Ac. Sc. Fenn. t. 18, 1922, no 5.
    PAINLEVÉ P.
1903. *Leçons sur la théorie analytique des équations différentielles professées à Stockholm*, Institut Mittag-Leffler.
    PETRINI H.
1908. *Les dérivées premières et secondes du potentiel*, Acta Math. 31, 1908, 127-332.
1917. *La densité d'une masse électrique en équilibre*, Arkiv 12, 1917, nr 8.
    PETROVSKI I.G.
1945. *On the diffusion of waves and the lacunas for hyperbolic equations*, Mat. Sbornik t. 17, 1945, no 3, 289-370.
    PHRAGMÉN E.
1891. a. *Le logarithme intégral et la fonction $f(x)$ de Riemann*, KVA Öfversigt 48, 1891, 599-616.
1891. b. *Über die Berechnung der einzelnen Glieder der Riemannschen Primzahlformel*, Öfversigt 48, 1891, 721-744.
    RIEMANN B.
1859. *Ueber die Anzahl der Primzahlen unter einer gegebenen Grenze*, Saml. Arbeten Dover Publ. 1953.
    RIESZ M.
1988. *Collected Works*, Springer 1988.
1906. *Framställning av den analytiska fortsättningen av en given potensserie*, I,II (på ungerska) Math. és Phys. Lapok XVI, 1906, 1-25, 96-10.
1909. *Sur les séries de Dirichlet et les séries entières*, C.R. Paris 149, 1909.
1911. *Über einen Sats des Herrn Fatou*, Journ. reine u. angew. Mathematik 140, 1911, 89-99.
1914. *Eine trigonometrische Interpolationsformel und einige Ungleichungen für Polynome*, Jahresber. Deutsch. Math. Ver. 23, 1914, 354-368.
1915. *The general theory of Dirichlet series*, Cambridge tracts 18, 1915, reprinted 1952 (with G.H. Hardy).
1916. a. *Neuer Beweis des Fatouschen Satzes*, Gött. Nachrichten 1916, 62-65.
1916. b. *Über einen Satz des Herrn Bernstein*, Acta Math. 40, 1916, 337-347.
1916. c. *Über die Randwerte einer analytischen Funktion*, 4. Skand. Matematikerkongressen 1916 (1920), 27-44.
1921. -1923 *Sur le problème des moments 1,2,3*, Arkiv 16, 1921, no 12, no 19; Arkiv 17, 1923, no 16.
1923. a. *Sur la sommation des séries de Fourier*, Acta Szeged Sect. math. I, 1923, 104-113.
1923. b. *Sur le problème des moments et le théorème de Parseval corréspondent*, Acta Szeged Sect. math.I, 1923, 209-225.
1924. *Neuere Untersuchungen über trigonometrische Reihen*, (med E. Hilb) Enz.d. Math. Wiss. II C 10, 1924, 1189-1228.
    SCHMIDT E.
1903. *Über die Anzahl der Primzahlen unter gegebener Grenzen*, Math. Ann. 57, 1903, 195-208.
    SHOHAT J.A.; TAMARKIN J.D.
1943. *The Problem of Moments*, Math. Surveys 1, Amer. Math. Soc. 1943.
    SIEGEL C. L.
1949. *Transcendental numbers*, Annals of Mathematics studies 16, 1949.
    STIELTJES J, *Recherches sur fractions continues*, Ann. Fac. SDc. Tolouse 1(8), 1894.
    STRIDSBERG E.
1910. *Sur quelques propriétés arithmétiques de certaines fonctions transcendentes*, Acta Math. 33, 1910, 233-292.
    WIGERT C.S.
1906. *Sur l'ordre de grandeur du nombre des diviseurs*, Arkiv 3, 1906-07, no 18.
1909. *Sur un théorème de la théorie des fonctions analytiques*, Arkiv 5, 1909, no 8.
1914. *Sur quelques fonctions aritmétiques*, Acta Math. 37, 1914, 113-140.
1916. *Sur un théorème concernant les fonctions entières*, Arkiv 11, 1916, no 21.

1918. *Sur la série de Lambert et son application à la théorie des nombres*, Acta Math. 41, 1918, 197-218.

VOLTERRA V.

1892. *Sur les vibrations dans les milieux birefringents*, Acta Math. 8, 1892.
1906. *Leçons sur l'intégration des équations différentielles aux dérivées partielles*, Uppsala 1906. Almquist and Wicksell.
1914. *Drei Vorlesungen über neuere Fortschritte der mathematischen Physik*, Archiv d. Mathematik und Physik. Vol. 22, 1914, 97-182.

ZEILON N.

1911. *Das Fundamentalintegral der allgeminen partiellen linearen Differentialgleichung mit konstanten Koeffizienten*, Arkiv 6, 1911, no 38.
1913. a. *Sur les intégrales fondamentales des équations à caractéristique réelle de la Physique Mathématique*, Arkiv 9, 1913, no 18.
1913. b. *Sur les dérivées d'ordre $n-1$ de l'intégrale fondamentale d'une équation différentielle elliptique*, Arkiv 9, 1913, no 19.
1919. -1921 *Sur les équations aux dérivées partielles et le problème optique des milieux birefringents*, I,II, Nova Acta Reg. Soc. Sc. Upsal. IV vol. 5.3 and 5.4, 1919, 1-59; 1921, 1-13.

ZYGMUND A.

1959. *Trigonometric series*, I,II, Cambridge University Press 1959.

CHAPTER 12

Uppsala 1900 - 1930

Wiman   Holmgren

At the turn of the century science in Sweden had attained international standard and could offer professional careers. The development from the middle of the nineteenth century had been very fast, both quantitatively and qualitatively. The ensuing period to the end of the second world war was one of consolidation for the universities. New chairs were created at long intervals and the student body grew so slowly that the existing organization was capable to cope with the influx of newcomers.

Mathematics followed the general trend. The two professors at Uppsala and Lund at the middle of the nineteenth century had increased to four and together with the two at Stockholm university there were six professors of mathematics with research as their main task. At the technological institutes in Stockholm and Gothenburg there were professors of mathematics, but they did in general not have any advanced students and the work of the two professors was dominated by teaching and examination.

The daily life of a professor at the philosophical faculties meant four hours lecturing a week, faculty meetings, work on various committees and the task of examining the students. There were three successive steps in every subject called one, two, three *betyg* which normally required full-time study during the corresponding multiples of half an academic year. In many cases the professor was the only teacher and examiner of his subject.

The students studied one subject at a time and could combine the various subjects of the philosophical faculty rather freely. At least seven *betyg* in properly chosen subjects gave a master's degree which qualified the student for teaching in the schools after some supervised teaching. The beginning of a scientific career was the *licentiate* degree which permitted the defense of a thesis.

The foremost product of the universities was the doctors. The system that the professor wrote the theses had been abandoned long ago. To defend a thesis now meant to have written a thesis with new results and to have paid for the printing. When the defense, called *a disputation*, was over and the thesis had been judged properly and got at least a pass, tradition prescribed that the defendant should invite everybody closely connected with the thesis to a dinner party.

For a doctor of philosophy life after the thesis meant as a rule one year of unpaid supervised teaching at some gymnasium, after that teaching in various insecure positions and, at the end, a tenured position at some gymnasium as a *lektor* with teaching duties in two subjects. If the thesis had received good marks, the new doctor could get an unpaid position at the university as a docent and, in that case, he could compete with others, also in other subjects, for those six-year scholarships for docents that happened to be open. After such a period the door was closed to a further career at the university unless a post as a professor was open, usually after the retirement of the previous holder. This meant a competition between five to ten applicants with one winner. Under these conditions, not many could count on a scientific career and those who did were driven by a burning interest and a wish to test their capacities. A side effect of the system was a concentration of research to the main subjects for teaching in the schools.

Towards the end of the nineteenth century Uppsala had two professorships in mathematics, one of them a somewhat insecure extraordinary position. With a view to Berger (see Chapter 6) this post was specialized to algebra and number theory. A committee of specialists, E.R. Neovius from Finland, L. Sylow from Norway, and Mittag-Leffler from Sweden judged the applicants, Berger, von Koch, and Wiman and put them in the reverse order with Wiman first. It was not without difficulties that the specialists could argue that von Koch's papers, which mostly were about analysis, had some merit in algebra and number theory. Berger's paper about the asymptotic distribution of primes, which the author with all probability had taken from Tjebyshef, was mentioned in passing — the problem had just got its solution by Hadamard and de la Vallée Poussin.

The science section of the philosophical faculty asked one of its members, Matths Falk, to judge the applicants. After a very stingy treatment of Wiman, Falk put Berger in the first position, but an expected promotion fight came to nothing through Berger's death and Wiman was nominated. Falk was succeeded in 1909 by Erik Holmgren.

Wiman

Anders Wiman (1865-1957) came from the southern province of Scania where his father was a prosperous farmer. Anders was the only one of the family to go to higher studies and he took his name from the rural district of Willie. After going to school in Lund he went to the university and took a first degree with, among other things, botany and Latin in the curriculum. He started his mathematical studies at a time when algebraic geometry was the most favoured subject and his thesis on ruled surfaces was the best result of this tradition (see Chapter 4). In the beginning of the new century he applied for the professorship in Lund after Björling and entrusted a friend with the task of delivering the necessary papers to the university authorities. But the friend was one day late and that is why Wiman, who had been the natural first choice, could not return to his alma mater. Instead he got a chair in Uppsala. He never married and led a quiet life of study and research in the circle of his students who gave him many signs of appreciation and friendship. Some years after his retirement Wiman went back to Lund.

His work on the classification of ruled surfaces of degree six (see pp. 47-49) had made Wiman at home in the algebraic geometry of the time but this was just one of his fields of research. The others were group theory, the theory of equations and entire functions. He did good and respected work in all these fields and therefore he may be said to be the most versatile of all Swedish mathematicians.

Group theory

After his 1892 thesis Wiman worked for ten years with the classification of finite geometrical groups, a fast growing field started by Felix Klein in his determination of all finite groups of Möbius (linear fractional) transformations $z \to (az+b)/cz+d$. Via a stereographic map, it turned out that each of them is group of congruence transformations of one of the classical perfect bodies or the dihedron. The most complicated and most interesting object was a group with 60 elements named after the icosahedron.

The Möbius group is the same as the group of projective bijections of the complex projective line. After Klein C. Jordan tried to determine the finite groups of projective bijections of the complex projective plane. His list had two gaps which were soon filled. One of the forgotten groups with 360 elements was analyzed in detail by Wiman (1896) who proved that it is isomorphic to the group of even permutations of 6 elements. The geometric feature of such research is the investigation of objects which are invariant under a given group, for instance polynomials or linear spaces spanned by polynomials.

Together with the projective maps there are in algebraic geometry also the birational ones. In homogeneous coordinates a rational map has the form

$$x = (x_0, x_1, x_2) \to (f_0(x), f_1(x), f_2(x)),$$

where f_0, f_1, f_2 are homogeneous polynomials of the same degree. This map is birational if its inverse has the same form. The simplest example is

$$(x_0, x_1, x_2) \to (x_1 x_2, x_2 x_0, x_0 x_1).$$

An iteration gives (x_0, x_1, x_2) back, multiplied by $x_0 x_1 x_2$. The map is a projective involution. Birational maps are also called Cremona transformations after the Italian mathematician who introduced them. In his paper (1897) Wiman used his talent and his knowledge of algebraic geometry to complete a paper by Kantor where the goal was to determine all finite groups of birational transformation of the complex projective plane. The main work had been done by Kantor but Wiman's methods were simpler and he found some gaps.

The paper (1899) is a determination of all groups of linear fractional bijections of the Galois fields $GF(p^n)$, i.e., the zeros of the equations

$$x^{p^n} - x \equiv 0 \mod p,$$

where p is a prime.

The beginning of Wiman's group theoretical work was two papers (1895 a, b) where the author classified all algebraic curves of genus > 2 and < 7 which possess non-trivial algebraic automorphisms. The point of departure was a paper by Hurwitz (1888) where it is shown that curves of genus $g > 1$ (in contrast to those of genus 1) only have discrete algebraic automorphisms and that these have at most the period $10(g-1)$. In his paper Wiman sharpened this majorant to $2(2g+1)$ and made a classification of the groups which may occur, completed with the corresponding normal forms of the curves.

Wiman's work in group theory gave him the task of writing the section on finite groups of linear maps in the mathematical encyclopedia (1900). A large part of this paper deals with a theory of equations which has now lost its novelty, for instance Hermite's solution of the general equation of degree five by the modular functions and that of Klein through the so-called icosahedral irrationality.

In connection with his work on groups Wiman wrote (1900) about fields with abelian Galois group over a given field K. When K is the rational numbers, Hilbert had proved that such a field is generated by roots of unity, a result partly extended by Wiman to more general cases.

The papers sketched above gave Wiman the professorship in Uppsala. It is not probable that he submitted a paper on probability written in a polemic with a docent Brodén in Lund. The curious origin of this affair was a remark by Gyldén (see Chapter 9) where he argues that the troublesome small denominators which appear in astronomy do this with probability zero. The discussion with Brodén touched upon the difference beween mean and median and the existence of measures on everywhere dense sets. Wiman defended Borel measure, a notion which Brodén found doubtful against his own experience of point sets.

Wiman on solvable equations

When Wiman wrote about solvable equations his source was a section in Weber's algebra (1912) on the structure of the Galois groups of solvable primitive equations, necessarily of primary degree. This result, completed with Abel's formulas for the structure of the zeros, is the subject of an Appendix (pp. 27-28) of Chapter 2. In Abel's formula (l.c. (3)) the zeros are written as sums of pth roots of elements in the last field L before the splitting field. The formula has the flaw that the arguments of the roots cannot be chosen arbitrarily. In a letter to Crelle Abel had succeeded in writing the zeros of a solvable equation of degree 5 as a polynomial in four fifth roots with arbitrary arguments. This gave first Weber and then Wiman the impulse to look for a similar formula in the general case. In Weber's solution $p^n = 8$, and the zeros x_j and Lagrange's resolvents (ω^j, x) have indices $j = (j_1, j_2, j_3)$ with integral components modulo 2. Weber shows that the resolvents can be written as

$$(\omega^j, x) = a_j \sqrt{\prod_{(j,k)=1} (\omega^{j,k}, x)}$$

where $a_j \in L$ and $j = (j_1, j_2, j_3) \mod 2$. The signs of the seven square roots of this formula are then arbitrary. The same formula where j, k each have two components modulo 2 holds for equations of degree four. Wiman (1907) managed to carry out a similar construction for primitive solvable equations of degree p^2. Wiman's student Bucht (1908) deduced the formula (2) of the appendix. His proof used the Galois group and he was also able to extend Weber's formula to certain Galois groups. In *Acta Mathematica*'s memorial volume for Abel (1903 a) Wiman returned to Abel's paper on solvable equations of prime degree. Abel assumed that the ground field contains all necessary roots of unity. Without making this assumption Wiman studied how large a part of the Galois group is generated by the Galois group of the roots of unity of prime order which is cyclic of order $p - 1$. This group produces permutations of the zeros induced by $\omega \to \omega^m$ where m is an integer and ω is a primitive p^{th} root of unity.

From time to time Wiman wrote about the theory of equations. He proved for instance in (1927) in connection with a paper by Hilbert that every general equation of degree $n > 9$ can be transformed into

$$x^n + ax^{n-5} + .. = 0$$

by algebraic manipulations involving solutions of equations of degree at most three.

Entire functions

If the polynomial $G_n(u) = \sum_1^n z^k/k$ is the sum of the first n terms of the power series of $\log(1-u)$, then

$$E_n(u) = (1-u)e^{-G_n(u)} = O\left(|u|^{n+1}\right)$$

for small u. If $z_1, z_2, ...$ is a sequence of complex numbers $\neq 0$ which tends to infinity, there are always positive numbers m_n such that the series

$$\sum |z/z_n|^{-m_n-1}$$

converges for all z. This in turn means that Weierstrass's canonical product

$$f(z) = \prod_1^\infty E_{m_n}(z/z_n)$$

is an entire function with the zeros z_n, each with the multiplicity m_n. If the numbers m_n can be chosen to be bounded, this means that the zeros have a finite exponent of convergence, i.e., the lower bound ϱ of numbers r such that $\sum |z_n|^{-r} < \infty$. Hadamard proved that the number ϱ also equals the *order* of the product, defined as the lower bound of numbers r such that $\log |f(z)| = O(|z|^r)$, and that every entire function of finite order ϱ is the product of a Weierstrass product $f(z)$ with an exponent of convergence at most equal to ϱ and a function $e^{g(z)}$ where $g(z)$ is a polynomial of degree at most equal to ϱ.

When an entire function $f(z) \neq 0$ has the order $\varrho < 1$, the exponential factor is a constant and the function is a product

(4) $$a_0 z^m \prod (1 - z/z_n)$$

where $\sum |z_n|^{-r} < \infty$ for every $r > \varrho$.

After Hadamard's papers the interest in entire functions grew. Borel's series of lectures (1900) and (1902) enriched the subject by the introduction of some important notions which have a sense for every convergent power series,

(5) $$f(z) = a_0 + a_1 z + a_2 z^2 + \ldots,$$

in the first place the maximum modulus

$$M(r) = \max_{|z|=r} |f(z)|,$$

and the minimal modulus

$$m(r) = \min_{|z|=r} |f(z)|,$$

which of course vanishes when the circle $|z| = r$ contains a zero of $f(z)$. There is also the *maximal term* $\mu(r)$ of (5), attained for one or several *central indices* $n = n(r)$. When $f(z)$ is not a polynomial, it is easy to see that the central index tends to infinity when r tends to the radius of convergence.

Borel's main results for functions of finite order are, briefly, that the maximal term is not much less than the maximal modulus and the minimal modulus is not much less than the maximal modulus on a sequence of radii tending to infinity.

Wiman's first papers about entire functions deal with estimates of the logarithm of a canonical product away from the zeros and similar estimates of Mittag-Leffler's function

$$E_\alpha(z) = \sum z^n / \Gamma(\alpha n + 1)$$

for different values of α. Among other things he established a careful estimate (1903 b) of the size of (4) when $z_k = n^{1/\varrho}$ and $0 < \varrho < 1$, which means that the product $f(z)$ has the order ϱ. Wiman found that

$$\log |f(z)| = \frac{\pi}{\sin \varrho \pi} \operatorname{Re} z^\varrho + o(r^\varrho)$$

outside small circles around the zeros. Here the maximal modulus $M(r)$ occurs for $z = r$ and the minimal modulus $m(r)$ för $z = -r$ and one gets, approximatively,

$$m(r) \sim M(r)^{\cos \pi \varrho}$$

away from the zeros. The natural problem now arose to prove the same connection for every entire function of order $0 < \varrho < 1$. For Wiman this problem was the more interesting since Hadamard had shown that

(6) $$m(r) > e^{-r^{\varrho+\varepsilon}}$$

for a sequence of radii $\to \infty$ when ϱ is the order of the function. At least for order < 1 it ought to be possible to do away with the minus sign of the exponential!

It follows from (4) that we then may assume that all $z_k > 0$ are positive which means that the zeros $-z_k$ are on the negative axis. If $\varrho < 1/2$ we can use the Phragmén-Lindelöf principle in the sector $|\arg z| \le \pi$ where the critical growth is $e^{\sqrt{r}}$. If $\log m(r) \le (\cos \pi \sigma) r^\sigma + a$ for all r and some $\sigma < \varrho$ the Phragmén-Lindelöf principle with the comparison function

$$g(z) = e^{z^\sigma + a}, \quad |g(-r)| = e^{\cos \pi \sigma \, r^\sigma + a},$$

shows that $f(z)$ has the order $\le \sigma$, which is a contradiction. Hence it follows that

$$m(r) > e^{r^\sigma}$$

for every $\sigma < \varrho$ and some sequence $r \to \infty$. This result was proved by Wiman (1905) using Phragmén's preliminary paper (1904) (see p. 109) and soon attracted attention. The complete version of (6) in the form

(7) $$m(r) > M(r)^{\cos \pi \varrho - \varepsilon}$$

for a sequence $r \to \infty$ was obtained by Wiman in (1915) with a rather involved proof, which has been simplified later, for instance in Kjellberg (1948).

Wiman returned to the minimal modulus for entire functions in (1916). He first proved that

$$M_\theta(r) > M(r)(1 - \varepsilon)$$

for a sequence of radii tending to ∞ where the left side means the maximum of $|f(z)|$ when $|z| = r$ and $\arg z = \theta$. For entire functions without zeros,

$$f(z) = e^{F(z)}$$

this result shows that

$$M_f(r)^{-(1+\varepsilon)} < m_f(r) < M_f(r)^{(1+\varepsilon)}$$

for a sequence of radii $\to \infty$, which in this case is a considerable sharpening of Hadamard's inequality

$$\log \frac{1}{m(r)} < (\log M(r))^{1+\varepsilon},$$

true for all entire functions. Wiman conjectured that this inequality could be sharpened to

$$m(r) > M(r)^{-(1+\varepsilon)}$$

for all entire functions. Both inequalities are of course meant to be true for some sequence of radii $\to \infty$. Wiman's conjecture stood till 1951, when Walter Hayman constructed a counterexample with zeros of strongly increasing density and also showed that there is a best constant $A > 0$ in the inequality

$$\log m(r) > -A \log M(r) \log \log M(r)$$

for entire functions and sequences of radii.

Another theme of Wiman's work in function theory is majorants of the maximal modulus $M(r)$ of a power series $f(z)$ which only depend on the maximal term $\mu(r)$.

For entire functions $f(z)$ Wiman proved in (1914) that to every $\varepsilon > 0$ there are arbitrarily large r such that

$$M(r) \leq \mu(r)(\log \mu(r))^{1/2+\varepsilon}.$$

This result can be shown to be sharp except for the parameter ε and sharpens an inequality by Borel which has $\mu(r)^{1+\varepsilon}$ on the right side.

Wiman's elegant proof, which is presented in Pólya-Szegö II (1925) (see p.9), depends on the following fact: if

$$g(z) = 1 + \sum_1^\infty b_n z^n + ...$$

is a power series with positive coefficients ≥ 1 with radius of convergence equal to 1 and such that b_{n-1}/b_n increases with n for large values of n, then every $b_n|z|^n$ is a maximal term of some $g(z)$ and for some $|z|$ when n is sufficiently large. The simple proof can be left to the reader.

With g as above and $\varphi(r) > 0$ an increasing function such that

$$M_g(r) \leq \mu_g(r)\varphi(\mu_g(r)),$$

Wiman's theorem depends on the circumstance that this inequality can be transferred to any entire function (5) with maximum modulus $M(r)$ and largest term $\mu(r)$. In the proof of this statement we may assume that $a_0 = 1$. If $n = n(r)$ is the central index of the entire function $\sum a_k z^k / b_k$, then

$$|a_k|r^k/b_k \leq |a_n|r^n/b_n$$

for all k. With $r = x/y$ this means that

(8) $$|a_k|x^k/|a_n|x^n \leq b_k y^k / b_n y^n$$

for $k = 0, 1, ...$ For large r we may now choose $y = y(r)$ in such a way that $b_n y^n = \mu_g(y)$ and a summation gives

$$M(x)/a_n x^n \leq M_g(y)/\mu_g(y)$$

where x, y, n tend to infinity with r. Since $|a_n|x^n \leq \mu(x)$ it follows that

$$M(x)/\mu(x) \leq M_g(y)/\mu_g(y) \leq \varphi(\mu_g(y)).$$

Now $k = 0$ in (8) shows that $\mu_g(y) \leq \mu(x)$ and this proves that the argument of φ can be replaced by $\mu(x)$. Hence we get the desired result that $M(x) \leq \mu(x)\varphi(\mu(x))$ for a sequence of positive numbers $x \to \infty$.

Different choices of $g(x)$ produce different functions φ. Wiman's function

$$g(z) = \sum e^{n^\alpha/\alpha} z^n, \quad 0 < \alpha < 1$$

turns out to give $\varphi(u) \sim (\log u)^{1+\varepsilon}$ while for instance $\sum n^k z^n$ gives $\varphi(u) = u^{1/k}$.

In his work on entire functions, which went from explicit calculations to general results, Wiman showed that he could compete with the best analysts. It is also remarkable that he could do this work after he had established himself as a specialist in another field.

Differential equations

Wiman wrote a couple of papers on oscillatory solutions of second order differential equations of the form

$$u'' + f(x)u = 0, \quad f(x) > 0, \text{ continuous.}$$

By comparing with the case $f(x) = a^2/x^2$, which has two solutions x^α such that $\alpha(\alpha - 1) + a^2 = 0$, he proved among other things in (1917) that the bound for oscillation is $a = 1/2$. If $x^2 f(x)$ is strictly greater than $1/4$ far away, all real solutions oscillate between positive and negative values. But if $x^2 f(x)$ is strictly less than $1/4$ for all large x, the solutions have a constant sign for all large x.

A solution of the equation above is said to be stable if it is bounded for all positive x. The origin of (1935) was a false statement by Fatou that all solutions are stable when $f(x) > 0$ and $f(x)$ and $1/f(x)$ are bounded. Wiman proved that stability occurs under the additional condition that $\log f(x)$ has bounded variation.

We shall sketch the solution. Since f has a positive lower bound, the classical theory of Sturm and Liouville shows that every solution u oscillates for large $x > 0$. Multiplication of the equation by u', an integration between two bounds $x < x + h$ and an integration by parts show that

$$u'^2(x+h) - u'^2(x) + u^2(x+h)f(x+h) - u^2(x)f(x) = \int_x^{x+h} u^2(t)df(t).$$

If $x, x + h$ are two consecutive maxima of $|u|$, the two first terms disappear. Hence, if $df(t) \geq 0$ it follows that $u^2 f$ does not decrease between these maxima. On the contrary, the function $v = u^2$ has the opposite property. In fact, rewriting the left side when the two first terms vanishes gives the equalities

$$v(x+h)\Delta f(x) + f(x)\Delta v(x) = v(x)\Delta f(x) + f(x+h)\Delta v(x) = \int_x^{x+h} u^2(t)df(t)$$

where $\Delta f(x) = f(x+h) - f(x)$ and analogously for $v(x)$. If $df(t) \geq 0$ and, for instance, $v(x+h) \geq v(x)$, the integral is at most $v(x+h)\Delta f(x)$ and it follows that $\Delta v(x) \leq 0$ and the same conclusion follows when $v(x+h) \leq v(x)$. Hence $\Delta v(x) \leq 0$, i.e., the successive maxima of $|u|$ do not increase so that u is bounded. If $df(t) \leq 0$ an analogous argument gives the same result.

When $df(t)$ changes its sign, the situation is more delicate. Wiman's proof that u is bounded when $\log f(x)$ has bounded variation cannot be given here.

Wiman also investigated the case $f(x) = a^2 + b^2 \cos 2x$, $a^2 > b^2$, which was important to Gyldén, and proved that the equation then has unbounded solutions. As in the earlier paper, he uses a careful and sharp argumentation based on elementary facts.

Algebraic geometry

After his analytical period, Wiman returned to his original interests and wrote about finite groups and algebraic geometry till long after retirement.

Wiman did not share his interest in ruled surfaces with many mathematicians, but in 1931 the English mathematician W.L. Edge published a book on the subject with an extended analysis for the orders four, five, and six. Wiman's thesis (1892), written in Swedish, was not quoted. As Wiman before him, Edge had chosen the nature of the double curve as a point of departure but he had also used the corresponding double envelope which had resulted in a better classification. The book by Edge made Wiman write a number of papers about ruled surfaces, followed by papers on W-curves, projective lines contained in algebraic surfaces, and papers about rational points on curves of degree three. This was also the theme of Wiman's last paper (1956).

Obituary

In his obituary (1960) of Wiman, Pleijel wrote among other things:

> Wiman was by nature unassuming but with a firm self-confidence. His vivid interest in people was a decisive factor in his behavior towards the many students whom he took care of during his time as a professor. He was not a fluent talker, his speech was slow and came in spurts accompanied by vivid movements of his hands and fingers. His judgement of people and events was sharp and to the point. Wiman was the only member of his rural family who went to study. One reason was that he was not considered suited for work in the fields. Wiman often talked about this with a shade of disappointment. Otherwise Wiman looked on the world with a positive eye — he was a composed and harmonic human being of a kind rarely met.

Wiman's students

A mathematician wanting to write a thesis had two possibilities, either to find his way through mathematics and find a workable subject by himself or to use the teaching, research, and help of the professor to find a suitable thesis subject. The second, more secure possibility was used by a majority, especially since this road meant less risk of being involved in an unsuccessful venture.

During his time in Uppsala, Wiman was generous with ideas for theses, but in the final result the dependence on the teacher varied. In the beginning the favorite field was entire functions of finite order. R. Mattson (1905) and G.P. Persson (1908) wrote on estimates of entire functions outside the zeros. Both found an opportunity to use the Phragmén-Lindelöf principle. Persson remarks about the application to functions $f(z)$ of exponential type in a half-plane that the desired result can be obtained via $\log|f(z)|$, an idea later carried out by Carleman. A. Wahlund (1915) improved a result by Valiron and Littlewood which says that the maximal module $M(r)$ and the minimal module $m(r)$ of a function of order zero satisfy the inequality $m(r) > M(r)^c$ for every $c < 1$ and some sequence of radii $r \to \infty$. Wiman inspired A. Carlson's thesis (1907) by asking about the convergence of the iterates of the function $f(x) = a^x$. In his studies of iteration of polynomials of the second degree, Carlson touched upon their chaotic bahavior. In special cases M. Ålander (1914) investigated the distribution in the plane of zeros of repeated derivatives of entire functions. When the order is > 1 and the function has real zeros, it was found that these zeros wander out to infinity along rays where the function is small.

One of Wiman's students, Fritz Carlson, who became a professor in Stockholm (see Chapters 4 and 14), was perhaps not so dependent on Wiman. In his thesis (1914) he studied analytical continuation of power series

$$\sum g(n) z^n,$$

in the right half-plane, $g(z)$ being entire analytic of exponential type. This thesis contains a well-known theorem: if a function $g(z)$ is analytic in the right half-plane, bounded on the imaginary axis and if $\log|g(z)| = o(\operatorname{Re} z)$, then the function cannot vanish on the integers $0, 1, 2, \ldots$ without vanishing identically. After division by $\sin \pi z$ the theorem reduces to another one which has been treated in an earlier chapter (p. 114) in connection with the Phragmén-Lindelöf principle. One year later Matts Essén defended a thesis with new criteria for the convergence and divergence of series. An entirely new subject, Dirichlet series and their zeros, appeared in Sven Wennberg's thesis (1920). Among other things the author improved on results by Bohr and Landau on upper bounds of the zeros of Riemann's ζ-function in the critical strip, a result which shortly afterwards was improved by F. Carlson.

For many theses the subject was group theory. Josef Andersson (1913) treated a special problem for Abelian groups, Bolinder (1909), Bucht (1909) and Lindwall (1919) wrote about metacyclic groups. Lindwall modernized Jordan's big treatise. Malmrot (1925) catalogued groups whose order is a product of six primes. E.L. Petterson (1935) wrote about criteria of irreducibility of polynomials with integral coefficients which extend Eisenstein's well-known criterion.

Algebraic geometry is represented by Stenström (1925) who wrote about surfaces of degree three in a synthetical vein, by Brundin (1927), Sjöstedt (1929) and Jonzon (1930) who wrote about maps of algebraic curves and by Lunell (1940), who combined Wiman's and Edge's classifications of ruled surfaces of degree six.

Holmgren

Wiman's younger colleague in Uppsala was Erik Holmgren (1873-1943), professor in 1909 after a thesis in 1898. Many members of his family devoted themselves to science. His father was the mathematician Hjalmar Holmgren (see Chapter 2).

Almost all of Holmgren's papers deal with the theory of differential equations, especially the equation of heat conduction.

To describe these papers we need some definitions. A linear differential operator $P(x, D)$ in n variables x of order m is a polynomial of degree m in the gradient operator $D = \partial_x$. The corresponding linear differential equation is

$$P(x, D)u = v.$$

It is said to be homogeneous when the right side v vanishes. The principal part P_m of the operator is the part of the operator which is homogeneous of degree m in D. A surface $f(x) = 0$ is said to be characteristic for P at the places where $P_m(x, \operatorname{grad} f(x)) = 0$. In two dimensions a characteristic curve is also said to be a characteristic. The Cauchy data of a solution at a surface are the derivatives of the solution of order $< m$ at the surface. These notions, suitably formulated, extend to systems of differential equations with as many equations as there are unknowns.

Before the turn of the century the theory of systems of linear partial differential equations contained one very general result, the Cauchy-Kovalevski theorem, valid for analytical differential equations and analytic Cauchy data on noncharacteristic analytic surfaces. The theorem says that these equations have unique locally analytic solutions with given analytic Cauchy data. It was proved at a time when power series were the universal tool in the study of partial differential equations. But in physics and especially in wave propagation it is in the nature of things that nonanalytic solutions occur which are only differentiable or even possess singularities.

Very early Holmgren took an interest in nonanalytic solutions of differential equations and systems in two variables. His thesis (1897) deals with differential operators of the form

$$L = D_x^p D_y^q + \sum_{j<p, k<q} a_{jk}(x,y) D_x^k D_y^j$$

with suitably regular coefficients. The characteristics are here lines parallel to the coordinate axis. For the homogeneous equation $Lu = 0$ in a rectangle $0 < x < a$, $0 <$

$y < b$ it is natural to consider a boundary value problem where the x-derivatives of order $< p$ of u are given on the y-axis and the y-derivatives of order $< q$ on the x-axis. This problem, considered before Holmgren by Delassus (1897), is easy to solve by successive approximations. Delassus had worked with analytic functions, but Holmgren proved existence and uniqueness for sufficiently differentiable ones. Holmgren is very careful to use to concepts of strict analysis: continuity, existence of derivatives, uniform convergence, etc. This is very clear in the second part of his paper, which otherwise treats a somewhat uninteresting problem.

Holmgren's uniqueness theorem

Holmgren is known for a simple but useful theorem, Holmgren's uniqueness theorem. He got the good idea to ask himself what remains of the Cauchy-Kovalevski theorem for linear equations when the boundary problem has a solution which need not be analytic. His uniqueness theorem gives the answer: if such a solution exists, it is unique. Holmgren's simple proof has also survived and is easy to reproduce.

For simplicity let us work with two variables x, y and a linear differential operator

$$L = A(x,y)D_x + B(x,y)D_y + C(x,y)$$

where the coefficients are square matrices of order m which are analytic functions of x, y close to the point $O = (0,0)$. A curve $f(x) = 0$ is then characteristic at points where $A(x,y)f_x + B(x,y)f_y$ has no inverse. The operator has an adjoint

$$L' = -D_x A'(x,y) - D_y B'(x,y) + C'(x,y)$$

(where ' means transposition), characterized by the property that

(1) $$(Lu, v) = (u, L'v) + D_x(Au, v) + D_y(Bu, v).$$

Here $u = u(x, y), v = v(x, y)$ are matrices of the type $n \times 1$ and $(u, v) = u'v$. We can choose coordinates such that $f(x, y) = y$ close to the origin. That the corresponding surface it is not characteristic there means that $B(x, y)$ is invertible and in this case the curves $y = \varepsilon x^2$ and $y = \delta$ are not characteristic near O when ε, δ are small. Consider a lens-shaped region

$$T = T(\varepsilon, \delta) : \varepsilon x^2 \leq y \leq \delta$$

where $\varepsilon, \delta > 0$ are small.

Suppose now that $u(x, y)$ is continuously differentiable, that $Lu = 0$ in D, and that $u = 0$ when $y = \varepsilon x^2$ and that $L'v = 0$ in a neighborhood of T. If we integrate (1) over T it then follows that

$$\int_{T'} (Bu, v) dx = 0$$

where T' is the part of the boundary of T where $y = \delta$. Here B is not singular and, according to the Cauchy-Kovalevski theorem, v can be chosen arbitrary analytic on T'. It follows that $u = 0$ on T'. If we let δ and then ε tend to zero, it follows that $u = 0$ when $y \geq 0$ in a neighborhood of O. The same reasoning when $\varepsilon < 0$ shows that $u = 0$ in a neighborhood of O, which is Holmgren's uniqueness theorem. When Holmgren had to quote this result he called it 'un théorème connu', but this does not mean that he did not realize its importance.

Repeated applications of Holmgren's uniqueness theorem produce global results. According to the theorem, a smooth solution of a differential equation with analytic coefficients which vanishes in some region has to vanish in every region which can be fibered by connected pieces of non-characteristic surfaces with boundaries in the first region. Laplace's equation $\Delta u = 0$ in R^n has no real characteristic surfaces. Hence a twice continuously differentiable solution in a connected region cannot vanish of second order on a piece of a surface without vanishing everywhere. If a solution u of the wave equation $u_{tt} = \Delta u$ vanishes of the second order when $t = 0$, $x \neq 0$, it vanishes outside the double cone $|t| = |x|$. For all surfaces $|t| = a|x|$ are non-characteristic when $0 \leq a < 1$.

Holmgren's uniqueness theorem and its extensions play an important part in the modern theory of general differential operators where, in combination with functional analysis, they give important theorems of existence (see Hörmander (1963, Chapter III)).

Other papers

Holmgren published frequently in the beginning of the century. A theme which was present already in his thesis is Volterra's integral equations

$$(2) \qquad g(x) = \int_0^x H(x,y)f(y)dy, \quad x, y \geq 0,$$

where $H(x,y)$ is continuously differentiable. Differentiation gives an integral equation

$$H(x,x)f(x) + \int_0^x H_x(x,y)f(y)dy = g'(x),$$

which is easy solve when $H(x,x) \neq 0$. Under the heading *On the inversion of definite integrals* Volterra wrote in (1896) four notes about what happens when this condition is not satisfied. The most intricate situation occurs when both sides of (2) vanish at $x = 0$ as x^{n+1} and $H(x,y) = P(x,y) + Q(x,y)$, where all derivatives of Q of order n vanish for $x = y = 0$, and the principal part of $P(y,x)$ is a homogeneous polynomial of degree n,

$$P(y,x) = a_0 x^n + a_1 x^{n-1} y + ... + a_n y^n.$$

If $H = P$, differentiation n times of (2) gives an Eulerian differential equation for $f(y)$ with solutions y^λ where

$$\frac{a_0}{\lambda + 1} + ... + \frac{a_n}{\lambda + n + 1} = 0.$$

Volterra required the zeros of this equation to be separate and to be situated in in the left half-plane. This means that the homogeneous equation only has bad, unbounded solutions. Under this condition, Volterra gave a rather complicated proof that the inhomogeneous equation has a unique, bounded solution. (It could be said that Volterra had a variant of Fredholm's alternative.) Already in his thesis Holmgren considered Volterra's problem and he returned to it in (1899 a, c) where he investigated the case when the equation above has zeros in both the right and the left half-plane. This gives good solutions of the homogeneous equations and conditions on the right side for the inhomogeneous equation to have good solutions. When Holmgren gave his seminar talk in Göttingen in 1900 about Fredholm's integral equations he was well prepared.

In order to qualify for the professorship he got in 1909, Holmgren wrote in *Arkiv* about integral equations and boundary problems in the plane. These papers are well done but of varying interest. One of Hilbert's problems was to show that all solutions of analytic elliptic equations are themselves analytic. Holmgren gave a simple contribution to the problem. In (1904) he constructed a fundamental solution of the operator $\Delta + p(x, y, z)$ where p is analytic. He also wrote about boundary problems for hyperbolic systems. In his perhaps most interesting paper (1908 a,b), which is about a boundary problem for the heat equation $u_t - u_{xx} = 0$, Holmgren proves that every continuously differentiable solution $u(t, x)$ is infinitely differentiable and locally analytic in x. Hence a solution, defined for small x and for t in an interval, can be written as a convergent power series,

$$(3) \qquad u(t,x) = \sum_0^\infty \left(\frac{a_n(t)x^{2n}}{(2n)!} + \frac{b_n(t)x^{2n+1}}{(2n+1)!} \right)$$

where

$$a_n(t) = D_t^n u(t,0), \quad b_n(t) = D_t^n D_x u(t,0)$$

since u solves the heat equation. If we let the local Gevrey class $G(\alpha)$ be the set of infinitely differentiable functions $h(x)$, defined in some interval and satisfying the inequalities

$$(4) \qquad |h^{(n)}(t)| \leq M^n \Gamma((1+\alpha)n), \quad n = 0, 1, \ldots$$

where M may depend on h, it follows that $f(t) = u(t,0)$, $g = u_x(t,0)$ belong to the class $G(1)$. In a note Holmgren treats also the case $f, g \in G(\alpha)$, $0 < \alpha < 1$. Then the right side of (3) is an entire function of x and $\log|u(t,x)|$ has the order of magnitude $O(|x|^{1/(1-\alpha)})$.

It is implicit in Holmgren's paper that the functions $f(t)$ and $g(t)$ may have zeros of infinite order without being identically zero. This means that there are solutions of the heat equation which vanish at $t = 0$ without being identically zero.

After he got his professorship Holmgren's intense activity waned and he wrote very few papers. The most remarkable one is the paper (1924) in which he returned to his paper (1908 a,b). He proved that if $u(t,x)$ solves the heat equation in a band $a > t > 0$ then

$$\log|u(t,x)| = O(|x|(\log|x|)^\alpha), \quad 0 \leq \alpha \leq 1$$

if and only if $u(t,0), u_x(t,0)$ satisfy the inequality (4) with the right side replaced by $M^n n^n (\log n)^{\alpha n}$. Now Denjoy had proved some years earlier that this class is quasi-analytic, i.e., no member can have a zero of infinite order without being zero itself. It follows that the boundary problem for $x = 0$ with data as above must have a unique solution.

Holmgren's work on regularity problems for solutions of partial differential equations has been greatly extended and found a prominent place in the general theory of partial differential operators with constant coefficients (Hörmander 1963). Therefore Holmgren's results— except the uniqueness theorem — have only a historical interest.

In the beginning of the 1920's Tricomi initiated a period of interest in boundary problems for the differential operators

$$L_m = y^m D_x^2 + D_y^2,$$

and especially for $L = L_1$, which is elliptic for $y > 0$ and hyperbolic for $y < 0$. The theory of these operators was close to Holmgren's interests. His contribution (1927) solves a certain boundary problem for the equation $L_m u = 0$ in the upper half-plane with data on a part of the x-axis.

Obituary

In a collection of biographies, *Swedish men and women* there is also one of Erik Holmgren from which the following is quoted.

> H. had many interests outside his own profession. He was well read in art history and he had a love for the esthetical side of mathematics, something which could be seen in the problems he composed for his teaching. Ever since his youth he had a warm interest in France and French culture. Italy and France were frequent goals for his summer travel in which he combined study and pleasure. During his first years as a professor, Holmgren took a vivid interest in the student debates on politics. With special fervor he threw himself into the debate on defense which raged in the years before the first world war. His strong temperament could also occasionally show itself in his seminars.

Holmgren's students

For simplicity, mathematicians at Uppsala, who wrote theses connected to Holmgren's work, are classified here as Holmgren's students although some of them could have worked entirely by themselves.

Holmgren's most successful student was Torsten Carleman, whose 1917 thesis was the beginning of a brilliant career in analysis (see Chapter 14). Gustaf Mikaelsson (1920) extended Holmgren's regularity theorem for the wave equation to certain equations of higher order. Olof Sjöstrand, who worked at the Chalmers Technological Institute in Gothenburg, and was entirely independent, wrote in (1929) about boundary problems for differential equations in the plane with data on two intersecting lines. His results are also published in two papers (1930, 1934). In the first of these papers, the solutions are analytic and the problem is to find suitable majorant series. The second one has connections with Goursat's problem, in the simplest case the problem of finding a solution of the equation $u_{xy} = 0$ with given values on two lines in the first quadrant intersecting at the origin. The proof depends on the fact that every solution is a sum of a function of x and another function of y. Similar statements hold for homogeneous equations with constant coefficients and real characteristics. Sjöstrand's two solid papers improve results by others.

Sven Gellerstedt (1935) treated Tricomi's boundary value problem for the equation $yu_{xx} + u_{yy} = 0$, extended to the equation $y^m u_{xx} + u_{yy} - cu = f(x,y)$ where $m > 0$ is odd and $c < 0$. Sven Täcklind (1936) extended Holmgren's paper (1924) to general quasi-analytic classes.

BIBLIOGRAPHY

Wiman.
BOREL E.
1900. *Leçons sur les fonctions entières*, Paris 1900.
1902. *Leçons sur séries à termes positives*, Paris 1902.
BUCHT G.
1908. *Über die Wurzeln der primitiven metazyklischen Geichungen von Primzahlpotenzgrad*, Arkiv 4, 1908, no 27.
EDGE W. L.
1931. *The Theory of Ruled Surfaces*, Cambridge 1931.
GÅRDING L.
1992. *Abel och lösbara ekvationer av primtalsgrad*, NORMAT 40.1, 1992.
HAYMAN W.K.
1951. *Sur le module des fonctions entières*, C.R. Paris 252, 591-593.
HURWITZ A.
1888. *Über diejenigen algebraischen Gebilde welche eindeutige Transformationen in sich zulassen*, Math. Ann. 32, 1888, 290-308.

KJELLBERG B.
1948. *On certain integral and harmonic functions*, Akad. avh. Uppsala 1948.
PLEIJEL Å.
1960. *Anders Wiman*, K. Fysiogr. Sällsk. Förh. 1960.
POLYA G., SZEGÖ G.
1925. *Aufgaben und Lehrsätze der Analysis*, I,II, Springer 1925, Dover 1945.
WEBER H.
1912. *Lehrbuch der Algebra*, I,II, zw. Aufl. Braunschweig, Vieweg u. Sohn, 1912,1899.
WIMAN A.
1892. *Klassifikation af regelytor af sjette graden*, Akad. avh. Lund 1892.
1895. a. *Über die hyperelliptischen Curven und diejenigen vom Geschlecht P = 3 welche eindeutige Transformationen in sich zulassen*, KVA Bihang, 1895, bd 21 Afd 1 no 1.
1895. b. *Über die algebrischen Curven von den Geschlechtern p = 4, 5 und 6 welche eindeutige Transformationen in sich zulassen*, KVA Bihang, 1895, bd 21 Afd 1 no 3.
1896. *Über eine einfache Gruppe von 360 ebene Collineationen*, Math. Ann. 1896, 531-556.
1900. *Endliche Gruppen linearer Substitutionen*, Enz. Math. Wiss, 1900, IB 3f.
1903. a. *Über die metazyklischen Gleichungen vom Primzahlsgrad*, Acta Math. 27, 1903, 163-175.
1903. b. *Über die angenäherte Darstellung von ganzen Funktionen*, Arkiv 1, 1903, 105-111.
1905. *Sur une extension d'un théorème de M. Hadamard*, Arkiv 2, 1905.
1907. *Ueber metazyklische Gleichungen vom Grade p^2*, Arkiv 3, 1907, no 27.
1914. *Über den Zusammenhang zwischen dem Maximalbetrage einer analytischen Funktion und dem grössten Gliede der zugehörigen Taylorschen Reihe*, Acta Math. 37, 1914,1-28.
1915. *Über eine Eigenschaft der ganzen Funktionen von der Höhe Null*, Math. Ann. 76, 1915, 197-221.
1916. *Über den Zusammenhang zwischen dem Maximalbetrage einer analytischen Funktion und dem grössten Betrage bei gegebenem Argumente der Funktion*, Acta Math 41, 1916, 1-28.
1917. *Über die reellen Lösungen der linearen Differentialgleichung zweiter Ordnung*, Arkiv 12, 1917, no 14.
1931. *Über die Regelflächen mit einem Leitgeraden*, Acta Math. 57, 1931, 343-387.
1932. *Über die Regelflächen sechsten Grades ohne Leitgerade*, Acta Mathematica 59, 1932, 1-62.
1935. *Über eine Stabilitätsfrage in der Theorie der linearen Differentialgleichungen*, Acta Math. 66, 1935, 121-145.
1937. *Über die Cayleysche Regelfläche dritten Grades*, Math. Annalen 113, 1937, 161-198.
1944. a. *Über Regelflächen von beliebigem hohen Grade mit vollständig zerfallenden Doppelkurven*, Acta Math. 76, 1944, 1-30.
1944. b. *Über den Rang von Kurven $y^2 = (x(x + a(x + b)$*, Acta Math. 76, 1944, 225-251.
1945. *Über rationale Punkte auf Kurven $y^2 = x(x^2 - c^2)$*, Acta Math. 77, 1945, 281-320.
1948. *Über rationale Punkte auf Kurven dritter Ordnung vom Geschlechte eins*, Acta Math. 80, 1948, 223-257.

Theses by Wiman's students.

MATTSON R., *Contributions à la théorie des fonctions entières*, Uppsala 1905.
PERSSON G.P., *Recherches sur une classe de fonctions entières*, Uppsala 1908.
BOLINDER E.G., *Über die Strukturverhältnisse bei einer besonderer Klasse vollkommener Gruppen*, Ak. avh. Uppsala 1909.
BUCHT G., *Über die metazyklischen Gleichungen vom Grade p^n*, Uppsala 1909.
ANDERSSON J., *Über eine Klasse Untergruppen einer Abelschen G_N^m*, Uppsala 1913.
ÅLANDER M., *Sur le déplacement des zéros des fonctions entières*, Uppsala 1914.
CARLSON F.D., *Sur une classe de séries de Taylor*, Uppsala 1914.
ESSÉN M., *Över några svagt konvergerande eller divergerande serier med regelbundet avtagande termer*, Uppsala 1915.
WAHLUND A., *Sur quelques propériétés des fonctions entières de genre zéro*, Uppsala 1915.
LINDWALL B.A.G, *Jordans teori över primitiva metacykliska ekvationer*, Uppsala 1919.
WENNBERG S., *Zur Theorie der Dirichlet'schen Reihen*, Uppsala 1920.
MALMROT R., *Studien über Gruppen deren Ordnung ein Produkt von sechs Primzahlen ist.*, Uppsala 1925.
BRUNDIN C.G., *Realitätsdiskussion von Curven mit eindeutigen Transformationen in sich nach der zyklischen und der Abel'schen {2,2,...} Gruppe*, Uppsala 1927.
SJÖSTEDT C.E., *(α, β)-Korrespondenzen aud elliptische und hyperelliptische Kurven*, Uppsala 1929.
JONZON B.A.S., *Über die Gruppen birationaler Transformationen der elliptischen und der hyperelliptischen Kurven in sich*, Uppsala 1930.

STENSTRÖM O., *Synthetische Untersuchungen des Systems von 27 Geraden einer Fläche dritter Ordnung*, Uppsala 1925.

PETTERSON E.L., *Über die Irreduzibilität Ganzzahliger Polynome*, Stockholm 1936.

LUNELL E., *Liniengeometrische Studien mit besonderer Rücksicht aud Regelflächen mit vollständigen Zerfalle der Doppelkurve*, Uppsala 1940.

Holmgren.

CARLEMAN T.

1917. *Über das Neumann-Poincarésche Problem für ein Gebiet mit Ecken*, Ak. Avh. Uppsala 1917.

GELLERSTEDT S.

1935. *Sur un problème aux limites pour une équation linéaire aux dérivées partielles du second ordre de type mixte*, Ak. Avh. Uppsala 1935.

HOLMGREN E.

1898. *Om differentialekvationen* $\sum A_{ik}\partial_x^i \partial_y^k = 0$, Uppsala Årsskrift 1897.

1899. a *Sur l'inversion des intégrales définies*, KVA Bihang 25 Sep. tryck 1898.

1899. b *Sur les intégrales des équations différentielles considerées comme fonctions de leurs valeurs initiales*, KVA Bihang 25, 1899.

1899. c *Recherches sur l'inversion des intégrales définies*, Nova Acta 1899.

1901. a *Über Systeme von linearen partiellen Differentialgleichungen*, KVA Öfversigt 58, 1901, 91-103.

1904. *Über die Existens der Grundlösung bei einer linearen partiellen Differential gleichung der zweiten Ordnung vom elliptischen Typus*, Arkiv 1, 1904, 209-224.

1908. a,b *Sur l'équation de la propagation de la chaleur*, Arkiv 4, 1908, no 14,18.

1924. *Sur les solutions quasi-analytiques de l'équation de la chaleur*, Arkiv 18, 1924-25, no 9.

1925. *Sur un problème aux limites pour l'équation* $y^m u_{xx} + u_{yy} = 0$, Arkiv 19B, 1925-27, no 14.

HÖRMANDER L.

1963. *Linear Partial Differential Operators*, Springer Grundlehren 116, 1963.

MIKAELSSON G.

1920. *Sur la nature analytique des solutions des équations linéaires aux dérivées partielles à caractéristique multiple*, Ak. Avh. Uppsala 1920.

SJÖSTRAND O.

1929. *Sur le problème de M. Goursat pour les equations aux dérivées partielles du second ordre ou de l'ordre supérieur*, Ak. Avh. Göteborg 1929.

1930. *Sur l'emploi de la méthode de M. Gunther pour les equations aux dérivées partielles du troisième et quatrième ordre*, Arkiv 22A, 1930, no 27.

1934. *Sur un probléme aux limites pour les équations aux dérivées partielles du troisième ordre du type hyperbolique*, Arkiv 24A, 1934, no 18.

TÄCKLIND S.

1936. *Sur les classes quasianalytiques des solutions des équation aux dérivées partielles du type parbolique*, Ak. Avh. Uppsala 1936.

VOLTERRA V.

1896. *Sulla inversione degli integrali definiti*, I-IV, Atti Torino vol XXI, 1896, 217-275.

CHAPTER 13

Lund 1900 - 1925

Brodén   Nörlund   Block

In 1906 Torsten Brodén succeeded C.F.E. Björling as professor of mathematics at Lund University. At about the same time a second professorship was created which went to the young Danish mathematician Niels-Erik Nörlund. In addition to these two, the astronomer Henrik Block wrote mathematical papers which earned him a professorship at Chalmers Technological Institute in Gothenburg. Only about half of the mathematical theses written in Lund during the first quarter of this century were inspired by the local professors.

Brodén

After a short time in Uppsala, Torsten Brodén (1857-1931) studied in Lund. At first he worked in the local tradition of algebraic geometry. The subject of his thesis was transformations of surfaces of rotation, other papers dealt with elliptic functions and algebraic correspondences, including Zeuthen's theorem (see p. 42). In the middle of the 1890's Brodén switched to the construction of curious continuous functions of different kinds. He also wrote about summation of trigonometric series and a problem of Riemann in connection with the ramification of solutions of Fuchsian differential equations. Brodén, who wrote reviews in Jahrbuch, was an able but somewhat superficial writer. In his later years he devoted himself to the foundations of mathematics.

Like many others at the same time Brodén preferred publication in academic journals, for instance Öfversigt. One exception is his paper (1898) in *Mathematische Annalen* which may serve as an example of Brodén's constructions of curious functions. Consider an infinte product

$$f(x) = \prod_1^\infty \varphi(\cos 2\pi [K_n x]), \quad \varphi(x) = xe^{x-1}$$

where $[y]$ means the integral part of y and K_n is the product of the first n number in a bounded sequence of odd integers $k_j \geq 3$, for instance $K_n = 3^n$. When the product diverges it equals zero and, when it converges, all $[K_n x]$ with large indices n are close to 1. We see that all the factors of the product are finally zero when $x = $ odd integer$/4K_n$ for some n and equal to 1 when $x = $ integer$/K_n$. This means that the zeros of $f(x)$ are dense but that the function does not vanish in any interval. The author shows that the function is continuous at the zeros and assumes positive and negative values in every interval. A modern reader may remember the classical function $g(x)$ which equals zero when x is irrational and $(-1)^p/q$ when $x = p/q$ and p,q are coprime. The qualitative properties are the same as those of the product above.

In the same journal and a year later, in (1899), Brodén wrote a review article about the convergence of Dirichlet's integral, i.e.,

$$\int_0^a f(x) \frac{\sin \omega x}{\sin x} dx$$

when $\omega \to \infty$. Convergence of the integral means that the partial sums of the Fourier series of f converge to $\pi f(0)/2$ at the origin. This happens for instance when $f(x)$ is continuously differentiable. When $f(x)$ is monotone or of bounded variation, an integration by parts shows that the limit as $\omega \to \infty$ is $\pi f(+0)/2$. Difficulties arise when $f(x)$ oscillates strongly for small x. Brodén's long paper is a review of work by Dirichlet, Riemann, Lipschitz, du Bois-Reymond, Dini and Kronecker but contains nothing new. Often the author takes a superior attitude towards others for which he himself has no collateral. Certain results become sharper if integration is taken in the Lebesgue sense, but Brodén could not know this. Lebesgue's paper appeared at about the same time as his paper and it took some time for the new theory to be accepted.

In his fifties Brodén began taking part in the fight between idealists and intuitionists about the foundations of mathematics. In his book (1924) on the subject he considered the problem from a dialectical point of view.

Brodén came to influence Swedish literature in an indirect way. In the 1880's Strindberg lived for a short time in Lund and frequented a circle of local notables who used to meet in a well-known pub. Brodén, who was an able pianist, could then entertain the company with classical music by Bach, Mozart and Beethoven. According to a notation by Strindberg, it was Brodén's rendering of Beethoven's 'ghost' sonata (no. 17 in d minor) which inspired the atmosphere of the play 'Crime and Crime'.

Differential equations of Fuchsian type and Riemann's problem

One of Brodén's interests was a problem by Riemann: to construct a Fuchsian equation with a given monodromy group. Brodén's contributions are not remarkable, but a section on the problem and its history may give an insight into a circle of problems which once was important and has not yet lost its interest. The next section is a short overview of a related subject, the theory of discrete subgroups of the Möbius group.

Let
$$L = a_n(z)(d/dz)^n + ... + a_0(z)$$
be a linear differential operator of order n in the complex plane with rational coefficients. Suppose that all solutions u of the homogeneous equation $Lu = 0$ are analytic in the extended complex plane C outside m singular points $Z = (z_1, ..., z_m)$. When γ is a loop from a base point z_0 not in Z, analytic continuation along γ produces a linear map $u \to S(\gamma)u$ of the n-dimensional space of solutions of the equation $Lu = 0$ at the point z_0. In this way one gets a representation $\gamma \to S(\gamma)$ of the fundamental group Γ of the region $C \backslash Z$. A change of base point gives an equivalent representation. The image group $S(\Gamma)$ is called the monodromy group of the equation. Its elements can be seen as matrices of type $n \times n$.

The fundamental group Γ has $m - 1$ independent generators corresponding to $m - 1$ of m elements $\gamma_1, ..., \gamma_m$ where γ_k is a loop from z_0 around z_k. If these loops are equally oriented it is clear that their product is homologous to a point. Hence the product $S(\gamma_1)...S(\gamma_m)$ is the identity.

If the operator L is Fuchsian (see p. 116), every solution u of $Lu = 0$ grows at z_k at most as an inverse power of $|z - z_k|$. A classical example of this situation is the hypergeometric operator
$$x(1-x)(d/dz)^2 + (c - (a + b + 1)x)d/dz - abu = 0,$$
with singular points at $0, 1, \infty$.

In a posthumous paper (see (1892) p. 389) Riemann posed the following inverse problem. A representation $\gamma \to S(\gamma)$ of the fundamental group Γ of the complement in C of m different points $Z = (z_1, ..., z_m)$ is given, the $S(\gamma)$ being matrices of order $n \times n$. Construct a Fuchsian equation of order n with the monodromy group $G = S(\Gamma)$. By the above, it is natural to give G in the form of $m-1$ invertible matrices, corresponding to $m-1$ independent generators of Γ.

In the first part of his paper Riemann proved that every system of n functions $u_1, ..., u_n$ which are analytic outside of Z and transform according to G at the points of Z are solutions of a Fuchsian differential equation with rational coefficients. In the proof he considers the matrix

$$u^{(n)}, ..., u$$

of type $(n+1) \times n$ which consists of n successive derivatives of the column $u = (u_1, ..., u_n)^t$. If S_k is the given generator at z_k, one sees that all subdeterminants of n must be eigenvectors of S_k with the eigenvalue $\det S_k$. Additional arguments show that they are proportional to rational functions which are the coefficients of the differential equation sought for. Riemann's paper can be seen as a program with the following points, the second of which was proved by Riemann.

1. Construct a system of n functions, analytical outside Z, which transform linearly under $m-1$ generators of the monodromy group.
2. Show that the system satisfies a Fuchsian differential equation.
3. Analyze the class of differential equations with a given monodromy group.

The problem 1 was solved for second order equations by Hilbert (1900). His solution was simplified by Plemelj (1908) who also extended the solution to equations of higher order. In that case the system reduces to a system of integral equations of the second kind with solutions given by Fredholm's theory. An earlier solution by Schlesinger (1905) was attacked by Plemelj in a furious polemic. Without having contributed anything of importance Brodén (1905 b) joined in the polemic against Schlesinger.

In the present version of Riemann's problem, all these names have disappeared. Algebraic geometry has contributed new methods and the theory of hypergeometric functions has been extended to several variables. The booklet (1987) by Yoshida gives an overview and contains among other things solutions of the second and third points of Riemann's program.

Discrete subgroups of the Möbius group

The elements of the Möbius group are fractional linear bijections

$$z \to S(z) = (az+b)/(cz+d), \quad ad - bc = 1,$$

of the complex plane, ∞ included. The theory of Fuchsian differential equations led Poincaré to a general theory of discrete subgroups of the Möbius group, i.e., such subgroups where orbits from a point do not return arbitrarily close to the point. The connection is the following one. When u_1, u_2 form a basis of the solutions of a second order Fuchsian equation, the quotient $d(z) = u_1(z)/u_2(z)$ is transformed as above when z runs through a loop from a base point around a singular point. The inverse function $\varphi(z)$ defined by $\varphi(d(z)) = z$ then has the property that $\varphi(S(z)) = \varphi(z)$ for every element S of the monodromy group G of the equation. When G leaves a circle invariant (we can let it be the unit circle), Poincaré named these functions Fuchsian (against the advice of Klein) and in other cases Kleinean. Poincaré's great papers about automorphic functions were the main attraction that Mittag-Leffler secured for the first volumes of *Acta Mathematica*.

Briefly, Poincaré's general theory runs as follows. When G is a discrete subgroup of the Möbius group, two points are said to be equivalent if they are images of each other under the group. A fundamental domain of G is a maximal part of the plane which only contains inequivalent points. Poincaré proved that there exist fundamental domains bounded by circular arcs. When the group is Fuchsian, there is a fundamental domain in the unit disk bounded by circular arcs orthogonal to the unit circle. Such a fundamental domain is also the analytical and topological image of a Riemannian surface whose universal covering surface is bijective to the unit disk (the other possibilities are the entire plane or the plane minus one point). A Riemann surface with finitely many leaves gives a fundamental region strictly contained in the unit disk. The analytic functions on the Riemann surface correspond to functions which are automorphic under a Fuchsian group.

In general a discrete group G has infinitely many elements and it is therefore not possible to construct automorphic functions by just taking the mean value over the group of a rational function $f(z)$. Instead Poincaré constructed so-called thetafuchsian functions $F(z)$ with the property that

$$F(S(z)) = F(z)(S'(z))^m, \quad S'(z) = (ac - bd)/(cz + d)^2$$

where m is a positive or negative integer, called the weight of F. Since $T'(Sz) = (TS)'(z)/S'(z)$, the function

$$F(z) = \sum_{T \in G} f(T(z))(T'(z))^{-m}, \quad f(z) \text{ rational}$$

is thetafuchsian of weight m when the series converges. If m is large positive, depending on G, we may get uniform convergence and then construct automorphic functions as quotients of two thetafuchsian ones with the same weight.

Another one of Poincaré's constructions is the zetafuchsian functions. We meet here an abstract version of the first part of Riemann's problem: construct a column vector

$$f(z) = (f_1(z), ..., f_n(z))^t$$

of rational functions such that

$$f(S(z)) = f(z)A(S)$$

for every S in a discrete subgroup G of the Möbius group, where $S \to A(S)$ is a representation of G as invertible matrices of order n. A vector which behaves like this under a thetafuchsian group G is given, at least formally, by

$$F(z) = \sum_{T \in G} A(T)^{-1} g(T(z))(T'(z))^{-m}$$

where the elements of $g(z)$ are rational functions. It is immediately clear that $F(S(z)) = A(S)F(z)(S'(z))^{-m}$. For large positive m we can hope for convergence and division by a scalar thetafuchsian function of weight m then gives the desired result. A remaining problem is to show that that the elements of $F(z)$ are linearly independent.

Nörlund

Niels Erik Nörlund (1885-1980) made his debut in 1908 with papers about linear difference equations. After a doctor's degree in Copenhagen in 1910 he became professor in Lund in 1912. He left in 1923 to be director of the National Danish Geodetic Institute in Copenhagen.

Mittag-Leffler, who became impressed by the young and distinguished Dane, had made plans for Nörlund to succeed him as director of the Mittag-Leffler Institute. But when Nörlund got an offer to be director of the Geodetic Institute in Copenhagen he accepted. One of the reasons may have been that Mittag-Leffler at that time had lost so much of his wealth that the future of the Institute was at stake.

Difference equations and interpolation

Nörlund's first position was as assistent at the astronomical observatory in Copenhagen. The director Thorvald Thiele (1838-1910) had taken an interest in interpolation and difference equations and constructed a method of interpolation by 'reciprocal differences' which used continued fractions. A function $\varphi(x)$ with given values φ_k for $x_0 < x_1 < ... < x_n$ is interpolated by the continued fraction

$$A_0 + /A_1 + (x - x_1)/A_2 + (x - x_2)/A_3 + ...$$

where $A_0, ...$ depend on $\varphi_0, ..., \varphi_n$ by a rather complicated rule. In his first major paper (1911) Nörlund worked with the convergence of these formulas when the points x_k come together. This problem is rather loosely connected with difference equations.

In (1913) Nörlund began his systematic investigations of difference equations. The name difference equation comes from an analogy with differential equations. The differentials are replaced by differences,

$$\Delta_\pm f(x) = (f(x \pm h) - f(x))/h,$$

and repeated differences of the same kind $\Delta_\pm^n$. For simplicity we shall choose $h = 1$ and $\Delta = \Delta_+$ in the sequel.

A linear homogeneous difference equation for a function u can also be written as

(1) $$\sum_{j=0}^{k} a_j(n) u(j + n) = 0, \quad n = 0, \pm 1, \pm 2, \ldots,$$

where the coefficients are defined on the integers. If $a_k(n)$ does not vanish for any n, a solution of (1) is uniquely determined by its values on n integers in a sequence, $z, z+1, ..., z+n-1$ and these values can be given arbitrarily. In particular, the space of solutions is linear with dimension n. In the simplest case, when the coefficients $a_j(n)$ in (1) are constant, the equation has exponential solutions $u = e^{\lambda n}$ where

$$\sum_{0}^{k} a_j e^{j\lambda} = 0.$$

In the older literature the solutions u were only defined on the integers. Later the unknown function and the coefficients acquired arbitrary real or complex arguments and then it was possible to use the theory of analytic functions. Around 1910 this step was taken by Nörlund and other mathematicians.

Nörlund's main theme was the asymptotic behavior of the solutions of difference equations in analogy with the well-developed theory for linear differential equations with analytic coefficients, in the first place those of Fuchsian type. For a difference equation this limitation is interesting only when the equation is of Fuchsian type at infinity in which case it has the form

$$\sum_{0}^{k} P_j(x) \Delta^j u(x) = 0,$$

where the coefficients have decreasing degrees with the consequence that the solutions increase at most exponentially. The behavior at infinity is determined as for differential equations. Instead of power series, Nörlund used factorial series

$$\sum \frac{a_n n!}{x(x+1)...(x+n)}.$$

in his papers (1911,1913). They are better adjusted to difference equations and have natural convergence domains $\operatorname{Re} x > \text{const}$.

In the next major paper (1916) the problem is the behavior at infinity of solutions of difference equations which are not necessarily of Fuchsian type. The impulse came from Poincaré's treatment (1885) of the same problem for differential equations. Instead of power series the result may then be asymptotical series of the size $e^{Q(x)}$ where Q is a polynomial. When all coefficients of the equation have the same degree, $Q(x)$ has degree one. It is in the first place this case which Nörlund transplants to difference equations. As Poincaré before him, Nörlund uses the Laplace transform which means that solutions are sought of the form

$$u(z) = \int_\gamma t^{z-1} v(t) dt,$$

where $v(t)$ is locally analytic and γ is a loop in the complex plane which does not go around 0. One sees that $v(t) \to t^k v(t)$ corresponds to $u(z) \to u(z+k)$ and that $v(t) \to -tv'(t)$ corresponds to $u(z) \to zu(z)$ if an integration by parts without boundary terms is possible. From this follows that a difference equation (1) with polynomial coefficients gives a differential equation in $v(t)$ with polynomial coefficients

$$\sum \theta_i(t) v^{(i)}(t) = 0,$$

whose order is the maximal degree of the coefficients $a_k(z)$.

In order to elucidate Nörlund's methods and results we shall choose his own example, namely the case when the polynomials $a_k(z) = a_k + b_k z$ are linear. This gives the equation

$$\sum_0^n a_k t^k v(t) - \sum_0^n b_k t^{k+1} v'(t) = 0.$$

for $v(t)$. We now assume that the polynomial $\sum b_k t^{k+1}$ has separate zeros $B_0 = 0, B_1, ..., B_n$ so that

$$v'(t)/v(t) = \frac{A_0}{t} + \sum_1^n \frac{A_k}{t - B_k}.$$

If we further assume that no number A_k is a negative integer, the result is that

(2) $$v(t) = \text{const } t^{A_0} \prod (t - B_k)^{A_k}.$$

Hence, if γ_k is a loop from the origin around the point $B_k \neq 0$ and $v(t)$ is defined by (2), we get n solutions of the difference equation (1), namely

$$u_k(z) = \int_{\gamma_k} t^{z-1} v(t) dt.$$

To avoid boundary terms at 0 it suffices to assume that $\operatorname{Re} z > 0$. Then it is not difficult to see that the solutions $u_1, ..., u_n$ constitute a basis of the solution space of (1).

It is now easy to see the asymptotic behavior for large Re z of the function $u_k(z)$. As Nörlund shows, it has an asymptotic series with a main term const $A_k^z z^{-B_k-1}$. Concurrent with the solutions u_k there are solutions

$$u_k^*(z) = \int_{\gamma_k^*} t^{z-1} v(t) dt$$

where Re z is assumed to be large negative and γ_k^* is a loop from ∞ around B_k. These functions also constitute a basis. Nörlund investigates the linear dependence with the previous base and gets asymptotic solutions for Re $z \to \infty$.

Nörlund wrote many more substantial papers (1913, 1914, 1922 a, 1923 b) on difference equations. The last one is devoted to the simplest equation

$$f(x+1) \pm f(x) = 2g(x).$$

The series

$$-\sum_0^\infty g(x+s), \quad 2\sum(-1)^s g(x+s)$$

are formal solutions but diverge in general. Nörlund gave conditions for their summability.

Interpolation was another of Nörlund's interests. One of the problems was to extend classical interpolation formulas to infinitely many points, for instance Newton's interpolation

$$f(z) = f(1) + \sum_1^\infty f(n+1)(z-1)(z-2)...(z-n)/(n+1)!$$

where $f(1), f(2), ...$ are given. In (1922 b) Nörlund reworked Fritz Carlson's result that this interpolation is correct for functions which are analytic in the right half-plane and there grow less than $e^{\operatorname{Re} z}$.

Very soon Nörlund became a known expert in his field, the modern theory of difference equations. A review (1923 b) of the field in *Enzyklopädie der Mathematischen Wissenschaften* can be seen as the end of a successful career.

Block

Henrik Block (1882-1929) studied astronomy in Lund and became a docent in 1909. His main interest was mathematics and he wrote about planetary motion and partial differential equations. His papers rendered him a professorship at Chalmers Technological Institute in Gothenburg in 1915.

In (1907) the Finnish astronomer proved the following important result about the three body problem: If the three bodies collide in one point, the future motion occurs in a plane in such a way that all bodies move on straight lines or at the corners of equilateral triangles. These modes of motion had been found earlier by Lagrange as explicit solutions of the equations of motions. In a paper (1920) Block used perturbation calculus to study motions which are infinitely close to these. A later paper (1920) is a careful study of stability and existence of periodic solutions for the much studied model equation

$$\eta'' + \cos\eta \sin\eta = -\mu b^2 \cos bt, \quad \eta = \eta(t).$$

It describes the motion of a pendulum under the influence of a small periodic force whose direction forms the angle η with a fixed direction.

Block's second theme was differential operators, mainly parabolic ones. The paper (1910) is more of an exercise, but the following ones, (1912-13), were probably decisive for the Chalmers professorship. They deal with a class of differential equations with no physical meaning but with properties which can be found in the equation of heat conduction. This choice of subject may have been judged to be fresh and audacious. Block's results disappear into the modern theory of differential operators (see Hörmander 1963) but they deserve a short review for which an appendix about fundamental solutions at the end of Chapter 10 may serve as a background.

When Block constructed his fundamental solutions in 1912-14 the subject had been opened by Fredholm's work. In addition, Zeilon's paper (1911) might have given Block the idea to use the Fourier transform. These references are missing in Block's papers. Instead he refers to a paper by E.E. Levi (1909) which deals with diffferent types of differential equations and their characteristics.

Block's differential operators have the form

$$(1) \qquad P(\partial_t, \partial_x) = \partial_t^n - \partial_x^m + \sum a_{jk} \partial_x^j \partial_t^k, \quad j < m, k < n,$$

where $n < m$ and $a \neq 0$. The principal part is $P_m = \partial_x^m$ and the characteristics are straight lines parallel to the t-axis.

A special case of (1) is the operator of heat conduction $\partial_t - \partial_x^2$ with a fundamental solution

$$H(t) \frac{1}{\sqrt{2\pi t}} e^{-x^2/2t}$$

where $H(t)$ is Heaviside's function, equal to 1 when $t > 0$ and 0 when $t < 0$.

In Block's first paper all a_{jk} vanish and m, n are coprime. Later he considers the general case. Block's text is strict and carefully written, but the general scheme is hidden under the details. What he does may be briefly described as follows. Block tried to write his fundamental solution as a Fourier transform

$$(2) \qquad E(t, x) = \frac{1}{2\pi} \int_{-\infty}^{\infty} F(t, \xi) e^{i(x, \xi)} d\xi,$$

which means that $F(t, \xi)$ ought to be a fundamental solution of the differential operator $P(\partial_t, i\xi)$ so that

$$P(\partial_t, i\xi) F(t, \xi) = \delta(t).$$

(Precisely this point is not very clear in Block's papers).

The problem is now reduced to an elementary determination of $F(t, \xi)$ chosen in such a way that the integral (2) converges. In Block's situation it is almost always possible to arrange that $F(t, i\xi)$ decreases exponentially to zero in ξ when $t \neq 0$. This gives a fundamental solution which, precisely like that of the heat operator, is infinitely differentiable in x when $t \neq 0$.

Theses

During the time from 1901 till 1922 seven theses with very varied subjects were written in Lund. One thesis can be traced to Brodén, one to Block, and two to Nörlund.

Brodén's student Lundberg wrote in (1912) about pairs of applicable surfaces for which there exists a bijection which transforms the line element of one the line element of the other. In this theory there is a group of linear bijections of R^6 which produces new applicable pairs from a given one.

Block's student G. Stadler (1916) followed Block in the construction of fundamental solutions of the differential operator

$$P(\partial_t, \partial_x) = \partial_t^p + \Delta^m, \quad \Delta = \partial_{x_1}^2 + \partial_{x_2}^2.$$

As in Block's case it is possible to write down a fundamental solution directly,

$$E(t,x) = \frac{1}{4\pi^2} \int F(t,\xi) e^{i(x,\xi)} d\xi$$

where $\xi = (\xi_1, \xi_2)$ are real, $(x,\xi) = x_1\xi_1 + x_2\xi_2$ and $F(t,\xi)$ is a fundamental solution of $P(\partial_t, i\xi)$, composed of exponential solutions which are small or at most oscillating when $|\xi|$ is large.

In addition to Block's papers. G. Stadler also quotes Zeilon's method (see Chapter 11) to construct fundamental solutions by summation of a divergent Fourier integral, but this method would just have given him unnecessary diffficulties. Stadler's thesis seethes with youthful enthusiasm and joy of computing, but it shares the fate of Block's papers to have been trivialized by a later development.

S. Stadler (1918) and Ryde (1921) got their subjects from Nörlund. S. Stadler wrote about a system of difference equations corresponding to a system of ordinary differential equations of the form $du/dt = Au$ where A is a square matrix. Ryde treated geometric difference equations where the elementary difference is

$$Qu(x) = (u(qx) - u(x))/(qx - x)$$

with a fixed $q \neq 1$. The factorial series must then be replaced by other objects. As Nörlund had done, Ryde treated his subject in the framework of analytic functions. After his time in Lund he wrote a series of papers about finite repeated fractions of the form

$$a + b/b + c/c + d/ \ldots$$

where $a, b, c, \ldots$ is a monotone sequence of positive integers. The problem was to construct an algorithm for the representation of rational numbers in this form (see (1944)).

If $F(z, w)$ is an analytic function of two complex variables and γ is a suitable contour in the complex plane, the function

$$f(z) = \int_\gamma F(z,w) dw$$

is analytic, and in many cases a change of the contour γ gives an analytic continuation of the function $f(z)$. This possibility is the main theme of Ernst Lindelöf's classical book (1905). It is also the mainstay of Bernhard Svensson's thesis (1908) where Lindelöf is quoted. The problem is to continue an analytic function $f(z)$, given in the unit disk by its power series, for the case when the coefficients $a_n = \varphi(n), n > 0$ are values of another analytic function $\varphi(z)$. We may for instance assume that $\varphi(z)$ is analytic when $\text{Re } z > 0$ and such that $\log |\varphi(z)| = o(|z|^\varepsilon)$ for every $\varepsilon > 0$. The point of departure of Svensson's thesis is the formula

$$f(z) - a_0 = \frac{1}{2\pi i} \int_\gamma \frac{\varphi(w) z^w dw}{e^{2\pi i w} - 1}, \quad 0 < z < 1,$$

where γ is a loop in positive direction around the interval $(1, \infty)$. The residue sum of the right side is $\sum a_n z^n$ for $n > 0$. If $z = re^{i\theta}$ and $w = \varrho e^{i\alpha}$ with $|\alpha| < \pi/2$ then

$$|z^w| = e^{\varrho(\log r \cos \alpha - \theta \sin \alpha)}$$

is exponentially decreasing in ϱ when z belongs to a kidney-shaped region S_α with the cusp at 1 defined by $\log r < |\theta \tan \alpha|$. The region S_α contains the unit disk and increases when $|\alpha| \to \pi/2$ to the entire plane except for the positive real axis.

As long as z belongs to S_α we can modify the loop to the boundary of the region $|\arg w - 1/2| < \alpha$ giving an integral which is an analytic function of $z \in S_\alpha$. This a first elementary result of Svensson's very carefully written thesis. His results were taken up later and extended by Fritz Carlson (1914).

Svensson became lektor at a gymnasium in Lund. He was reluctant to fail students and for this he was nicknamed *B-minus*, the name of a mark meaning 'passed, but not quite'. The 'minus' must here be interpreted as a 'plus' for the teacher.

A woman wrote a thesis during this period, Louise Petrén, who had many scientifically notable brothers. Her thesis (1911) was published in the yearbook of the university and treated a classical subject: the solution of certain partial differential equations by quadrature, i.e., simple integration. When the process succeeds, the result is a general solution containing as many independent arbitrary functions as the order of the equation. Petrén's contribution is part of an old tradition with a large literature.

The introduction states that already Euler remarked that the operator

$$L = \partial_x \partial_y + a \partial_x + b \partial_y + c, \quad a = a(x,y) \quad \text{etc.}$$

can be written as a product $(\partial_x + b)(\partial_y + a)$ precisely when the first one of the functions

$$h = a_x + ab - c, \quad k = b_y + ab - c$$

vanishes. When the second one vanishes, L equals the reversed product. If one of the functions h, k vanishes it is clear that the equation $Lu = v$ can be solved by two integrations which give a general solution containing two independent arbitrary functions of x and y exactly as in the case $L = \partial_x \partial_y$. But there are more possibilities. The functions $U(z) = z_x + az$ and $V(z) = z_y + bz$ satisfy equations of the type $Lu = 0$ above, and if these are solvable, the original equation is also solvable. This process may be iterated, which is an idea of Laplace in a paper to which the title of the thesis alludes. It is also presented in Darboux (1889).

The object of Petrén's thesis is to use Laplace's method in order to find conditions for the solvability of equations $Lu = v$ where

$$L = \sum_0^{n-1} A_k(x,y) \partial_x \partial_y^k + \sum_0^n b_k(x,y) \partial_y^k.$$

This program, which is carried out with care and clarity in 150 densely printed pages, defies a short review.

The problem to find general solution of partial differential equations with analytical coefficients can be said to have been solved by the solution of Cauchy's problem. The original point of view, where independent functions play the same part as arbitrary constants for ordinary differential equations, has been a blind alley in spite of some interesting results. Louise Petrén's talent and capacity for work was worthy of a better theme.

Two remaining theses that cannot be understood without large background presentations will only get a summary treatment. Carl Wilhelm Oseen's thesis (1902) lists a number of Lie algebras in six variables which have a number of given invariants. This is the first time that Sophus Lie's great theory of continuous groups appears in Sweden. The remarkable thing about Oseen's thesis is not the result — a list of Lie algebras — but rather the fact that it was written when the author was twenty-two years old and had had a very short time to penetrate a difficult and vast theory.

After the thesis which earned him the title of docent, Oseen left Lie's theory and wrote papers on different parts of mathematics. Among other things there is a paper

(1906) on a boundary value problem for the heat equation. But he devoted himself more and more to mathematical physics and in 1909 he was nominated professor of mathematical physics in Uppsala. In 1933 Oseen became director of the Nobel Institute of the Swedish Academy of Sciences.

Oseen was active in all parts of mathematical physics, pure mechanics, electricity, hydrodynamics and, at the end of his life, quantum mechanics and statistical mechanics with a contribution to the theory of fluid crystals. His main work on hydrodynamics is reviewed in Chapter 15. During a very long period Oseen had a leading position in his subject in Sweden. He was also a prolific writer in popular science and had historical interests. Apart from the book about Klingenstierna (see Chapter 1) he wrote penetrating obituaries of Swedish scientists, for instance Bäcklund and Gullstrand, both of which are quoted in this book.

Another independent thesis (1921) was written by Arvid Uhler. He refers to two papers by Ritter (1894,1897). In the first one the author constructs on a Riemann surface of genus g a many-valued analytic function with given ramifications at n given points and along the $2g$ fundamental cycles on the surface. In the second paper, which is not so complete, the problem is extended to systems of functions.

Uhler's first chapter contains computations on the basis of Ritter's results, while the two others constitute an effort to construct Ritter's systems using zetafuchsian functions and theory. Uhler presupposes that the reader has a thorough knowledge of Ritter's paper and this makes it very difficult to understand the details of his thesis. The reader encounters a kind of literary style with an awkward terminology, vague reasonings, and unclear results. The thesis got two tired lines in *Jahrbuch*, but Uhler seems to have worked all by himself. He made a brave try in a difficult and not very easily accessible area.

BIBLIOGRAPHY

Brodén.

BRODÉN T.
1886. *Om rotationsytors deformation till nya rotationsytor med särskilt avseende på algebraiska ytor*, Lund 1886.
1889. *Über die durch Abel'sche Integrale erster Gattung rectificierbare Curven*, KVA Bihang bd 15, Afd 1, no 5..
1894. a. *Zur Transfomation elliptischer Functionen. Erste Mitteilung*, Lunds Univ. Årsskr. 1894, no 30.
1894. b. *Ueber Zeuthens Correspondenzsats*, KVA Öfversigt 1894, 345-359.
1895. *Über unendlich oft oscillierende Funktionen*, KVA Öfversigt 52, 1895, 763-768.
1896. *Das Weierstrass-Cantorsche Condensationsverfahren*, KVA Öfversigt 53, 1896, 583-682.
1898. *Darstellungen von reellen Funktionen mit unendlich dicht liegenden Nullstellen durch unendliche Produkte*, Math. Ann. 51, 1898, 299-320.
1899. *Ueber das Dirichlet'sche Integral*, Math. Ann. 52, 1899, 177-227.
1904. *Ueber gewisse Arten linearer Differentialgleichungen zweiter Ordnung*, Arkiv Mat. Astr. Fysik 1, 1904, 419-447.
1905. a. *Über eine Verallgemeinerung des Riemann'schen Problems in der Theorie der linearen Differentialgleichungen*, Acta Math. 29, 1905, 273-294.
1905. b. *Bemerkungen über die Uniformisierung analytischer Funktionen*, Berling, Lund 1905.
1922. *Det nittonde århundradet*, Lund 1922.
1924. *Eine realistische Begründung der Mathematik*, Lunds Univ. Årsskr. 1924.
    HILBERT D.
1912. *Riemanns Probleme in der Theorie der Funktionen einer komplexen Veränderlichen*, in *Grundzüge einer Allgemeinen Theorie der linearen Integralgleichungen*, Leipzig 1912, s. 81.
    PLEMELJ J.
1908. *Riemannsche Funktionenscharen mit gegebener Monodromiegruppe*, Monatshefte d. Math. u. Phsyik 19, 1908.
    RIEMANN B.
1892. *Gesammelte Werke und Nachlass*, Herausgeg. H. Weber, zweite Aufl. 1892 Dover 1952.

SCHLESINGER I.
1905. *Zur Theorie der linearen Differentialgleichungen im Anschlusse an das Riemannsche Problem*, J. f. Math. 130, 1905.
YOSHIDA M.
1987. *Fuchsian Differential Equations*, Aspects of Mathematics E 11, Braunschweig 1987.

Nörlund.

NÖRLUND N.E.
1911. *Fractions continues et différences réciproques*, Acta math. 34, 1911, 1-108.
1913. *Sur l'intégration des équations linéaires aux différences finies par des séries de facultés*, Rend. Palermo 35, 1913, 177-216.
1914. *Sur les séries des facultés*, Acta Math. 37, 1914, 327-387.
1916. *Sur les équations linéaires aux différences finies à coéfficients rationels*, Acta Math. 40, 1916, 191-249.
1922. a. *Mémoire sur les polynomes de Bernoulli*, Acta Math. 43, 1922, 121-205.
1922. b. *Sur les formules d'interpolation de Stirling et de Newton*, Ann. École Norm. (3) 39, 1922.
1923. a. *Mémoire sur le calcul de différences finies*, Acta Math. 44, 1923, 71-212.
1923. b. *Neuere Untersuchungen über Differenzengleichungen*, Enzyklopädie der Mathematischen Wissenschaften Band II.3.2, 1923-27.
POINCARÉ H.
1882. *Théorie des groupes fuchsiennes*, Acta Math. 1, 1882, 205-278.
1884. *Mémoire sur les fonctions zétafuchsiennes*, Acta Math. 5, 1884, 209-278.
1885. *Sur les équations aux différentielles ordinaires et aux différences finies*, Amer. J. Math. 7, 1885, 203-238.
1886. *Sur les intégrales irrégulières des équations linéaires*, Acta Math. 8, 1886, 295-344.

Block.

BLOCK H.
1909. *Sur les chocs dans le problème des trois corps*, Arkiv 5, 1909, no 9.
1910. *Sur les équations aux dérivées partielles du type parabolique*, Arkiv 6, 1910, no 31.
1914. *Sur les équations aux dérivées partielles à caractéristiques multiples*, Arkiv 7, 1912, no 13,21; Arkiv 8, 1913-14, no 23; Arkiv 9, 1913-14, no 8.
1920. *Sur une équation différentielle du problème de la rotation des corps célestes*, Arkiv 14, 1920.
HÖRMANDER L.
1963. *Linear partial differential operators*, Springer 1963.
LEVI E.E.
1909. *Caratteristiche multiple e problema di CAUCHY*, Annali di Matematica 1909.
SUNDMAN K.F.
1907. *Recherches sur le problème des trois corps*, Acta Soc. Sci. Fenn. 34, 1907, no 6.

Theses.

LUNDBERG E, *Om par av applicabla ytor*, Ak. avh. Lund 1912.
OSEEN C.W, *Über die endlichen, continuierlichen, reduciblen Berührungstransformationen im Raum.*, Ak. Afh. Lund 1901.
PETRÉN L, *Extension de la méthode de Laplace...*, Ak. avh. Lund 1911.
RYDE F, *A contribution to the theory of linear homogeneous geometric difference equations...*, Ak. avh. Lund 1921.
STADLER G, *Etudes sur l'équation...*, Ak. avh. Lund 1916.
STADLER S, *Sur les systèmes d'équations aux différences finies linéaires et homogènes*, Ak. avh. Lund 1918.
SVENSSON B, *Etudes sur les fonctions définies par une série de Taylor*, Ak. avh. Lund 1908.
UHLER H, *Sur les séries zétafuchsiennes*, Ak. avh. Lund 1921.

Bibliography.

CARLSON F.
1914. *Sur une classe de séries de Taylor*, Diss. Uppsala 1914.
KOCH J.
1945. *Carl Wilhelm Oseen*, K. Fysiogr. Sällsk. Förh. bd 15, 1945.
LIE S., ENGEL F.
1893. *Theorie der Transformationsgruppen*, I,II,III Leipzig 1888-1893.
LINDELÖF E.
1905. *Le calcul des résidues*, Paris 1905.
DARBOUX G.

1889. *Leçons sur la théorie générale des surfaces II*, Paris 1889.
   PISATI.
1905. *Sulla estensione del metodo di Laplace alle equazioni lineari di ordine qualunque con due variabli independenti*, Rend. Palermo 20, 1905, 344-374.
   RITTER E.
1894. *Die multiplicativen Formen auf algebraische Gebilde beliebigen Geschlechtes mit Anwendung auf die Theorie der automorfen Formen*, Math. Ann. 44, 1894, 261-374.
1896. *Ueber Riemann'sche Formenscharen auf einem beliebigen algebraischen Gebilde*, Math. Ann. 47, 1896, 157-221.
   RYDE F.
1944. *Der Algorithmus der monotonen nicht-wachsenden Kettenbrüche*, Arkiv 31 A, 1944, no 19.

CHAPTER 14

Stockholm 1925 - 1950

Carleman   Carlson

In the 1920's two new professors of mathematics were appointed at Stockholm University. Torsten Carleman succeeded von Koch who had died in 1924 at the age of fifty-four. Bendixson retired in 1926 and was replaced by Fritz Carlson, who came from a professorship of descriptive geometry at KTH.

In 1923 Carleman was for a short time professor in Lund, but on Mittag-Leffler's initiative he was invited to Stockholm University. This was against the wishes of the many students who had preferred Marcel Riesz as a professor. After Mittag-Leffler's death in 1927 Carleman became director of the Mittag-Leffler Institute in Djursholm, a post which he combined with his professorship at the university.

Both Carleman and Carlson had studied in Uppsala and did not belong to the Mittag-Leffler circle. But Mittag-Leffler's great initiative in mathematics still paid dividends when two of his third generation students, Nils Zeilon and Marcel Riesz became professors at Lund University.

Carleman

With Torsten Carleman (1892 - 1948) Sweden acquired its most gifted and powerful mathematician so far. He was born in a rural district in southern Sweden where his father was a school teacher and precentor of the local church. Carleman graduated from the gymnasium in Växjö where he had had a highly qualified teacher of mathematics and physics, Henrik Petrini (see Chapter 11). During his studies in Uppsala, Carleman devoted himself with great energy to mathematics and mathematical physics and the results soon became apparent. A thesis (1916) was followed by a sequence of papers in different parts of analysis: integral equations, Fourier series, analytic functions and mathematical physics, including a pioneering spectral analysis of the Schrödinger operator.

Carleman wrote important papers about integral equations (1923), quasianalytic functions (1926), harmonic analysis (1944) and Boltzmann's equation (1944). But his great talent is best seen in his short papers with concise proofs of important results.

Integral equations

Carleman's thesis (1916) is a thorough treatment of a problem in potential theory which reduces to the properties of the integral operator of the double layer potential,

$$Kf(s) = \int_C K(s,t)f(t)dt.$$

Here s and t denote arc-length on a simple closed curve $C : s \to x(s)$ in the plane and

$$K(s,t) = \frac{1}{\pi}D_t \log \frac{1}{|x(s) - x(t)|} = \frac{(n_t, x(s) - x(t))}{|x(s) - x(t)|^2},$$

where D_t denotes differentiation at the point $x(t)$ along the inner normal n_t at the same point. When C has a continuous curvature, $K(s,t)$ is a continuous and bounded function so that Fredholm's theory applies. The main purpose of Carleman's thesis is to extend Fredholm's theory to the case when C has a continuous curvature except at a number of corners. His results, formulated for the integral equation $f - \lambda K f = g$, will be rewritten below in terms of properties of the operator K.

At a corner of C the kernel $K(s,t)$ tends to infinity when s, t approach the corner from different directions, but $\int |K(s,t)| dt$ is finite. Carleman's main results are the following ones: when g is a continuous function, the equation $Kf - \lambda f = g$ has a unique integrable solution $f(x, \lambda)$ which is analytic in λ outside of the real axis and meromorphic with finitely many poles when

$$|\lambda| > R,$$

where R is the maximum of $|\pi - \alpha|/\pi$ for all interior corner angles α at the finitely many corners.

This result becomes more understandable by the fact that the kernel $K(s,t)$ is symmetrizable. If

$$L(s,t) = \log \frac{R}{\log |x(s) - x(t)|}, \quad R > 0,$$

the kernel

$$M(s,t) = \int_C K(s,u) L(u,t) du$$

is symmetric. This follows from Green's formula applied in the interior of C to the harmonic functions

$$\log \frac{1}{|x - x(s)|}, \quad \log \frac{1}{|x - x(t)|}.$$

It can also be shown that $L(s,t)$ is positive definite when R is large. As the author points out, his problem can be seen as a diagonalization of the kernel $M(s,t)$ relative to an interior product given by the kernel $L(s,t)$. The corresponding operator turns out to be bounded with a continuous spectrum in the interval $|\lambda| < R$ and a discrete spectrum outside. Carleman also gives an explicit spectral decomposition corresponding to Hilbert's spectral family for bounded and self-adjoint operators (see the appendix at the end of the chapter).

In a limited space it is not possible to sketch Carleman's proof which uses known results from the theory of integral equations, Fourier analysis and an explicitly solved model problem in which C consists of two circular arcs. With its tight analysis and almost 200 pages Carleman's thesis is both difficult to read and difficult to interpret. This circumstance caused Carleman's teacher Holmgren to write an abstract of the results for a broader audience (see (1960), p. 223).

Carleman kernels

Carleman's thesis was followed five years later by a treatise (1923 a) on singular integral equations. This work, which generalized most earlier papers in the subject, was written some years before John von Neumann in (1929) introduced abstract Hilbert space and extended Hilbert's spectral theory for bounded, self-adjoint operators to unbounded ones. Therefore Carleman's theory of integral equations ought to be seen in the light of the abstract theory and its prehistory (see the appendix of this chapter).

Carleman's treatise is about integral equations of the type

$$f(x) + \lambda \int_I K(x,y) f(y) dy = g(x)$$

where I is a bounded interval on the real axis and $K(x,y)$ is a real, symmetric kernel such that

$$K(x)^2 = \int_I K(x,y)^2 dy < \infty, \quad K(x) \geq 0, \tag{1}$$

when x is outside a countable set. The theory is equivalent to a spectral theory for operators K defined by

$$Kf(x) = \int_I K(x,y)f(y)dy \tag{2}$$

with f in some subspace of $L^2(I)$. The theory is not tied to an interval and one dimension. It can easily be extended to complex spaces $L^2(I)$ and hermitian symmetric kernels, $K(y,x) = \overline{K(x,y)}$, such that

$$\int_I |K(x,y)|^2 dy$$

is finite almost everywhere. These kernels are now called Carleman kernels.

To begin with, it is clear from (1) and Schwarz's inequality that the double integral

$$\int_{I \times I} |K(x,y)||f(x)||g(y)|dxdy \tag{3}$$

converges for instance when the functions f and g are square integrable and the integral $\int K(x)|f(x)|dx$ is finite. This last condition determines a linear domain $D = D(A)$ of an operator A defined by

$$Af(x) = \int K(x,y)f(y)dy \in L^2(I)$$

when $f \in D(A)$. It is easy to see that A is closed and, in addition, symmetric, i.e. $(Af, g) = (f, Ag)$ when $f, g \in D$. In fact, the integral (3) converges when $f, g \in D$ and hence $(Af, g) = (f, Ag)$ by Fubini's theorem. Carleman has the space D and the operator A, but the leading part is played by an operator $K \supset A$ whose domain of definition $D(K) \supset D(A)$ consists of all $f \in H = L^2(I)$ such that $Kf(x) = \int K(x,y)f(y)dy \in H = L^2(I)$.

Carleman chooses to approximate the kernel $K(x,y)$ by kernels $K_N(x,y)$ of the Hilbert-Schmidt class. We can define

$$K_N(x,y) = K(x,y) \quad \text{when} \quad K(x) < N, \quad K(y) < N$$

and $K_N(x,y) = 0$ otherwise. Then $K_N(x,y)$ is the kernel of a Hilbert-Schmidt operator and we have

$$\int K_N(x,y)f(x)g(y)dxdy = \int \lambda d(E_N(\lambda)f,g)$$

where $E_N(\Delta)$ is the corresponding spectral family (see the appendix). Every $E_N(\Delta)$ has a kernel $E_N(\Delta, x, y)$ of the Hilbert-Schmidt type. In fact, if $\varphi_k(x)$ is an orthonormal and complete system of eigenfunctions of K_N with corresponding eigenvalues then

$$E_N(\Delta, x, y) = \sum_{\lambda_k \in \Delta} \varphi_k(x)\varphi_k(y).$$

Carleman makes a passage to the limit $K_N \to K$. He starts with the equation

(4) $$K_N f_N - z f_N = g$$

with $g \in H$ and $z = x + iy$ with $y \neq 0$. This equation has the unique solution

(5) $$f_N = \int \frac{1}{\lambda - z} dE_N(\lambda) g.$$

It follows from (4) that $\|f_N\| \leq \|g\|/|y|$. Hence we may choose a sequence of functions f_N which converge weakly to some element $f \in H$ such that

$$\int_I K(x, y) f(y) dy$$

converges for almost all x, in which case

(6) $$Kf - zf = g$$

with the operator K given above. At the same time we may make a passage to the limit in the weak form of (4),

(7) $$(f_N, h) = \int \frac{1}{\lambda - z} d(E_N(\lambda) g, h).$$

This shows that there exists a family of increasing self-adjoint operators $B(\Delta)$, majorized by unity, such that

(8) $$(f, h) = \int \frac{1}{\lambda - z} d(B(\lambda) g, h).$$

In particular, $f = f(x, z)$ can be assumed to be analytic in z outside of the real axis.

The solutions of the homogeneous equation (6) with $\text{Im } z \neq 0$ constitute a linear space $L(z)$. Since the kernel $K(x, y)$ is real, we have $L(\bar{z}) = \overline{L(z)}$. In his book Carleman shows that $\dim L(z)$ is independent of z.

Carleman divides his kernels into two classes according as the dimension of $L(i)$ is zero (the class I) or > 0 (the class II). He verifies that if

$$(K - zE)f_1 = g_1, \quad (K - zE)f_2 = g_2$$

then the equality $(f_1, g_2) = (f_2, g_1)$ means that K is symmetric, i.e., $(Kf, g) = (f, Kg)$ when $f, g \in D(K)$. Since this holds when $K = K_N$ it follows by a passage to the limit that an operator K with a kernel of the class I is symmetric. The distinction between kernels of the classes I and II was present already in Carleman's note (1920).

Carleman also shows that the family $B(\Delta)$ of (8) is unique for open intervals Δ when K belongs to the class I and that it consists of commuting projections such that $B(\Delta) B(\Delta') = B(\Delta \cap \Delta')$.

We can find a simpler proof by remarking that if $\alpha(\lambda)$ is a function of bounded variation and if

$$F(z) = \int (\lambda - z)^{-1} d\alpha(\lambda),$$

where $\varphi(x)$ is continuous with compact support, then the limit

$$\lim_{y \searrow 0} \frac{1}{2\pi i} \int \varphi(x) (F(x + iy) - F(x - iy)) dx$$

equals
$$\int \varphi(\lambda)d\alpha(\lambda).$$

Hence the measure $d\alpha(\lambda)$ is uniquely determined by $F(z)$. If this is used in (8) one gets a map of functions $\varphi(\lambda) \to \varphi(A)$ defined by

$$(\varphi(A)f,g) = \int \varphi(\lambda)d(B(\lambda)f,g)$$

with the property that $\lambda\varphi(\lambda) \to A\varphi(A)$. This suffices for a functional calculus and identifies $B(\Delta)$ with a family of projections.

Carleman found the projections $E_N(\Delta)$ and their limits $B(\Delta)$ by the construction of the corresponding kernels $E_N(\Delta, x, y)$ and $B(\Delta, x, y)$, called spectral functions. They are defined by the integrals

$$E_N(\Delta)f(x) = \int E_N(\Delta, x, y)f(y)dy. \quad B(\Delta)f(x) = \int B(\Delta, x, y)f(y)dy.$$

Carleman and the abstract theory

It is easy to see that Carleman's operator K is the adjoint of his operator A. Carleman had also the correct condition for K to be self-adjoint, namely that $K + z$, considered as an operator from $D(K)$ to L^2, is invertible and surjective when $\operatorname{Im} z \neq 0$. The extension of Carleman's theory from real kernels to complex ones is only a formality.

When Carleman read von Neumann's paper (1929) (see the appendix), he did not see much that he did not know already and, since he did not attach much value to abstract versions of known facts, he felt passed over. In his major congress lecture (1932) he remarks that a complete orthonormal system in the domain of a selfadjoint operator is all he himself had needed to construct a spectral measure. In his paper (1935) von Neumann answered the criticism indirectly by proving that a selfadjoint operator is given by a Carleman kernel if and only if its spectrum contains 0. It is easy to imagine that von Neumann also could have said that Carleman had the essential feature of the theory, namely a correct definition of selfadjointness.

In the long run the abstract theory won because of its inherent advantages. It is clear, simple and closed, it provides an instant insight, and it can be applied to a great many different problems. An important feature of von Neumann's paper is the explicit definition of the graph of an operator. This notion, now universally used, took away all possible insecurities from the class of unbounded operators. Marshall Stone took Carleman more seriously than von Neumann. His book (1932) has a chapter where Carleman's theory is fitted into the abstract frame.

Generalized kernels

Not all interesting kernels are Carleman kernels, for instance not Green's functions $G(x, y)$ of classical boundary value problems for Laplace's operator

$$\Delta = \sum_1^n \partial^2/\partial x_k^2$$

when the dimension n exceeds two. This depends on the fact that the principal part, Newton's potential $\operatorname{const} |x - y|^{2-n}$, is then not square integrable in x or in y at $x = y$. This defect made Carleman, who took an interest in the physically important case $n = 3$, write a chapter on kernels $K(x, y)$, such that the functions $y \to K(x, y) - K(x', y)$ with fixed x, x' are square integrable.

In a collection of applications these kernels are used to investigate the propagation of light outside of a reflecting and smoothly bounded body $S \subset R^3$. In modern terminology Carleman shows that the pairs $(u, -\Delta u)$ such that $u, \Delta u \in L^2(R^3 \setminus S)$ and such that the outer normal derivative of u vanishes at the boundary of S, from the graph of a self-adjoint operator $A \geq 0$. Its inverse has a kernel of which Newton's potential is the principal part and Carleman could construct a spectral decomposition of A with kernels $E(\lambda, x, y)$.

It follows that the function

$$u(t, x) = \int \cos\sqrt{\lambda} t \, d \int E(\lambda, x, y) f(y) dy + \int \frac{\sin\sqrt{\lambda} t}{\sqrt{\lambda}} d \int E(\lambda, x, y) g(y) dy$$

solves the wave equation $u_{tt} - \Delta u = 0$ with Cauchy data f, g at $t = 0$. Carleman sketches a proof that the space derivatives of the solution tend to zero locally uniformly as $t \to \infty$. The meaning of this is that the initial energy is reflected out at infinity.

Applications

The applications of the theory that Carleman found occupy more than half of his book. The first one is a spectral decomposition of the operator $Af(x) = \int e^{-|x-y|} f(y) dy$. This result is perhaps not so interesting since A is diagonalized by the Fourier transform to multiplication by $\text{const}/1 + \xi^2$. In other examples Carleman's analytical skill is more visible. He proves for instance that a kernel of the form

$$H_1(|x - y|) + H_2(x - y)$$

belongs to the class I if H_1 and H_2 are square integrable and shows that it is easy to fit Weyl's spectral theory (1911) of second order differential operators into his own theory.

In Carleman's last application the theory of integral operators is devoted to Hamburger's moment problem: given real numbers $c_0 = 1, c_1, \ldots$ such that all quadratic forms

$$\sum_{i+k \leq m} c_{i+k} x_i x_k$$

are positive definite, find a measure $d\mu > 0$ on the real axis such that

$$c_k = \int_{-\infty}^{\infty} t^k d\mu(t), \quad k = 0, 1, 2, \ldots.$$

The connection with Carleman's theory is an expansion of the asymptotic series

(9) $$\int \frac{1}{t-z} d\mu(t) \sim -(z^{-1} + c_1 z^{-2} + c_2 z^{-3} + \ldots)$$

into a unique continued fraction

$$J(z) \sim 1/a_0 - z - b_0^2/a_1 - z - b_1^2/a_2 - z - b_2^2/\ldots$$

with positive not vanishing $b_0, b_1, \ldots$. This continued fraction determines in turn a symmetric matrix $(K(j, i))$ which has only three nonvanishing elements in the row numbered by j,

$$K_{j,j-1} = -b_j, \ K_{j,j} = a_j, \ K_{j,j+1} = -b_{j+1}.$$

Hence this matrix is a Carleman kernel of a linear transformation K operating on square summable column matrices

$$x = (x_0, x_1, ...)^t,$$

with an interior product $(x, y) = \sum x_k \overline{y_k}$. The integral (9) turns out to have the form

$$\int ((K - zI)^{-1} e_0, e_0) d(E(\lambda) e_0, e_0),$$

where $e_0 = (1, 0, 0, ...)^t$ and $E(\Delta)$ is the spectral family of a self-adjoint extension of K.

Carleman's main result is that the moment problem is uniquely solvable if and only if K is of the class I.

By a stepwise solution of the equation $(K - z)x = 0$ one finds that $x = 0$ if $x_0 = 0$. If $x_0 = 1$, the equation can be solved explicity with $x_1 = (a_1 - z)/b_1$ and, in general, $x_k = f_k(z)$ where the right side is a polynomial in z of degree k depending only on the numbers $K_{i,j}$ with $j, k \leq n$. This shows that K belongs to the class I precisely if

$$\sum |f_k(z)|^2 = \infty$$

for some (and hence all) z with $\operatorname{Im} z \neq 0$.

Carleman has also a more directly applicable criterion: the moment problem is uniquely solvable if

$$\sum \frac{1}{\sqrt[2n]{c_{2n}}}$$

diverges. This is in fact a proof of uniqueness (1922 a) for asymptotic series of the form (9) which uses what is called below the Jensen-Carleman formula (see (1923 b)).

Carleman's other papers on integral equations

The object of Carleman' first paper on integral equations (1917) is the power series

$$D(\lambda) = \sum_0^\infty c_n \lambda^n$$

of Fredholm's denominator for continuous kernels. Fredholm had shown that $|c_n| \leq C^n n^{n/2}$ which means that $D(\lambda)$ is an entire of function of order at most two. Carleman shows that the factor $e^{A\lambda^2}$ is missing in the canonical product and later, in (1918), that the order cannot be < 2 for all continuous kernels. The far from trivial proofs show Carleman's mastery of what has been called hard analysis. Some later papers on special integral equations complete the picture.

Two invited major lectures (1931) and (1932) confirmed Carleman's position as a leading mathematician and as Hilbert's heir in the theory of integral equations. Due to its many applications the subject has never lost its interest, but its theoretical importance, now replaced by various abstract notions in functional analysis, has decreased considerably after Carleman's time.

Jensen's formula in a half-plane

Jensen's formula (1909), which is one of the most important ones in complex analysis, says that

$$\log |f(0)| + \sum_{f(z)=0} \log(R/|z|) = \frac{1}{2\pi} \int_0^{2\pi} \log |f(Re^{i\theta})| d\theta$$

where $f(z)$ is analytic when $|z| \leq R$ and $\neq 0$ when $z = 0, |z| = R$. In the sum on the left side, z runs through all zeros of $f(z)$ in the region $|z| < R$. The formula became important because it relates the number and size of the zeros in the disk $|z| < R$ to the size of the function on the circle $|z| = R$.

The proof runs as follows. The function $u(z) = \log(|z|/R)$ is harmonic outside the origin and vanishes when $|z| = R$ and, if $f(z)$ is analytic for $|z| \leq R$, the function $v(z) = \log|f(z)|$ is harmonic when $f(z) \neq 0$. If we integrate the equality $0 = u\Delta v - v\Delta u$ over $|z| < R$ but take away small disks around the zeros with radii at most $\varepsilon > 0$, use Green's formula and let $\varepsilon \to 0$, the result is Jensen's formula.

The Jensen-Carleman formula

Many of Carleman's results around 1922 depend essentially on a variant of Jensen's formula in which the circle is replaced by a half-circle and the function $u(z) = \log|z|/R$ above by the function

$$u_R(z) = \sin\theta\left(\frac{1}{r} - \frac{r}{R^2}\right), \quad z = re^{i\theta},$$

which is harmonic in the upper half-plane and vanishes on the real axis and on the half-circle $r = R$. If the function $f(z) \not\equiv 0$ is meromorphic in the upper half-plane when $|z| \geq 1$ and continuous at the boundary and $v = \log|f(z)|$, an integration of $u_R\Delta v - v\Delta u_R$ over the part of the upper half-plane where $1 \leq |z| \leq R$, small disks around the zeros excepted, and a passage to the limit give the following result:

THE JENSEN-CARLEMAN FORMULA. *With assumptions as above, one has*

(1) $\displaystyle\sum_{f(z)=0} \varepsilon(z)u_R(z) =$

$$\frac{1}{\pi R}\int_0^\pi v(Re^{i\theta})\sin\theta\, d\theta + \frac{1}{2\pi}\int_1^R \left(\frac{1}{x^2} - \frac{1}{R^2}\right)(v(x) + v(-x))dx + O(1).$$

In this formula the term $O(1)$ represents a boundary integral over the upper half of the unit circle. The sum on the left side runs over all zeros ($\varepsilon(z) = 1$) and poles ($\varepsilon(z) = -1$)) of $f(z)$ in the upper half-plane such that $1 \leq |z| \leq R$. If there are zeros or poles on the boundary, the corresponding terms assume the factor $\varepsilon(z)/2$.

Carleman wrote down his formula when $f(z)$ has no poles and this is assumed in the sequel. It follows for instance from (1) that the sum of the two integrals on the right cannot tend to $-\infty$ as $R \to \infty$ without the function $f(z)$ vanishing identically. This remark produces several results which had been proved earlier by the Phragmén-Lindelöf principle, for instance the statements below about functions $f(z)$ which are analytic in the upper half-plane with the bound $\log|f(z)| = O(\text{Im}\,z) + O(1)$.

(i) If $\log|f(z)| \leq -c|z|, c > 0$, on the positive or negative part of the real axis, the last integral above tends to $-\infty$ so that $f(z)$ is identically zero (a theorem by F. Carlson).

(ii) $\sum \text{Im}\, z/|z|^2 < \infty$ where z runs through the zeros of $f(z)$ in the upper half-plane.

(iii) If $f(z)$ is analytic in the upper half-plane and if $v(z) = \log|f(z)|$ is ≤ 0 on the real axis and if the region where $v(z) > 0, |z| < R, R > 0, \text{Im}\,z > 0$ is not empty, it is bounded by curves γ on which $|z| < R$ and arcs Γ on which $|z| = R$. If we integrate $u_R\Delta v - v\Delta u_R$ over this region and observe that

$$du_R(e^{i\theta})/dr = -2\sin\theta/R^2$$

when $r = R$, we get

$$2 \int_\Gamma v(Re^{i\theta}) \sin\theta d\theta / R = \int_\gamma u_R(z) v_n(z) ds$$

where ds is the arc element on γ and $v_n(z) \geq 0$ is the inner normal derivative. The right side is a non-decreasing function of R. With $\log^+ c = \max(1, \log c)$, this gives the following attractive version of the Phragmén-Lindelöf principle: If $f(z)$ is analytic in the upper half-plane and $|f(z| \leq 1$ on the real axis, then

$$\frac{1}{R} \int_0^\pi \log^+ |f(Re^{i\theta})| \sin\theta d\theta$$

is a nondecreasing function of R.

Approximation by powers z^λ

Carleman's paper (1923 b) about approximation of analytic functions by linear combinations of prescribed powers is inspired by a theorem by Müntz which says that the linear combinations of a sequence of powers

$$1, x^{\lambda_k}, \quad \lambda_k > 0$$

are dense in $L^2(0,1)$ if and only if the series

$$\sum \lambda_k^{-1}$$

diverges. Carleman's powers z^{λ_k} are defined in a sector $|\arg z| < \alpha\pi$ and the condition

$$\limsup_{R \to \infty} \frac{1}{\log R} \sum_{\lambda_n < R} 1/\lambda_n = \alpha$$

suffices that an analytic function in the sector can be approximated locally uniformly by linear combinations of these powers. The proof is given by the Jensen-Carleman formula applied to the entire function

$$F(\lambda) = \int_\Gamma g(z) z^\lambda |dz|$$

where Γ is a closed curve in the sector and $g(z)$ is analytic there. If $F(\lambda_k) = 0$ for all k and the condition above is satisfied, g must vanish. This paper, which is an example of Carleman's terse and effective style, also has other results about approximation without a connection to the Jensen-Carleman formula.

Quasianalytic classes

In the beginning of the twentieth century the interest in analytic functions extended to infinitely differentiable functions. For these functions the connection between the function and its Taylor series disappears. For instance: all derivatives of the function

$$f(x) = e^{-1/x}, x \geq 0, \quad f(x) = 0, x \leq 0,$$

vanish at the origin. On the other hand there is great freedom. Borel proved that there exists an infinitely differentiable function whose derivatives at a point are equal to any numbers given in advance. Hence, if the connection with the Taylor series is to be preserved, this can only happen for subclasses of the class of infinitely differentiable functions. As proposed by Hadamard this can be achieved very simply and naturally by inequalities,

$$|f^{(n)}(x)| \leq C^n M_n$$

where x belongs to some interval, the number $C = C(f)$ may depend on f but not so the numbers of the increasing sequence $M = (M_0, M_1, ...)$. The corresponding class C_M of functions is linear and admits affine changes of the variable, $x \to ax+b$. When $M_n = n^n$ one gets the class of analytic functions from the interval. A class C_M is said to be quasianalytic if a function of the class vanishes when all its derivatives vanish at a point. Earlier examples had shown that $M_n = n^{cn}$ does not give a quasianalytic class when $c > 1$, and therefore the quasianalytic classes are rather close to each other and to the analytic functions.

Denjoy made the first progress in the field when he proved that a class C_M is quasianalytic if

$$(2) \qquad \sum \frac{1}{\sqrt[n]{M_n}} = \infty$$

and the numbers M_n have a certain regular growth. Shortly afterwards, Carleman proved that this auxiliary condition is superfluous and that a precise form of Denjoy's theorem holds. Thereafter Carleman proved the final result which in modern terms may be written as follows: a class C_M is quasianalytic if and only if the least not increasing majorant of the series above diverges. Later in the same year Carleman was invited to lecture at Collège de France about quasianalytic functions. The lectures were published in a rich book (1926 a) where Carleman also reconstructs a quasianalytic function by summation of its Taylor series. It was his work on quasianalytic functions rather than the book on integral equations which made Carleman's name.

Criteria for quasianalyticity

Simple estimates of the derivatives show that the class C_M is conserved under more transformations than just the affine ones, for instance $f(t) \to f(t(1-t))$ when the interval is $(0,1)$ or $f(t) \to f(t/(1-t))$ in which this interval is mapped to the positive real axis.

Carleman turned the quasianalyticity of a class C_M on the positive real axis to the following problem

(*) When is there an analytic function $F(z) \neq 0$ in the upper half-plane such that

$$(3) \qquad |F(z)| \leq C^n M_n/|z+i|^n, \quad n = 0, 1, 2, 3, ...?$$

The link between the two problems is the Fourier transform

$$F(z) = \int_0^\infty e^{i(z+i)t} f(t) dt.$$

If $f(t) \in C_M$ then $F(z)$ is analytic in the upper halfplane, and, if all derivatives of f vanish at the origin, then

$$i^{n+2}(z+i)^n F(z) = \int_0^\infty e^{i(z+i)t} f^{(n)}(t) dt$$

where the absolute value of the right side is at most $C^n M_n$. Conversely, if $F(z) \neq 0$ satisfies (3), then

$$f(t) = \frac{1}{\pi} \int_{-\infty}^{\infty} e^{-izt} F(z+i) dz/(z+i)^2, \quad z = x+iy, \ y \geq 0,$$

is infinitely differentiable when $t \geq 0$ and the differentiations can be performed under the integral sign. Since $e^{-ty} f(t)$ is the Fourier transform of the function $F(x+iy) \neq 0$, f does not vanish identically and it is easy to see that (3) entails

$$|f^{(n)}(t)| \leq C^n M_n$$

for all n. When $t = 0$ all derivatives of f vanish since the integral on the right side is independent of $y > 0$ and tends to zero when $y \to \infty$.

We can now limit ourselves to the problem (*). Carleman's answer (1926 a p. 50) is that a required function $F(z)$ exists or not according to whether the integral

(4) $$\int^{\infty} \log\left(\sum r^{2n}/M_n^2\right) dr/r^2$$

converges or diverges. The answer is at the same time a condition that an analytic function in the upper half-plane is or is not determined by its asymptotic series

$$F(z) \sim a_0 + a_1 z^{-1} + a_2 z^{-2} + ...$$

in the sense that $F(z)$ is approximated of order $O(C^n M_n/|z|^n)$ by the sum $S_n(z)$ of the first n terms of the series. If the terms of the series vanish, this means precisely that (3) holds.

We shall show a simpler condition: the function $F(z)$ vanishes if and only if

(4') $$\int^{\infty} T(r) dr/r^2 = -\infty$$

where

$$T(z) = \inf_{n>0} (\log M_n - n \log|z|)$$

majorizes $\log |F(z)|$. In fact, Carleman's formula shows at once that $F(z) = 0$ when (4') diverges. Conversely, if this integral converges then

$$\frac{1}{i\pi} \int \frac{y}{y^2 + (x-s)^2} T(s) ds/(s-z)$$

is harmonic in the upper halfplane and equal to $T(x) = T(-x)$ when $y = 0$. If $f(z)$ is a corresponding analytic function then $F(z) = e^{f(z)}$ has the property that

$$|F(z)| \leq M_n/|z|^n$$

for all n.

What remains is to transfer the conditions (4) and (4') to more direct conditions on the sequence M_n. From (4) Carleman deduced

THE DENJOY-CARLEMAN THEOREM. *The condition*

$$\sum_{n}^{\infty} \frac{1}{m_n} = \infty, \quad m_n = \inf_{k \geq n} \sqrt[k]{M_k}$$

is necessary and sufficient for the class C_M to be quasianalytic.

At the end of the Appendix which closes this chapter there is a simple proof of the Denjoy-Carleman theorem.

In his book (1926 a) Carleman constructed a method of summation for a quasianalytic class where numbers $c(k, n) > 0$ are given such that

$$f(x) = \lim_{n \to \infty} \sum_0^n f^{(k)}(a)(x-a)^k c(k,n).$$

Some twenty years after Carleman, H. Cartan (1941), S. Mandelbrojt (1942) and Th. Bang (1946) reworked the theory of quasianalytic classes in such a way that the theory is carried out without analytic functions.

Carleman's short papers

Carleman's thesis, his books on integral equations, quasianalytic functions and Fourier analysis and a posthumous book on Boltzmann's equation are all works on a large scale. In addition to these he wrote short papers on Fourier series, power series, analytic functions and, above all, on different subjects in mathematical physics. Here we find no padding and almost nothing indifferent. New problems are attacked and the results are stated and proved clearly and precisely. Carleman had a strict sense of quality and he required much of himself.

Analytic functions

Carleman's three papers 1919-1923 on Fourier series and power series show him in full command of the theory of his time, but these papers are peripheral in his production. On the other hand, his papers on analytic functions, in particular the use of harmonic majorants and harmonic measure, are seminal contributions to the Finnish-Swedish school of analytic functions.

Harmonic majorants and harmonic measure

If $f(z)$ is analytic, $\log |f(z)| = \text{Re} \log f(z)$ is harmonic at all points z where $f(z) \neq 0$. In the general case, $\log |f(z)|$ is *subharmonic* in the sense that this function, restricted to any bounded open set, is at most equal to the harmonic function with the same boundary values. Since $\log |f(z)| = -\infty$ when $f(z) = 0$, this follows immediately from the maximum principle for harmonic functions.

In principle, majorants of functions $v(z) = \log |f(z)|$ in regions B where f is analytic are obtained by solving Dirichlet's problem in B, i.e., by constructing harmonic functions $u(z)$ in B such that $u \geq v$ at the boundary. With equality one gets the least harmonic majorant. It is in this connection that the notion of harmonic measure has been used.

When D is a part of the boundary of a region B, let $\omega(z, D; B)$ be a harmonic function in B which equals 1 on D and is otherwise zero on the boundary. The function $\omega(z, D; B)$, which has values beween zero and 1 in B, is called the harmonic measure of D in B.

The harmonic measure of an interval I on the boundary of a half-plane is particularly simple. Its value at a point is the visual angle of I divided by π. In fact, the boundary measure on the real axis for Dirichlet's problem in the upper half-plane is

$$(1) \qquad \frac{y}{y^2 + (x-t)^2} dt/\pi = \frac{1}{\pi} d \arctan \frac{y}{x-t}.$$

If $B' \supset B$ and the boundaries of B, B' have a part D in common, the following inequality holds for the corresponding harmonic measures

$$(2) \qquad \omega(z, I; B) \leq \omega(z, I; B'), \quad z \in B, \quad I \subset D$$

where I is a part of D. In fact, the maximum principle shows that the right side majorizes the left side on the boundary of B. The harmonic measure and the use of the visual angle occur for the first time, but in rather implicit form, in the paper (1920 b) by Carleman. The end of this section has a simpler application of the same idea. In the paper (1926 b), which contains a refined application of harmonic measure, Carleman shows that an entire function $f(z)$ such that

$$\log \log |f(re^{i\theta})| < \varphi(\theta)$$

for all large r is a constant when $\varphi(\theta)$ is integrable. The motivation of this theorem, which is presented as an extension of Liouville's theorem, was probably Mittag-Leffler's use of entire functions which tend to zero in all directions except one.

Harmonic measure was to play a prominent part in the theory of analytic and meromorphic functions that was developed in Scandinavia by Finnish and Swedish mathematicians, for instance the brothers Nevanlinna and Lars Ahlfors in Finland and Arne Beurling in Sweden. The term 'subharmonic' was introduced by F. Riesz in a paper (1926) with a general definition and a systematic theory of subharmonic functions.[1]

As a simple application of the notion of harmonic majorant we shall now prove Hadamard's three lines theorem: if $u(z)$ is subharmonic and bounded from above by $o(|y|)$ in a strip $I : a \leq x \leq b$ then the function

$$M(x) = \sup_y u(x + iy)$$

is convex. In fact, let J be a substrip of I, suppose that $u \leq 0$ on the boundary of J and let $v(z)$ be the least harmonic majorant of $u(z)$ in the rectangle $z \in J, |y| \leq T$. At a given point, the harmonic measure of the horizontal sides of the rectangle is majorized by the visual angle of these sides times $o(T)$. Since the visual angle has the size $O(1/T)$, $v(z)$ has the upper bound $o(1)$ for any fixed z and it follows that $u \leq 0$ in J. This reasoning repeated with $M(x)$ replaced by $M(x) - cx - d$ shows that $M(x)$ is convex.

Asymptotic paths

The dominating problem in the theory of entire functions had been the growth at infinity and its angular distribution. Already the exponential function e^z shows that an entire function can tend to zero in certain directions and to infinity in others. Denjoy took up the question of asymptotic paths of an entire function, i.e., separate paths tending to infinity on which the function is bounded. Example: the entire function

$$f(z) = \int_0^z \sin u^{2n} du$$

---

[1] Riesz, who wrote in French, used the term sousharmonique.

has order $2n$, i.e., $\log|f(z)| = O(|z|^{2n+\varepsilon})$ for every $\varepsilon > 0$ and no $\varepsilon < 0$. It also has $4n$ asymptotic paths, namely straight lines through the origin whose angles to the real axis are multiples of $\pi/2n$. Denjoy guessed that this example is optimal, in other words that an entire function of order ϱ has at most 2ϱ asymptotic paths. If the paths are close to straight lines, this is a simple consequence of the Phragmén-Lindelöf principle about the least growth of analytic functions which are bounded on the boundary of sectors. In the general case the proof is more difficult.

In his paper (1920 b) Carleman used the harmonic majorant to show that the number of asymptotic paths of an entire function of order ϱ is at most 5ϱ. The correct result was proved by Ahlfors (1929) and again by Carleman (1933 b) as a consequence of a general inequality which shall be proved below. Carleman's own text can be recommended as an alternative.

An inequality by Carleman. Let S be a plane, open region whose projection on the x-axis contains an interval $x \geq$ const and suppose that every intersection S_x of S with a vertical line $\operatorname{Re} z = x \geq$ const has a finite positive measure $L(x)$. Also, let u be harmonic in S and vanish at the boundary. The problem is to estimate the growth of u as $x \to \infty$. Carleman shows that the function

$$(3) \qquad h(x) = \int_{S_x} u(x,y)^2 dy,$$

satisfies the inequality

$$(4) \qquad h''(x) \geq \frac{h'(x)^2}{2h(x)} + \frac{2\pi^2}{L(x)^2} h(x)$$

for large x. The proof is very direct. By differentiation under the sign of integration, by an integration by parts and use of the fact that u is harmonic, it follows that

$$h'(x) = 2\int_{S_x} uu_x dy, \quad h''(x) = 2\int_{S_x}(u_x^2 + u_y^2) dy.$$

Schwarz's inequality applied to the first equality shows that

$$\int u_x^2 dy \geq h'(x)^2/4h(x).$$

A vibrating string of length L fixed at its end points has the lowest frequency π/L. Hence

$$\int u_y^2 dy \geq h(x)\pi^2/L(x)^2,$$

and this proves (4). With $h(x) = e^{g(x)}$ it can be rewritten as

$$\left(\frac{2\pi}{L(x)}\right)^2 \leq g'^2 + 2g''.$$

It is proved below in a remark that there is a number a such that $g'(x) > 0$ when $x \geq a$. This means that

$$(5) \qquad \frac{2\pi}{L(x)} \leq g' + \frac{g''}{g'} = (g + \log g')'.$$

for $x \geq a$. We now observe that

$$\int_a^x (x-s) d(g(s) + \log g'(s)) = \int_a^x g(s) ds + \int_a^x \log g'(s) ds + O(x).$$

Since $\log g'(s) \leq g'(s)$, this gives an inequality

$$(6) \qquad 2\pi \int_a^x (x-s)ds/L(s) \leq \int_a^x g(s)ds + g(x) + O(x)$$

to be used below.

Proof of Ahlfors's theorem. All that is needed for this theorem is now some simple estimates of the right and left sides of (6). If $f(z)$ is entire of order ϱ with n asymptotic paths on which $|f(z)| \leq 1$, the function

$$v(z) = \log|f(e^z)|, \quad z = x + iy$$

is defined for all $x > 0$ and has the period 2π i y. Further, $v(z)$ vanishes on at least n paths in the region $\Omega : 0 \leq y \leq 2\pi$, $x >$ const, which do not cross and tend to infinity. If we let u be the least harmonic majorant of $v(z)$ in Ω which vanishes on these paths, we can use (6) in every strip between two neighboring paths. If $L_1(x), ..., L_n(x)$ are the corresponding breadths, their sum is at most 2π and hence

$$\sum \frac{1}{L_k(x)} \geq \frac{n^2}{2\pi}$$

for the sum is the least when all its terms are equal. Hence the left side of (6) has the minorant

$$(7) \qquad n^2 x^2/2 + O(x).$$

That $f(z)$ has the order ϱ means that

$$v(z) \leq e^{(\varrho+\varepsilon)x}.$$

The harmonic majorant $u(z)$ has the same upper bound so that

$$h(x) \leq L(x) e^{(2\varrho+\varepsilon)x}.$$

Since here $L(x) \leq 2\pi$ we get

$$\log h(x) = g(x) \leq 2(\varrho + \varepsilon)x + O(1).$$

It follows that the right side of (6) has the majorant

$$(8) \qquad nx^2(\varrho + \varepsilon) + O(x).$$

A comparison of (7) and (8) shows the desired result that $n \leq 2\varrho$.

Remark. It remains to prove that the function $h'(x)$ of (5) is positive in some interval $x > a$ if $u(z)$ is not bounded in the strip S. By (4), $h(x)$ is convex so that it suffices to show that u is bounded when h is bounded. Carleman uses this without a proof. The proof which follows can be seen as an instructive use of the inequality (2). If B is the part of the strip S which lies between a fixed S_r and S_t with $t > r$, then

$$u(z) = \int_{S_r} u(w)d\omega(z,w) + \int_{S_t} u(w)d\omega(z,w)$$

where $d\omega(z,w)$ is the harmonic boundary measure of B and both integrals are positive. The first term equals at most the maximum of u on S_r. By the principle (2), the second term does not diminish if we replace $d\omega$ by the boundary measure

$d\nu(z,w)$ of S_t in a half-plane to the left of S_t. The explicit form (1) of this measure and Schwarz's inequality show that the last integral is at most $O(\sqrt{h(t)})$ when z stays away from S_t. Hence, if $h(t)$ is bounded, the same is true of u, QED.

Approximation by entire functions, Lindelöf's function

In connection with the interest in asymptotic paths Gross proved that there is an entire function which is bounded on a given simple rectifiable curve which tends to infinity. In his paper (1927) Carleman proved that every continuous function on such a curve can be approximated uniformly by entire functions. The proof is presented for a straight line but is easily extended to the general case.

Carleman's sole contribution to analytic number theory is contained in a lecture (1931 b) in Copenhagen. The theme is the set of functions $f(\sigma+it)$ which like the ζ-function are analytic in strips $S : t \geq 0$, $\sigma \in I$, I an interval on the real axis. Such functions with the bounds $O(\varepsilon t)$ for every $\varepsilon > 0$ had been studied by E. Lindelöf, who introduced the growth function

$$\mu(\sigma) = \limsup_{t\to\infty} \frac{\log|f(\sigma+it)|}{\log t}$$

and proved that it is convex. In his lecture Carleman proved that any convex function can be obtained in this way. He also introduced a lower, concave growth function

$$\mu^*(\sigma) = \liminf_{t\to\infty} \frac{\log|f(\sigma+it)|}{\log t}.$$

The first part of the paper gives a proof that the curves $\sigma \to \mu(\sigma)$ and $\sigma \to \mu^*(\sigma)$ are equal and hence linear if they are tangent to each other at one point. The second part deals with other functions which describe growth, in particular the function $\kappa(\sigma,p)$ which is the lower bound of numbers k such that

$$\int^\infty |f(\sigma+it)|^p dt/t^k < \infty, \quad p > 0.$$

By a reasoning which cannot be reproduced in a few lines, Carleman proved that this function is convex in both its variables and he used this property to give a rather simple proof of a theorem by Hardy and Littlewood which says that the ζ-function has infinitely many zeros on the critical line $\operatorname{Re}\zeta = 1/2$.

Uniqueness

Holmgren's uniqueness theorem in its simplest form (see p. 165) says that a solution $u(x,y)$ of a square linear system

$$\partial_x u_j(x,y) + \sum_1^N a_{jk}\partial_y u_k(x,y) + \sum_1^N b_{jk}u_k(x,y) = c_j(x,y)$$

of partial differential equations with analytic coefficients is uniquely determined by its restriction $u(0,y)$ to $x=0$. That the coefficients are analytic is essential for the proof. In the paper (1938) Carleman extended uniqueness to such systems with real coefficients a_{jk} which are twice continuously differentiable, with b_{jk} continuous and the matrix $(a_{jk}(x,y))$ locally diagonalizable.

The proof, which defies a short review, uses a weight function which grows very fast from a point on the x-axis. Carleman's paper attracted attention in the 1950's when the problem of uniqueness for equations of higher order began to be interesting and counterexamples were produced. A large part of this area of research is accounted for in Hörmander's book (1985). The most important tool, called Carleman estimates, are inequalities with a weight function.

Mathematical physics

Carleman's thesis was followed by three papers (1918, 1919, 1921), in which he treated some variational problems in classical physics. In the 1920's Carleman's papers became more mathematical, but after 1930 mathematical physics returned as his main interest. Some of the most important papers are chosen for comment below.

Ergodic theorem

The most important differential equations of classical mechanics have the form $x_t = F(x)$ in n variables x. They determine orbits $t \to x(t)$ on a manifold M with an invariant volume measure ω. This means that

$$f(x) \to U(t)f(x) = f(x-t)$$

defines a unitary operator on the Hilbert space H of measurable complex functions from M which are square integrable for the measure ω. When A is a subset of M this means that $U(t)A = A(t) = (x(t); x(0) \in A)$.

Carleman's main result in the papers (1931 c, 1932) is that the differential operator $L = i(F(x), \partial_x)$, defined on continuously differentiable functions, has a self-adjoint extension A such that $U(t) = e^{itA}$. The proof, which uses the theory of integral equations, is now obsolete. The fact that every one-parameter group of unitary operators is generated by a self-adjoint operator is a well-known result by Stone. But the main idea of the current proof is present in Carleman's paper. If B is the projection onto the eigenspace of L with the eigenvalue zero, it follows from the spectral theorem that

$$\lim t^{-1} \int_0^t (U(s)f, g) ds = (Bf, Bg), \quad t \to \infty,$$

where (f, g) is the interior product of H. If the corresponding eigenspace of A only contains constants, then

$$(Bf, Bg) = \frac{1}{\omega(M)} \int f d\omega \int \bar{g} d\omega.$$

If f, g are the characteristic functions of two measurable parts $A, B \subset M$, the formula says that

$$\lim t^{-1} \int_0^t \omega(U(s)A \cap B) ds = \omega(A)\omega(B)/\omega(M).$$

This theorem, which was published also by von Neumann, is called the mean ergodic theorem. A sharper result with pointwise convergence was proved at the same time by G. D. Birkhoff. His theorem says that if T is a bijection of a measure space M such that the measures of A and TA are the same for every measurable subset A and no proper part of M is invariant under T, then the mean

$$\lim_{n \to \infty} \frac{1}{n} \sum_0^{n-1} f(T^k x)$$

converges almost everywhere to a constant for every integrable function f.

Carleman's paper got a large space in *Jahrbuch*, but later it was overshadowed by the more modern formulation by von Neumann who used Hilbert space and unitary operators.

The Schrödinger operator

A particle in space with mass m, position x and velocity v in a potential field $V(x)$ has the energy
$$\frac{m|v|^2}{2} + V(x).$$
The rule for getting the corresponding quantum-mechanical energy operator H is to exchange the operator mv, conjugate to x, for the differential operator
$$ih\partial/\partial x$$
where h is Planck's constant. With normalized constants of nature, the result is Schrödinger's energy operator for a particle in a potential field,
$$H = -\Delta + V(x).$$
For the potential $V(x) = -1/|x|$ the pioneers of the 1920's introduced polar coordinates and could establish square integrable eigenfunctions with negative eigenvalues representing the energy levels of a particle in centrally symmetric field, for instance the electron in hydrogen. This was a fantastic piece of progress in physics, but in the beginning no one bothered about the operator H which, in spite of encouraging experiences, could perhaps have a complex spectrum, an embarassment for quantum mechanics. Already in 1932, in his Zürich lecture, Carleman stated that H is self-adjoint. At that time he was probably the only one who could produce a proof of this statement. The proof came in the paper (1934) which was twenty years ahead of its time. Carleman introduced what is now known as weak solutions of the equation $Hf = g$, although not in the form this notion received later. He gave a very elegant proof that H is self-adjoint when $V(x)$ is locally square integrable and bounded from below. A general theory of the Schrödinger operator is now an essential part of quantum mechanics, documented for instance in Reed-Simon (1978).

Asymptotics of eigenvalues

Let $\lambda_0 \leq \lambda_1 \cdots \leq \lambda_{n-1}$ be the eigenvalues of a real quadratic form $Q(x)$ in n variables relative to a unit form $|x|^2$. Then, as is well known,
$$\lambda_k = \sup_{L_k} \inf_{x \in L_k} Q(x)/|x|^2$$
where L_k runs over all subspaces of R^n of codimension k. This formula, which goes under the name of the max-min principle, was found before the first world war by Herman Weyl, who used it in connection with integral operators. He could, for instance, show that all eigenfrequencies of a vibrating body are monotone functions of the size of the body. After the first world war, Richard Courant used the max-min principle to compute the asymptotic behavior of the eigenfrequencies of a vibrating membrane under different boundary conditions.

Carleman's method broke new ground. To describe it in the simplest case, let
$$\varphi_1(x), \varphi_2(x), \ldots$$

be a complete set of orthonormalized eigenfunctions of Laplace's operator $-\Delta$ in a bounded open set Ω in R^n and let $\lambda_1 \leq \lambda_2 \leq ...$ be the corresponding eigenvalues. If $N(\lambda)$ is the number of eigenvalues $< \lambda$, it is now known that

$$(10) \qquad N(\lambda) = \sum_{\lambda_k < \lambda} 1 \sim c_n |\Omega| \lambda^{n/2}$$

for large λ, a certain constant c_n and volume $|\Omega|$ of the region Ω. For $n = 2$ (with $c_2 = 1/4\pi$) this was proved by Weyl in (1911).

To the boundary value problem just described there is a self-adjoint extension A of $-\Delta$ with spectral deomposition

$$A = \int \lambda dE(\lambda),$$

where $E(\lambda)$ has the kernel

$$E(\lambda, x, y) = \sum_{\lambda_k \leq \lambda} \varphi_k(x) \varphi_k(y).$$

We observe that

$$N(\lambda) = \int_\Omega E(\lambda, x, x) dx.$$

In his paper (1934 b) Carleman proved in the case $n = 2$ that there is a constant $c > 0$ such that

$$(11) \qquad E(\lambda, x, x) = c\lambda^{n/2} + o(\lambda^{n/2})$$

is asymptotically independent of $x \in \Omega$ with a uniform remainder when x stays away from the boundary. He also proved similar estimates for the eigenfunctions and their derivatives. The formula (11) and a prudent integration over Ω gives the basic formula (10).

Carleman's method is analytic. He compared the operator A to a self-adjoint extension A_∞ of $-\Delta$ in all of space,

$$A_\infty = \int \lambda dE_\infty(\lambda),$$

and proved that the corresponding projection kernels $E(\lambda, x, y)$ and $E_\infty(\lambda, x, y)$ have the same asymptotics at the diagonal $x = y$,

$$(12) \qquad E(\lambda, x, x) \sim E_\infty(\lambda, x, x)$$

for large λ. This makes it easy to compute the constant c of (9) and (10). In fact, the operator A_∞ is diagonalized to multiplication by $|\xi|^2$ by the Fourier transform so that

$$E_\infty = (2\pi)^{-n} \int_{|\xi|^2 < \lambda} e^{i((x-y),\xi)} d\xi, \quad \xi = (\xi_1, ..., \xi_n).$$

When $x = y$, the right side equals $c\lambda^{n/2}$ where the constant is now explicit.

Carleman's indirect proof of (11) depends on a so-called Tauberian theorem where the asymptotic behavior of some transform of a measure $d\mu(x) \geq 0$ on the real axis, for instance

$$\int_0^\infty x^{-s} d\mu(x) \quad \text{or} \quad \int_0^\infty e^{-tx} d\mu(x),$$

when s decreases to a number s_0 or $t \to \infty$, determines the behavior of $\mu(x)$ at infinity. Carleman used the first method. The second one, which is connected with the equation of heat conduction, has been used later on a large scale in situations where it can even give two terms in an asymptotic expansion of the eigenvalues of elliptic operators defined on compact manifolds (Hörmander (1985), Chapter XXIX).

Carleman returned to spectral asymptotics in the paper (1938). The subject is eigenvalues of the Laplace operator with lower terms. This makes the eigenvalues complex, but Carleman proved that their real parts have the same asymptotics as in the self-adjoint case. To carry through the proof Carleman formulated a new Tauberian theorem (proved by Pleijel (1952)).

The kinetic theory of gases

In his papers (1932), (1933) on Boltzmann's equation, and the posthumous book (1957), Carleman encountered new problems requiring entirely new methods.

In kinetic gas theory, founded by Maxwell and Boltzmann in the 1870's, gases are considered to consist of many very small particles with different velocities colliding according to the laws of classical mechanics. The main object of the theory is the local velocity distribution F from which it is possible to determine velocity, pressure, temperature and density in various parts of the gas. Boltzmann's equation describes how F changes with time. A basic result of the theory for space-homogeneous velocity distributions F is the existence of a function, Boltzmann's H-function. It resembles entropy and, in spite of the time reversal symmetry of classical mechanics, it decreases with time.

The kinetic theory of gases is relevant for thin gases, for instance at low pressure or at high altitude.

Let $F(x, v, t)dxdv$ where $dx\,dv$ are euclidean volume elements in space and v is velocity, be the space and velocity distribution of the gas at the point x at the time t. If no outer forces influence the gas, the development in time of F is described by Boltzmann's equation
$$F_t + (v, F_x) = T(F), \quad t > 0,$$
where $T(F)$ is a collision term. A very approximate derivation of this term is the following one. The product $FF_* = F(t, x, v)F(t, x, v_*)$ represents the combined density of two colliding particles with the velocities v, v_* and the product $F'F'_* = F(t, x, v')F(t, x, v'_*)$ is the density of the reflected particles with the velocities v', v'_* after the collision. By the laws of mechanics,
$$v + v_* = v' + v'_*, \quad |v|^2 + |v_*|^2 = |v'|^2 + |v'_*|^2,$$
which limits a free choice of the velocities. But one gets
$$v - v' = v'_* - v_* = \lambda\omega$$
where ω is a unit vector and an insertion into the preceding formula gives $\lambda = (\omega, v_* - v)$. This means that the local net increase $F'F'_* - FF_*$ ought to have the density $|(v - v_*, \omega)|dv_*d\omega$ where $d\omega$ is the invariant measure on the unit sphere. Hence

(13) $$T(F) = \int (F'F'_* - FF_*)|(v - v_*, \omega)|dv_*d\omega,$$

possibly with a constant factor, and this makes $T(F)$ a quadratic function of F. To get an interesting result, we have to assume that
$$\int F\,dv < \infty.$$

For a gas in equilibrium, $T(F)$ must vanish which turns out to happen precisely if

$$F = \text{const} \cdot e^{-a|v|^2}$$

is a density characteristic of a Maxwellian gas. If $G = G(v)$ depends only on v, we can interchange both v and v_* and v, v_* and v', v'_* in the integral

$$\int G(v)F(v)F(v_*)dvdv_*|(v-v_*,\omega)|d\omega$$

with no effect on the volume element. Hence it follows from Boltzmann's equation that

(14) $\quad 2\partial_t \int G(v)F(v)dv = \int T(F)(G(v)+G(v_*)-G(v')-G(v'_*))dvdv_*|w|d\omega.$

With $G = v$ or $|v|^2$ the parenthesis on the right side vanishes so that

$$\int G(v)Fdv$$

is independent of time. Since $(a-b)\log a/b \leq 0$ when $a, b > 0$, the right side of (14) is ≤ 0 when $G(v) = \log F$. Hence Boltzmann's famous law follows: the integral

$$\int F\log F dv$$

is a not increasing function of time when F is independent of x.

In his paper (1932) Carleman proved that Boltzmann's equation with an initial function $f(|v|)$ such that

(15) $\qquad\qquad 0 \leq f(|v|) \leq \text{const}(1+|v|)^{-\kappa}, \quad \kappa > 6,$

has a unique solution $F \geq 0$ which is a function only of $|v|$ and t and satisfies the inequality above for all $t > 0$. In addition, the distribution $F(t, |v|)$ tends to a solution of the stationary problem for large t. As for most nonlinear equations, the proof uses estimates which hold for all time, successive approximations and, finally, a convergence argument. Carleman's rather long paper (1932) is not suited for a review, and the same is true of (1934) where Carleman studies solutions close to the equilibrium solution of Boltzmann's equation in a bounded set. The posthumous book does not go much further than these two papers, but it has some simplifications of Carleman's proofs due to Lennart Carleson.

In the beginning of his first paper, Carleman says that he envisaged a complete theory of Boltzmann's equation. His project was not realized but it has been taken over by others and work is still going on (ses Truesdell and Muncaster (1980) for a review). Carleman's strict existence proofs have been followed by others. His restrictive assumption (15) has recently been replaced by weaker ones which give existence but not uniqueness (ses Lions and DiPerna (1990)).

Late papers

At his last international conference (1947) Carleman talked about his lectures (1944) about Fourier analysis, partly inspired by Wiener's famous paper (1931) about Tauberian theorems. In the introduction Carleman says that he wanted to exhibit the important part played by analytic functions. His main result says that the Fourier transform

$$F(x) = \int_{-\infty}^{\infty} e^{ixt}f(t)dt$$

of an integrable function can be extended to more general functions. The idea is to divide the integral into a sum of two,

$$\text{(16)} \qquad \int_{-\infty}^{0} e^{(x+iy)t} f(t) dt + \int_{0}^{\infty} e^{(x+iy)t} f(t) dt,$$

where the first integral is analytic in the upper half-plane and the second one in the lower half-plane. Here it is immediate that we may permit polynomial growth of $f(t)$ or, more generally, that an integral of f has this property. Carleman worked this out including a generalization of the classical inversion formula to functions (16). But the result is not simple and his theory did not survive. For once the analytic functions had deserted him. The correct formulation was found in Laurent Schwartz's theory of distributions. The class S of all infinitely differentiable functions $f(t)$ for which every derivative is at most $O(|t|^{-N})$ for large t and every integer $N > 0$ is mapped onto itself by the classical Fourier transform. Hence the dual space, that of tempered distributions, has the same property. The Fourier transform operates on this class without problems.

Summary

From the beginning of the 1920's Carleman was considered the best mathematician in Sweden. International success came, but his spectral theory was overshadowed by the abstract theory and he had also bad luck with his mean ergodic theorem. It is certain that Carleman felt that he was the equal of the best mathematicians but also that he was not appreciated according to his merit. One reason was that many of his results, for instance the extension of Holmgren's uniqueness theorem, the analysis of the Schrödinger operator, and the existence theorem for Boltzmann's equation, were two decades ahead of their time and therefore not immediately appreciated.

In Fritz Carlson's obituary (1960), Carleman is described as retiring and taciturn. He regarded life and people with a bitter humour but he could also be kind and helpful to others, especially his students. Although not a natural athlete, he could perform amazing physical feats. Sometimes one got the impression of unbridled power both in his scientific and physical activity.

During the last period of his life Carleman lived alone in two rooms of the Mittag-Leffler Institute. His sole interests had been mathematics and mathematical physics, subjects which he had studied with such energy and fire during his student days. Towards the end the 1940's, when his health began to deteriorate, he sometimes remarked to his students that professors ought to be shot at the age of fifty. This statement reflects the severe discipline which is apparent in his mathematical work.

Carleman's students

Carleman had few students. The first one of them, Karl Persson (later Dagerholm), wrote a thesis (1938) in which he carried out a program sketched by Carleman in his congress lecture (1932). The point of departure is a quotient $f(z,u) = P(z,u)/Q(z,u)$ of two polynomials in u whose coefficients are analytic functions of z. The regular behavior of the function $n \to f(z,n)$ with integral arguments shows that if there is a nontrivial sequence of numbers $x_1, x_2, \ldots$ such that the series

$$\sum_{1}^{\infty} f(z,n) x_n$$

converges for $z = z_0$, then it represents an analytic function whenever the terms are well defined, grad $P(z,u)$ is maximal and no denominators vanish.

The paper deals mostly with a special case in the form of an infinite system of equations

$$\sum_1^\infty \frac{x_q}{p-aq} = c_q, \quad p,q = 1,2,\ldots$$

where a is real. If $a > 1$, the homogeneous system has only the trivial solution. In fact, the analytic function

$$\sum \frac{x_q}{z-aq}$$

has the zeros $z = 1, 2, \ldots$ and poles $z = a, 2a, 3a, \ldots$ in the right half-plane and we may apply the Jensen-Carleman formula ((1) p. 192) in the right half-plane. Then the left side of the formula tends to ∞ but the right side is bounded from above on semi-circles which avoid the poles. When $0 < a < 1$ the author constructs solutions of the homogeneous system and investigates solutions of the inhomogeneous one. The tool is always the theory of analytic functions.

In his thesis (1947), Ulf Hellsten determined the spectrum of integral operators with kernels $K(s,t)$ of a special form, for instance those which assume the value 1 in a polygonal subset of the square $0 \leq s, t \leq 1$ and vanish in the complement.

Pleijel

Carleman's student Åke Pleijel (1913-1989) became professor at KTH in 1947, later in Lund and finally in Uppsala. In his thesis (1941) he extended Carleman's work on the asymptotics of eigenvalues and eigenfunctions of certain vibration problems to other ones, for instance for clamped plates. In all cases there is a formally self-adjoint differential operator P with constant coefficients, two corresponding self-adjoint operators A and B, both bounded from below, where A comes from a boundary problem in a bounded region and B from a boundary problem in all of space with vanishing boundary data at infinity. Let $G(t,x,y)$ be the kernel of the resolvent $(A+tE)^{-1}$ for large $t > 0$. The kernel $H(t,x,y)$ of $(B+tE)^{-1}$ can be obtained explicitly by the Fourier transform. The problem, for which Pleijel developed a number of methods, is to show that

$$G(t,x,x) \sim H(t,x,x), \quad x \in \Omega$$

locally in Ω with some estimate at the boundary so that also

$$\int_\Omega G(t,x,x)dx \sim \int_\Omega H(t,x,x)dx$$

for large positive t. The degree of difficulty depends on how complicated the boundary problem is. In a series of later papers Pleijel worked out a spectral theory for problems of polar type. In these the resolvent $(\Delta + tE)^{-1}$ of Laplace's operator Δ is replaced by $(\Delta + tK)^{-1}$ where the operator K means multiplication by a smooth function $k(x)$ which may change its sign. The corresponding spectrum then has infinitely many positive and negative eigenvalues. The paper (1948) computes their asymptotic behavior.

In a well-known paper (1949, with Minakshisundaram) the authors studied the asymptotics of eigenfunctions and eigenvalues λ_n of the Laplace operator on a compact Riemannian manifold by determining the asymptotic behavior of the ζ-function

$$\sum \lambda_n^{-s},$$

which, as the usual one, is meromorphic with a single pole.

Carlson

Several of Wiman's students devoted themselves to the theory of analytic functions, among them Fritz Carlson (1888-1952). He became known for a theorem connected with the Phragmén-Lindelöf principle, for a theorem on the zeros of the ζ-function and several theorems about power series with integral coefficients. In addition, Carlson wrote a solid textbook of geometry.

The thesis

Many current power series, for instance

$$\sum e^{-n} z^n, \quad e^z = \sum z^n / \Gamma(n+1),$$

have the property that their coefficients have the form $g(n)$ for some entire function $g(z)$, in the two cases above, e^{-z} and $1/\Gamma(z+1)$. Since there are entire functions which vanish on the integers, the connection is not bijective. If $f(z) = \sum a_n z^n$ is a power series which converges when $|z| \leq R$, the integral

$$g(z) = g_r(z) = \frac{1}{2\pi i} \int_{|w|=r} f(w) w^{-1-z} dw, \quad |\arg w| \leq \pi,$$

where $r < R$ is arbitrary, is an entire function of z with the property that $g(n) = a_n$. In addition, g vanishes on the negative integers and simple estimates show that

$$|g(x+iy)| \leq Mr^{-x}(e^{\pi|y|} - e^{-\pi|y|})/|y|,$$

when $|f(z)| \leq M, |z| = r$. The converse relation has solutions as well, for instance

$$f(w) = \frac{1}{2\pi i} \int_V g(z) w^z dz / (e^{2\pi i z} - 1),$$

where V is a loop enclosing the real axis in a negative direction and $g(z)$ is analytic and suitably bounded in the right half-plane, which gives the residue sum $\sum g(n) w^n$.

Carlson's thesis is about classes F of power series $f = \sum a_n z^n$ and classes G of function for which the condition $g(n) = a_n$ gives a bijection. His classes G consist of functions $g(z)$ which are analytic in the right half-space and do not vanish on the integers without being identically zero. To every class G there is a corresponding class F of functions $f(w)$. The results are not simple and the author's way of writing does not contribute to clarity, but taken in its entirety, the thesis is an impressive piece of work.

Carlson's examples of the Phragmén-Lindelöf principle aroused interest. The best known ones follow from the Jensen-Carleman formula (see pp. 115, 192).

Theorems on power series

After the thesis Carlson was especially interested in improving the results about power series by Fatou, Jentzsch, and Pólya. In two of these papers, (1919) and (1921b), he was able to use results from the thesis. The problem of the first one is to characterize analytic functions of the type

$$f(z) = \frac{P(z)}{1-z^m} + g(z)$$

where $P(z)$ is a polynomial, $m > 0$ and integer and $g(z)$ is analytic when $|z| \leq 1$. Carlson proved that it is sufficient that $f(z)$ be analytic in the unit disk with finitely many singularities at the boundary and that the first $n(1 - o(1))$ coefficients of

$$f(z) = \sum a_n z^n$$

have values in a fixed set. If this condition is satisfied with $o(1) = 0$, the term $g(z)$ disappears, which is a theorem by Jentzsch. The theme of (1921b) is the minimal growth in different directions of entire functions which assume integral values if the integers, for instance

$$f(z) = \sum c_k^z$$

where $c_1, ..., c_m \neq 0$ are the zeros of a polynomial $z^m + ...$ of degree m with integral coefficients.

The most important paper of this series is (1921a), which answers a question by Pólya: What can one say about the functions in the class of analytic functions represented by power series with entire coefficients? There are such functions, for instance

$$\sum z^{n!},$$

which has the unit circle as a natural boundary and others, for instance

$$\sqrt{\frac{1}{1-4z}} = \sum \binom{2n}{n} z^n$$

which has a point of ramification in the unit disk. Carlson proved that a function in the class which has finitely many singularities in the unit disk and is analytic at one point of the boundary, has to be rational and then of the form

$$P(z)/(1-z^p)z^q, \quad p, q \text{ integers} > 0.$$

Hence, if a function in the class is not rational, it has a ramification point in the disk or else it has the unit circle as a natural boundary. Earlier, Fatou had proved this result for functions whose power series converge when $|z| < 1$.

Carlson's proof, which is technically complicated, depends roughly on the following circumstances. If $f(z) = \sum a_n z^n$ is a power series and $q > 1$ is a fixed integer, let S_n be the vector $(a_n, ..., a_{n+q-1})$. If $P(z) = 1 + az + ...$ is a polynomial of degree m and S'_n are the corresponding vectors for $g(z) = P(z)f(z) = \sum b_n z^n$, then

$$S'_n = S_n + aS_{n-1} + ..., \quad m \text{ terms}.$$

In advantageous cases it is possible to choose the coefficients b_n to be small for large n. This means that the vectors S'_n and the determinants

$$\det(S_n, ..., S_{n+q-1}).$$

are small for large n. Now, if $f(z)$ has entire coefficients, these determinants are integers and have to be zero when they are small enough. By a criterion by Borel this means that $g(z)$ is rational. In all this q is a fixed integer which may be large.

The central point of Carlson's proof is the existence of a polynomial $P(z)$ with $P(0) = 1$ whose absolute value is < 1 on the unit circle except in a small interval. The polynomial is constructed as an approximation of $1/(1 + 2z)$ outside a small sector around the negative axis, which puts the interval around -1. If $f(z) =$

$\sum a_n z^n$ is the given function and $e_1, ..., e_m$ its singular points in the unit disk, it is shown that multiplication by a high power of

$$P(z) \prod \left(1 - \frac{z}{e_k}\right)$$

gives a power series which converges in the unit disk and has such small coefficients that Borel's criterion can be used.

During his active period, Carlson contributed a number of very carefully written papers on power series in *Arkiv*. Some examples are given below.

As is well known, the exponential function e^z has a rational value for a rational argument only for $z = 0$. The paper (1935a) extends this property to functions of the type

$$f(z) = \sum_0^\infty \frac{z^n}{Q(0)...Q(n)}$$

where $Q(z)$ is a polynomial with rational coefficients which does not vanish on the integers. The unusually short proof uses results by Hermite and others. The subject of (1938) is Hadamard's three circle theorem which says that if $M(r)$ is the maximum of an analytic function on a circle of radius R, then $\log M(r)$ is a convex function of $\log r$. Equality is attained for fractional powers const z^λ. For single-valued functions the inequality improves. The paper (1940) is about functions

$$f(z) = \sum_0^\infty a_n z^n$$

which are analytic and of absolute value at most one in the unit disk. By reworking a criterium by Schur, Carlson proved that it is necessary that the absolute values $b_n = |a_n|$ of the coefficients satisfy the inequalities

$$b_{2n+1} \leq 1 - b_0^2 - ... - b_n^2$$

and

$$b_{2n} \leq 1 - b_0^2 - ... - b_{n-1}^2 - \frac{b_n^2}{1 + b_0}$$

and he investigates when equality occurs.

In this series of papers the following inequality (1935b) occurs

$$|\sum_1^\infty a_n|^4 \leq \pi^2 \sum |a_n|^2 \sum |a_n|^2 n^2.$$

In the proof a_n is divided and multiplied by the square root of $c + c^{-1}n^2$, $c > 0$. Then the left side to the power two is equals at most

$$\sum (c|a_n|^2 + c^{-1}|a_n|^2 n^2) \sum \frac{1}{c + c^{-1}n^2}, \quad c > 0,$$

where the last factor is majorized by the integral

$$\int_0^\infty dx/(c + c^{-1}x^2) = \pi/2.$$

Since $cA + c^{-1}B \geq 2\sqrt{AB}$, the inequality follows. It has also a variant in terms of integrals. Carlson verified that π^2 is the best constant.

Theorems on Dirichlet series

Dirichlet series have the form

$$f(s) = \sum_0^\infty a_n e^{-\lambda_n s}, \quad s = \sigma + it, \tag{1}$$

where λ_n is a sequence of real numbers tending to infinity with n. When $a_n = 1$ and $\lambda_n = \log n$, $f(s)$ reduces to Riemann's ζ-function

$$\zeta(s) = \sum n^{-s} = \frac{1}{\prod(1-p^{-s})}, \quad \sigma > 1,$$

where $s = \sigma + it$ and p runs through all prime numbers. Note also that

$$1/\zeta(s) = \prod(1-p^{-s}) = \sum_1^\infty \mu(n) n^{-s}, \quad \sigma > 1, \tag{2}$$

where $\mu(n)$ is the Möbius function, equal to $(-1)^k$ when n is a product of k different primes and $\mu(n) = 0$ otherwise.

It is well known (see for instance the section on analytic number theory p. 133) that the ζ-function is analytic in the entire plane apart from a pole at $\zeta = 1$. According to Riemann's as yet unproven hypothesis, all zeros in the critical strip $0 < \sigma < 1$ lie on the line $\sigma = 1/2$.

The number of zeros $\sigma + it$ of the ζ-function with $0 \leq t \leq T$ has the rather precise order of growth

$$\frac{T}{2\pi} \log \frac{T}{2\pi}.$$

In the absence of a proof of Riemann's hypothesis, attempts were made to estimate the size of the number $N(\sigma, T)$ of zeros whose real parts are $> \sigma > 1/2$ and whose imaginary parts are positive and $< T$. When Carlson wrote his paper (1921) on Dirichlet series, Bohr and Landau had shown that

$$N(\sigma, T) = o(T),$$

a result improved by Wiman's student Wennberg (1920). In principle their method uses Jensen's formula to estimate the number of zeros of a product

$$F(s) = \zeta(s) f_T(s) = 1 + \sum_N^\infty c_n n^{-s}$$

where $N = N(T)$ is suitably large, $f_T(s)$ has few zeros and the coefficients c_n are manageable. Bohr and Landau chose $f_T(s)$ as a partial product of the product (2), Carlson chose a partial sum of the sum of the same formula and succeeded in finding a majorant

$$N(\sigma, T) = O(T^a), \quad a = 1 - 4(\sigma - 1/2)^2 + \varepsilon,$$

for every $\varepsilon > 0$. With a similar method Selberg (1946) found the majorant

$$O(T^a \log T), \quad a = 1 - (\sigma - 1/2)/4.$$

In all cases Jensen's formula was used to estimate the number of zeros with $\log |u|$ majorized by $(1 + |u-1|^2)/2$. Many choices and a lot of computation led to the problem of estimating integrals

$$\int_{-T}^{+T} |F(s) - 1|^2 d\tau. \tag{3}$$

But Carlson's result for the ζ-function is not his main result. In the beginning of his paper he remarked that the product formula for the ζ-function is unnecessary in the scheme of Bohr and Landau. Using partial sums instead, Carlson carried through their scheme with a similar result for every Dirichlet series (1) which does not vanish in a half-plane $\sigma >$ const, and for which $1/f(s)$ can be represented in the same way. His results were then applied to the ζ-function.

Carlson's paper was a success also for himself as a mathematician. At a mature age he could remark that every mathematician ought to do some work with the ζ-function.

Afterwards Carlson wrote a series of papers in *Arkiv* with the title *Contributions à la théorie des séries de Dirichlet* where the main theme is the existence of the limit

$$\lim_{T\to\infty} (2T)^{-1} \int_{-T}^{T} |f(\sigma+it)|^2 dt$$

for different values of σ where $f(s)$ is given by (1). If the sum converges and $f(s)$ is analytic and bounded for $\sigma > \gamma$, the limit equals

$$\sum |a_n|^2 e^{-2\lambda_n \sigma}.$$

This is just the simplest result. The others cannot be treated in a simple review. On one occasion Carlson was able to prove an extension of the result by Hardy and Littlewood that

$$\lim \ (2T)^{-1} \int_{-T}^{T} |\zeta(\sigma+it)|^4 dt$$

converges when $\sigma > 1/2$.

Geometry

From 1920 to 1928 Carlson was professor at KTH in the old-fashioned subject of descriptive geometry. In two papers from 1920 he improved his standing in this discipline in which the main problem is to describe an object by means of two projections onto a plane. In the first paper (1920a) he gave an independent characterization of a projective transformation which maps a ray through a fixed point to a ray through another fixed point in such a way that the rays meet in a plane. Carlson's Geometry (1943) in two volumes is the result of his teaching at KTH, and it is intended to cover first year geometry at the university. It is a solid and rich work but was soon left unused for two reasons. The strict treatment of the subject was too much for most students and in the 1950's analytic geometry changed content and shape and turned into linear algebra, a subject which was more useful and more adapted to the general development of mathematics and physics.

Summary

Carlson was one of those who with refined methods continued Mittag-Leffler's effort in the theory of analytic functions. In his daily life he personified the correct professor. Carlson had a past as a high school teacher and served for a long time as censor at the student examination (see p. 33) This now historic activity started in the 1860's when the universities ceased to have entrance examinations. The replacement was a flying inspection in which teams of university professors went to the gymnasiums as auditors of the oral examinations and censors when the grades were decided. The following excerpt from Frostman's obituary (1953) catches the atmosphere of the work of a censor.

'In this task he found ample use not only of his experience of pure mathematics but also of his vast knowledge of literature, history, geography and French literature. Many high school teachers, whose teaching he supervised, keep the memory of a demanding censor with a certain stern sense of humour but also a man of superior comprehension and unfailing judgement.'

Appendix

The spectral theorem for self-adjoint operators

The first part of this appendix provides a historical background of Carleman's work on symmetric integral equations and, at the same time, it fixes the terminology used in the commenting text pp. 186-189. The second part completes the proof of the Carleman-Denjoy theorem p. 195.

Hilbert

Hilbert collected his research on integral equations in a famous book (1912). His main problem was to extend the diagonalization of quadratic forms to infinitely many variables.

A quadratic form can be written as (Af, f) where A is a symmetric, linear operator on a euclidean space with a distance $\| f \|$ whose square is the sum of the squares of the coordinates of f. The diagonalization means that there is a complete, orthonormal set $f_1, \ldots$ of eigenvectors of A with real eigenvalues $\lambda_1, \ldots$, such that

$$Af_k = \lambda_k f_k, \quad k = 1, 2, \ldots.$$

Hilbert proved that there are two objects which survive a passage to infinitely many coordinates, the resolvent

$$(A - z)^{-1} f = \int \frac{1}{\lambda - z} dE(\lambda) f, \quad \operatorname{Im} z \neq 0,$$

and the spectral family or spectral measure

$$E(\Delta) f = \sum_{\lambda_k \in \Delta} (f, f_k) f_k$$

of commuting, symmetric projections with the property that

$$E(\Delta) E(\Delta') = E(\Delta \cap \Delta')$$

for intervals Δ, Δ'. Simultaneously, the formula

$$\varphi(A) f = \int \varphi(\lambda) dE(\lambda) f$$

defines functions $\varphi(A)$ of A such that $\varphi\psi(A) = \varphi(A)\psi(A)$.

In his computations Hilbert had to assume that his operators were completely continuous in the modern sense and, of course, bounded so that their spectra are bounded from above and below. Among other things the theory provides diagonalizations of integral operators

$$Af(x) = \int_I K(x, y) f(y) dx$$

on the space of square integrable functions f from an interval I with the norm

$$\| f \| = \left(\int_I f(x)^2 dx \right)^{1/2}.$$

The bound on A means that $\| Af \| \leq$ const $\| f \|$. For kernels of the Hilbert-Schmidt type, such that

$$\int_{I \times I} K(x,y)^2 dx dy < \infty,$$

the diagonalization was particularly simple since also the projections $E(\lambda)$ have such kernels. By a reduction to kernels of the Hilbert-Schmidt type, Carleman (1923b) found a class of unbounded kernels for which he could prove the existence of a family of projections and of a resolvent.

von Neumann

Hilbert space, now one of the standard objects of analysis, was introduced by von Neumann (1929) as an abstract version of the complex space l^2 of sequences. Hilbert space is a linear space $H = (f, g, ...)$ over the complex numbers, provided with a complex inner product (f, g) which is sesqui-linear, i.e., linear in f and antilinear in g, which vanishes for $f = g$ only when $f = 0$. This makes (f, f) the square of a norm $\| f \|$. In addition it is required that the space is complete with respect to this norm. It is further assumed that the space H is separable in the sense that it has a countable everywhere dense subset. It then has a countable, complete, and orthonormal basis.

Since Hilbert's theory of self-adjoint bounded operators was formulated for the space of square summable sequences, his spectral theory for bounded, self-adjoint operators can at once be transplanted to Hilbert space. von Neumann's problem was to find a maximal class of unbounded operators which were diagonalizable in the sense of having spectral families.

In order to analyze unbounded operators, von Neumann used their graphs, which was a novelty at the time. The graph $G(A)$ of a linear operator A is the set of pairs f, Af, considered as a part of the direct sum $K = H \oplus H$ with elements $f = f_1 \oplus f_2$ and interior product $(f_1, g_1) + (f_2, g_2)$ which makes K a Hilbert space.

Every linear subspace L of $H \oplus H$ has an adjoint L^* consisting of all pairs g, g_1 such that

$$(f_1, g) = (f, g_1)$$

for all pairs f, f_1 of L. If L is closed, it follows that $L^{**} = L$. That L defines a closed, densely defined operator $A : f_1 = Af$ means that $f = 0 \Rightarrow f_1 = 0$ and that the projection $f, f_1 \to f$ has an everywhere dense range $D = D(A)$, called the domain of A. Then L is the graph $G(A)$ of A. We see immediately that A is closed and densely defined if and only if its adjoint A^*, defined by $G(A^*) = L^*$, has the same property.

In the sequel we shall restrict ourselves to densely defined and closed operators A with domain $D(A)$ and range $R(A) = AD(A)$. That $A \subset A^*$ means now that $G(A) \subset G(A^*)$ so that both $D(A) \subset D(A^*)$ and $A = A^*$ as operators on $D(A)$. This means that $(Af, g) = (f, Ag)$ when $f, g \in D(A)$ and we say that A is symmetric. If $A = A^*$, A is said to be self-adjoint. That A is symmetric means precisely that (Af, f) is real for all $f \in D(A)$.

von Neumann proved that every self-adjoint operator A has a spectral measure $E(\Delta)$, defined for intervals Δ, such that

$$\int \lambda^2 d(E(\lambda)f, f) < \infty$$

if and only if $f \in D(A)$. As above this spectral measure defines functions of A.

von Neumann's proof uses the fact that the maps $f \to (A \pm iE)f$ with $f \in D(A)$ are surjective, which makes the map

$$U : (A + iE)f \to (A - iE)f$$

unitary, i.e., U is a bijection of H such that $(Uf, Ug) = (f, g)$ for all f, g. The spectrum of U is then contained in the unit circle and it is not difficult, by limits of polynomials in U, to construct a family of projections of the form $\varphi(U)$ where $\varphi(t)$ is the characteristic function of an interval on the unit circle. Since

$$A = \frac{i}{2}(U - E)^{-1}$$

this construction gives a spectral measure of A.

von Neumann's paper also contains the well-known theory of defect indices and maximal symmetric operators.

A proof of the Denjoy-Carleman theorem

If $M_0, M_1, \ldots$ are positive numbers, let C_M be the class of infinitely differentiable functions $f(x)$ on the real axis such that

$$|f^{(n)}(x)| \leq C^n M_n$$

for some $C = C(f)$ and all $n \geq 0$. The Denjoy-Carleman theorem says the following

THEOREM. *The class C_M is quasi-analytic if and only if*

$$\sum \frac{1}{\sqrt[n]{M_n^*}} = \infty \tag{1}$$

where the sequence $\log M_n^*$ is the largest convex minorant of the sequence $\log M_n$.

BEVIS: We may assume that $M_0 = 1$. By the text of p. 194 there is a function

$$T(r) = \inf_{n \geq 0}(\log M_n - n \log r) \tag{2}$$

such that the condition

$$\int^\infty T(r)dr/r^2 = -\infty \tag{3}$$

means that the class is quasi-analytic. We must prove the equivalence of (1) and (3).

The first step is to introduce a convex nondecreasing function $S(u) \geq 0$ defined by

$$S(u) = -T(e^u) = \sup_{n \geq 0}(un - \log M_n) \quad u \geq 0, \tag{4}$$

and satisfying

$$\int^\infty e^{-u} S(u) du = \infty.$$

Then $S(u)$ grows faster than any power of u and hence the dual function

$$R(t) = \sup_{u \geq 0}(ut - S(u)), \quad t \geq 0, \tag{5}$$

is convex and finite everywhere. By (4) and the theory of convex functions, $R(n)$ is the largest convex minorant of $n \to \log M_n$.

The two integrals which have to equal infinity simultaneously are now

$$(6) \qquad \int^\infty e^{-u} S(u) du, \quad \int^\infty e^{-R(t)/t} dt,$$

where the first one corresponds to (3) and the second one to (1) and the lower bounds of integration are of no importance.

For a start, an easy computation shows that

$$(7) \qquad R(t) \leq t(\log t/\varepsilon - 1) \quad \Leftrightarrow \quad S(u) \geq \varepsilon e^u$$

for every $\varepsilon > 0$ and for large t and u. In this case both integrals (6) diverge. Hence we may assume from the beginning that

$$(8) \qquad \liminf e^{-u} S'(u) = 0, \quad \liminf e^{-u} S(u) = 0, \quad \sup(R(t)/t - \log t) = \infty,$$

for it is easy to see that $S(u)e^{-u}$ has a positive lower bound for large u if the same holds for $S'(u)e^{-u}$.

Now $S(u) + R(t) - tu \geq 0$ by the definition of $S(u), R(t)$ and the lower bound 0 is attained for pairs u, t such that $t = S'(u)$, $u = R'(t)$ except at the countably many points where $S(u)$ and $R(t)$ have different derivatives from left and right. Hence it follows from (8) and integrations by parts that if the first one of the integrals

$$\text{(a)} \int^\infty e^{-u} S(u) du, \quad \text{(b)} \int^\infty e^{-R'(t)} dt.$$

diverges so does the second, i.e., (a) $= \infty \implies$ (b) $= \infty$. Consider then the integral

$$\text{(c)} \int^\infty e^{-R(t)/t} dt.$$

Since $-R(t)/t \geq -R'(t)$ it follows that (b) $= \infty \implies$ (c) $= \infty$. Finally, if we choose t in the definition of $S(u)$ such that $u = (R(t) - t)/t$, then $S(u) \geq t$ so that

$$\int^\infty e^{-u} S(u) du \geq e^{-1} \int^\infty e^{-R(t)/t} t dR(t)/t = -e^{-1} \int^\infty t de^{-R(t)/t}$$

modulo constants. According to (8) we can integrate by parts in the third integral. It then follows that (c) $= \infty \implies$ (a) $= \infty$ so that the three integrals (a), (b), (c) diverge simultaneously.

BIBLIOGRAPHY

Carleman.

AHLFORS L.
1929. Über die asymptotischen Werte der ganzen Funktion von endlicher Ordnung, Ann. Acad. Sci. Fenn. Ser. A 32, 1929.
BANG TH.
1946. Om quasi-analytiske Funktioner, Ak. Afh. Köpenhamn 1946.
CARLEMAN T.
1960. Pleijel Å. ed., Édition complète des articles de Torsten Carleman, Malmö 1960, Institut Mittag-Leffler.
1916. Über das Neumann-Poincarésche Problem für ein Gebiet mit Ecken, Ak. Avh. Uppsala 1916.
1917. Sur le genre du dénominateur de Fredholm, Arkiv bd 12, 1917, no 6.

1918. *Über ein Minimalproblem der mathematischen Physik*, Math. Zeitschr. 1, 1918, 23-37.
1919. *Über eine isoperimetrische Aufgabe und ihre physikalischen Anwendungen*, Math. Zeitschr. 3, 1919, 1-7.
1920. a. *Sur les équations intégrales à noyau réel et symétrique*, C.R. Paris 171, 1920, 383-386.
1920. b. *Sur les fonctions inverses des fonctions entières d'ordre fini*, Arkiv 15, 1920, no 15.
1921. a. *Zur Theorie der Minimalflächen*, Math. Zeitschr. 9, 1921, 71-77.
1921. b. *Über eine nicht-lineare Randwertaufgabe bei der Gleichung* $\Delta u = 0$, Math. Zeitschr. 9, 1921, 5-13.
1922. a. *Sur un théorème de M. Denjoy*, CR Paris 174, 1922, 373-376.
1922. b. *Démonstration d'un théorème de Borel*, CR Paris 174, 1922, 129-997.
1923. a. *Sur les équations intégrales singulières à noyau réel et symétrique*, Uppsala U. Årsskrift 1923, matematik o. naturv. 3, 1-223.
1923. b. *Über die Approximation analytischer Funktionen durch lineare Aggregate von vorgegebenen Potenzen*, Arkiv 17, 1923, no 2.
1923. c. *Sur les fonctions indéfiniment dérivables*, CR Paris 177, 1923, 422-424.
1926. a. *Les fonctions quasi analytiques*, Leçons professées au Collège de France, Gauthiers-Villars 1926.
1926. b. *Extension d'un théorème de Liouville*, Acta Math. 48, 1926, 271-274.
1927. *Sur un théorème de Weierstrass*, Arkiv 20B, 1927, no 4.
1930. *Sur quelques problèmes dans la théorie mathématique de la diffraction des ondes electromagnétiques*, Arkiv 22, 1930, no 5.
1931. a. *La théorie des équations intégrales singulières et ses applications*, Ann. Institut Henri Poincaré 1, 1931.
1931. b. *Sur la croissance de certaines classes de fonctions*, Mat. Tidskr.B., 1931, no 4.
1931. c. *Application de la théorie des équations intégrales singulières aux équations différentielles de la dynamique*, Acta Math. 59, 1932, 63-87.
1932. a. *Sur la théorie de l'équation intégrodifférentielle de Boltzmann*, Acta Math. 60, 1932, 91-146.
1932. b. *Sur les caractéristiques du tore*, CR Paris 195, 1932, 478-481.
1932. c. *Sur la théorie des équations intégrales et ses applications*, Verh. internat. Kongr. math. Zürich 1932.
1933. a. *Sur la théorie de l'équation intégrodifférentielle de Boltzmann*, Arkiv 23A, 1933, no 22.
1933. b. *Sur une inégalité différentielle dans la théorie des fonctions analytiques*, CR Paris 196, 1933, 995-997.
1933. c. *Sur les systèmes linéaires aux dérivées partielles du premier ordre à deux variables*, CR Paris 197, 1933, 471-74.
1934. a. *Sur la théorie mathématique de l'équation de Schrödinger*, Arkiv 24B, 1934, no 11.
1934. b. *Propriétés asymptotiques des fonctions fondamentales des membranes vibrantes*, CR du VIII Congres de mathématiciens Scandinaves, Stockholm 1934.
1936. *Über die asymptotische Verteilung der Eigenwerte partieller Differentialgleichungen*, Ber. math.-phys. Klasse Ak. Wiss. Leipzig 88, 1936, 34-44.
1938. *Sur un problème d'unicité pour les systèmes d'équations aux dérivées partielles à deux variables indépendentes*, Arkiv 26B, 1938, no17.
1944. *L'intégrale de Fourier et questions qui s'y rattachent*, Publ. Sci. de l'institut Mittag-Leffler, Uppsala 1944.
1947. *Sur l'application de la théorie des fonctions analytiques dans la théorie des transformées de Fourier*, Colloque CNRS de l'analyse harmonique Nancy, 1947.
1957. *Problèmes mathématiques dans la théorie cinétique des gaz*, Publ. sci. de l'Institut Mittag-Leffler 2, éds O. Frostman, L. Carleson, Uppsala 1957.

CARTAN H.
1941. *Sur les classes de fonctions définies par des inégalités portant sur leurs dérivées*, Act. Scient. Paris 1941.

HELLSTEN U.
1947. *Determination of the denominator of Fredholm in some types of integral equations*, Acta Math. 79, 1947, 102-152.

HILBERT D.
1921. *Grundzüge einer allgemeinen Theorie der linearen Integralgleichungen*, B.G. Teubner 1912.

HÖRMANDER L.
1985. *The Analysis of Linear Partial Differential Operators IV*, Springer 1985.

LIONS P.L., DiPERNA R.J.
1990. *Global solutions of Boltzmann's equations and the entropy inequality*, Arch. Rat. Mech. Anal. 114, 1991, 47-55.

MANDELBROJT S.

1942. *Analytic functions and classes of infinitely differentiable functions*, Rice Inst. Pamphlet XXIX, 1, 1942.

PERSSON K.

1938. *Sur une classe de systèmes d'équations linéaires à une infinité d'inconnus*, Ak. Avh, Uppsala 1938.

PLEIJEL Å.

1941. *Propriétés asymptotiques des fonctions et valeurs propres de certains problèmes de vibration*, Arkiv 27A,1941, no 13 1941.

1943. *Quelques problèmes de vibration et les méthodes directes du calcul des vibrations*, Arkiv 29A, 1943, no 23.

1944. *Le problème spectral de certaines équations aux dérivées partielles*, Arkiv 30A, 1944, no 21.

1946. *Sur les opérateurs différentielles de type elliptique*, Arkiv 32, 1946, no 14.

1948. *Asymptotic relations for the eigenfunctions of certain boundary problems of polar type*, Am. J. Math. 70, 1948, 892-907.

1949. *Some properties of the eigenfunctions of the Laplace-operator on Riemannian manifolds*, Can. J. Math. 1, 1949, 242-256 (with Minakshisundaram).

1952. *On a theorem by Carleman*, Mat. Tidskr B, 1952.

REED M., SIMON B.

1978. *Methods of modern mathematical analysis IV*, Acad. Press 1978.

RIESZ F.

1926. *Sur les fonctions subharmoniques et leur rapport avec la théorie du potentiel I*, Acta Math. 48, 1926, 329-343.

TRUESDELL T., MUNCASTER R.G.

1980. *Fundamentals of Maxwell's kinetic theory of a simple monatomic gas*, Academic Press 1980.

STONE M.

1932. *Linear Transformations in Hilbert Space*, Amer. Math. Soc Coll. Publ. XV, 1932.

VON NEUMANN J.

1929. *Allgemeine Eigenwerttheorie Hermitischer Funktionaloperatoren*, Math. Ann. 102, 1929, 49-131.

1935. *Charakterisierung des Spektrums eines Integraloperators*, Act. Scient. Industr. 229, 1935.

WEYL H.

1911. *Über die asymptotische Verteiligung der Eigenwerte*, Nachr. mat.-phys. Klasse Göttingen 1911, 110-117.

Carlson.

CARLSON F.

1914. *Sur une classe de séries de Taylor*, Diss Uppsala 1914.

1919. *Über Potenzreihen mit endlich vielen verschiedenen Koeffizienten*, Math. Ann. 97, 1919, 237-245.

1920. a. *Über den Satz von Pohlke*, Arkiv 14, 1920, no 2.

1920. b. *Ein Satz über Kegelschnitte ...* , Math. Zeitschr. 8, 1920.

1921. a. *Über Potenzreihen mit ganzzahligen Koeffizienten*, Math. Zeitschr, 9, 1921, 1-13.

1921. b. *Über ganzwertige Funktionen*, Math. Zeitschr. 11, 1921, 1-23.

1921. c. *Über die Nullstellen der Dirichletschen Reihen und der Riemannschen ζ-Funktion.*, Arkiv 1, 1921, no 20.

1922. *Contributions à la théorie des séries de Dirichlet Note I*, Arkiv 16, 1922, no 18.

1926. *Contributions à la théorie des séries de Dirichlet Note II*, Arkiv 19A, 1926, no 25.

1933. *Contributions à la théorie des séries de Dirichlet Note III*, Arkiv 23A, 1926, no 19.

1935. a. *Sur une propriété arithmétique de quelques fonctions entières*, Arkiv 25A, 1938, no 7.

1935. b. *Une inégalité*, Arkiv 25B, 1935, no 1.

1938. *Sur le module maximum d'une fonction analytique uniforme. Note I*, Arkiv 26A, 1938, no9.

1940. *Sur les coéfficients d'une fonction bornée dans le cercle unité*, Arkiv 27A, 1940, no 1.

1943. *Lärobok i Geometri I,II*, Gleerup, Lund, 1943,47.

1952. *Contributions à la théorie des séries de Dirichlet Note IV*, Arkiv f. Mat. 2, 1952.

FROSTMAN O..

1955. *Fritz Carlson in memoriam*, Acta Math. 90, 1953, IX-XII.

CHAPTER 15

Lund 1925 - 1950

Riesz   Zeilon

In 1923 Brodén was succeeded by Carleman who, after a short time, left for Stockholm. At about the same time Nörlund decided to accept the post as director of the Danish Geodetical Survey in Copenhagen. With this both professorships of mathematics in Lund were open. The committee of experts suggested Marcel Riesz and Nils Zeilon for the first and second places but they were not able to decide which one to put first. In their judgments they had remarked among other things that Riesz had done very good work, often in collaboration with others, but that Zeilon's work, especially the paper on double refraction in crystals, was more innovative.

Anyway, because two chairs were available, who was put first by the University Council was irrelevant. But the rules required each member to make a definite choice and to state his motive. The retiring professor Brodén wrote that since the two were of equal scientific merit, other qualities must decide. From what the experts had written, he saw in Riesz a mathematician who improved the work of others and in Zeilon a man of action and initiative and this decided his vote. This judgment was flatly contradicted by the later developments. Zeilon turned out to be a less than enthusiastic lecturer and teacher. His lectures were rather dull and mostly restricted to analytical geometry and differential equations. His family and artistic interests came first. But Riesz, who lived alone and worked constantly with mathematics, was an enthusiastic lecturer with a large repertory from number theory to relativity theory. As a docent at Stockholm University Riesz had been so well appreciated that the students protested when Carleman was invited to the professorship. In Lund Riesz came to dominate the small group whose members devoted themselves to mathematics.

Riesz

When Riesz arrived in Lund in 1926, he had announced an inequality for conjugate functions and was in the process of publishing the proof (1927 b) including a theorem on the maxima of bilinear forms (1927 a). These two theorems, which are central in analysis, were to be his best known results. Work by Thorin (1938, 1948) has made it possible to give simple proofs of both.

Interpolation between inequalities, the Riesz-Thorin theorem

The theorem about bilinear forms was enunciated for spaces of finite dimension, but its importance is greatly enhanced by simple passages to the limit . To prepare for these the reader must be reminded of abstract measure theory. Let M be a set equipped with a class C of subsets which contains M and the empty set and is closed under countable intersections and unions. Such a space becomes a measure space M_μ when it is provided with a measure, i.e., a function μ from

C to numbers ≥ 0 which is countably additive, which means that

$$\mu(\cup_1^\infty X_k) = \sum_1^\infty \mu(X_k)$$

when the sets $X_k \in C$ are pairwise disjoint. Sets contained in a set of C of measure zero are said to be null sets. A set X which differs from a set Y in C by a null set is said to be measurable with measure $\mu(X) = \mu(Y)$.

Real functions $f(x)$ from M are said to be measurable if all sets where $f(x) = a$ or $f(x) \geq a$ are measurable. Complex functions are measurable when their real and imaginary parts have this property. A measurable function is said to be finite when it assumes finitely many values and vanishes outside a set of finite measure. Its integral $\int f(x)d\mu(x)$ is then a linear combination of these values with coefficients ≥ 0.

When M_μ is a measure space with measure $\mu \neq 0$, let $L_0^p(M)$ with $1 \leq p \leq \infty$ be the space of measurable functions from M such that

$$\| f \|_p = \left(\int |f(x)|^p d\mu(x) \right)^{1/p} < \infty.$$

It is well known that the space $L_0^p(M)$ modulo functions with $\| f \|_p = 0$ constitutes a Banach space $L^p(M)$. If $p < \infty$, every function f in $L^p(M)$ can be approximated by a sequence of finite functions g such that $\| f - g \|_p \to 0$. When $p = \infty$, a different approximation is possible.

Thorin's proof (1938, 1948) for what is now called the Riesz-Thorin theorem uses the following three line theorem by Hadamard that was proved on p. 197 under weaker assumptions: If $h(z)$ is analytic and bounded when $0 < \operatorname{Re} z < 1$ and

$$h^*(t) = \sup |h(z)|, \quad \operatorname{Re} z = t,$$

then $\log h^*(t)$ is a convex function of t. We can now formulate and prove

THE RIESZ-THORIN THEOREM. *Let M_μ, N_ν be two measure spaces and suppose that $B(f,g)$ is bilinear for finite functions f, g in the respective spaces. With real numbers α, β satisfying (2) below, put*

(1) $$M(f,g;\alpha,\beta) = B(f,g)/\| f \|_{1/\alpha} \| g \|_{1/\beta}$$

when the denominators do not vanish, and

$$M(\alpha,\beta) = \sup_{f,g} |M(f,g;\alpha,\beta)|.$$

Then $\log M(\alpha,\beta)$ is a convex function of α, β in the region

(2) $$0 \leq \alpha \leq 1, \quad 0 \leq \beta \leq 1.$$

Remark. Riesz's proof (1927 b) consisted of more or less explicit calculations and required that $\alpha + \beta \geq 1$.

Remark. The usual description of the proof says that the inequality $M(\alpha,\beta) <$ const is proved by interpolation between the inequalities at the endpoints of the interval. The Riesz-Thorin inequality is just one of many inequalities which are proved in a similar fashion.

PROOF: Let $\alpha(0), \alpha(1)$ and $\beta(0), \beta(1)$ be numbers between 0 and 1. With s a complex number with real part between 0 and 1, put $\alpha(s) = (1-s)\alpha(0) + s\alpha(1)$

and analogously for $\beta(s)$. Fix non-vanishing finite functions $F(x), G(x) \geq 0$ and real functions u, v and put

(3) $$f_s(x) = F(x)^{\alpha(s)} e^{iu(F(x))}, \quad g_s(y) = G(y)^{\beta(s)} e^{iv(G(y))}.$$

It follows that

$$\| f_s \|_{\alpha(s)} = \left(\int F(x) d\mu(x) \right)^{1/s}, \quad \| g_s \|_{\beta(s)} = \left(\int G(y) d\nu(y) \right)^{1/s}$$

when s is real. When s is complex, the right sides are taken as definitions of the left sides. This means that (1) with the arguments $f_s, g_s, \alpha(s), \beta(s)$ is a bounded analytic function of s when $t = \operatorname{Re} s$ is between 0 and 1. Hence, if

$$M(f, g, t) = \sup |M(f_s, g_s; \alpha(s), \beta(s))|, \quad \operatorname{Re} s = t,$$

then $\log M(f, g, t)$ is a convex function of t so that

$$M(f, g, t) \leq M(f, g, 0)^{1-t} M(f, g, 1)^t.$$

The supremum $M(t)$ of the left side for all possible functions (3) with t fixed includes the supremum above over $\operatorname{Re} s = t$. The corresponding supremum over the right side is not larger than the right side below and we get

$$M(t) \leq M(0)^{1-t} M(1)^t.$$

This proves the theorem.

Applications

The Riesz-Thorin theorem has a number of corollaries, earlier proved in other ways, for example Hölder's inequality

(1) $$\left| \int_M f(x) g(x) d\mu(x) \right| \leq \| f \|_p \| g \|_q .$$

This inequality holds since it is trivially true when $p = 1, q = \infty$ or $p = \infty, q = 1$. This theorem has a converse,

$$\| f \|_p = \sup_g \left| \int f(x) g(x) d\mu(x) \right| / \| g \|_q .$$

If we view $B(f, g)$ as (f, Tg) where T is a linear operator, the Riesz-Thorin theorem says that the set of pairs $1/p, 1/q$ with elements between 0 and 1 for which

$$\| Tf \|_q \leq c_{p,q} \| f \|_p, \quad c_{p,q} < \infty,$$

is convex.

An inequality of this type is the Hausdorff-Young theorem, the inequality

(4) $$\| Tf \|_q \leq c_p \| f \|_p, \quad c_p > 0,$$

in which $p \geq 2$ and $1/p + 1/q = 1$, the measure spaces are the real axis with Lebesgue measure and

$$Tf(t) = \int e^{-ixt} f(x) dx$$

is the Fourier transform of f. The inequality holds for $1/p$, $1/q$ equal to $1/2, 1/2$ by Parseval's theorem, it is trivial for the pair $1, 0$ and hence correct as stated.

Conjugate functions

The real and imaginary parts of the boundary values of a function analytic in a region, for instance the unit disk, are said to be conjugate functions. Between such functions there is a linear correspondence which we shall deduce for the upper half-plane and the real axis R.

When $g(x)$ is a real, infinitely differentiable function with compact support then the function

$$f(z) = \frac{i}{\pi} \int_R \frac{1}{z-t} g(t) dt, \quad y = \operatorname{Im} z > 0,$$

is analytic for $y > 0$ with boundary values

$$g(x) + iTg(x), \quad Tg(x) = \frac{1}{\pi} \int \frac{g(t)}{x-t} dt,$$

where the integral is a principal value, pv. In fact it is well known that

$$\lim_{0<y\to 0} \frac{1}{z-t} = \operatorname{pv} \frac{1}{x-t} - i\pi\delta(x-t).$$

In this case Riesz's theorem (1927 b) on conjugate functions says that

(6) $$\|Tg\|_p \leq c_p \|g\|_p, \quad c_p < \infty,$$

for all $p > 1$. It suffices to prove the theorem when g is restricted as above. The first step uses the fact that, with $f(z)$ as above,

(7) $$\int f(x+i0)^{2k} dx = 0, \quad k = 1, 2, \ldots,$$

for it is easy to see that the corresponding integral with the argument $x + iy$ is independent of y and tends to zero when $y \to \infty$. The formula (7) implies that

$$\int h^{2k} dx < \sum_1^k \binom{k}{j} \int_R h(x)^{2k-2j} g(x)^{2j} dx$$

where $h = Tg$. Now, if we put

$$X = \|g\|_{2k}, \quad Y = \|h\|_{2k},$$

Hölder's inequality with $p = 2k/2(k-j)$, $q = 2k/2j$ shows that

$$\int h^{2k-2j} g^{2j} dx \leq Y^{2k-2j} X^{2j}$$

so that

$$Y^{2k} < \sum_1^k \binom{k}{j} Y^{2k-2j} X^{2-2j}.$$

Hence $Y \leq \text{const } X$. This means that Riesz's inequality (6) is correct when $p = 2, 4, \ldots$. By an interpolation according to Riesz-Thorin, it follows when $\infty > p \geq 2$. Further, if g_1, g_2 are functions similar to g, the integral

$$(g_1, Tg_2) = \int g_1(x) Tg_2(x) dx$$

is symmetric by virtue of Fubini's theorem. Hence, from

$$|(Tg_1, g_2)| = |(g_1, Tg_2)| \leq \text{const } \| g_1 \|_p \| g_2 \|_q,$$

where $1/p + 1/q = 1$ and $p \geq 2$, it follows that

$$\| Tg_2 \|_q \leq \text{const } \| g_2 \|_q$$

when $1 < q \leq 2$ and this finishes the proof of Riesz's theorem on conjugate functions.

The same proof works for conjugate functions on the unit circle provided that the corresponding analytic function vanishes at the origin. Then

(8) $$\int_0^{2\pi} f(e^{i\theta})^{2k} d\theta = 0, \quad k = 1, 2, \ldots,$$

whence, as above,

$$\int_0^{2\pi} |v(\theta)|^p d\theta \leq c_p^p \int_0^{2\pi} |u(\theta)|^p d\theta, \quad p > 1,$$

when $f(e^{i\theta}) = u(\theta) + iv(\theta)$ with the same constants c_p as above. In his paper (1927 b) Riesz proved this inequality when v vanishes at the origin. At that time he did not have the Riesz-Thorin theorem and had to find a proof from integrals (8) with $p = 2k$ an even integer. This proof was only possible under some complicated restrictions.

Short papers, Medd. Lunds Univ. Mat. Sem.

During his first years in Lund, Riesz wrote three short papers in analysis. Later his interest in relativity theory and quantum physics led him to constructions which occupied him and his students for ten years. These papers will be analyzed below under the headings of *Fractional potentials* and *Spinors*.

Already in 1931, Marcel Riesz was elected 'promotor'[1] at the 'promotion', a graduation ceremony which marks the end of the academic year. For this occasion he wrote a paper (1931), founded on the fact that a disk C in the complex plane with radius R and center z_0 has a metric

$$|dz|_C = R|dz|/(R^2 - |z - z_0|^2),$$

which is covariant under Möbius maps, i.e., $|dz|_C = |dw|_D$ when $z \to w$ maps C to D. In addition,

$$|df(z)|_D \leq |dz|_C$$

when f is meromorphic and maps C into D. Many known facts in function theory follow from this fact. The paper has no news value, but it is characteristic of Riesz's personal style and love of the proper phrase.

---

[1] A master of ceremonies who distributes the insignia of the degree of doctor.

Very soon after his arrival in Lund Riesz founded a mathematical journal, *Meddelanden från Lunds Universitets Matematiska Seminarium*[2], here shortened to LMS. This journal, which published almost all the papers of the mathematicians in Lund, was the mathematical face of Lund University. Like many similar journals on the continent it was a parasitical enterprise. Each volume consisted of extra copies of doctoral theses paid for by their authors or of collections of reprints from a newly founded series[3] run by the Lund Physiographical Society and a similar series[4] from the university in Szeged, Hungary, where Riesz's brother Fredric was professor of mathematics. This system survived into the 1960's.

Two papers by Riesz dominated the first volume of LMS. In the first one he gives necessary and sufficient conditions that a subset M of $L^p(R)$ be compact. The conditions are the same as in the theorem by Arzelà-Ascoli with the supremum norm replaced by $\| f \|_p$, and the proof uses the fact that the means

$$f_d(x) = \int_{-d}^{d} f(x+h)dh/2d, \quad d > 0,$$

satisfy the conditions of this theorem.

The Danish mathematician Harald Bohr's great contribution to analysis was the theory of almost periodic functions. These functions are defined by the property that they can be approximated uniformly by finite sums

$$\sum a_k e^{ib_k x}$$

of exponential functions with real b_k. The main theorem says that every almost periodic function has countably many real frequencies $\lambda_1, \lambda_2, \ldots$, uniquely determined by the property that the limit

$$\lim_{T \to \infty} \frac{1}{T} \int_0^T f(x) e^{-i\lambda x} dx$$

does not vanish if and only if $\lambda = \lambda_k$ for some k. When Bohr wrote a detailed account (1932) of his theory, his original proof had already been simplified by, among others, Herman Weyl and Solomon Bochner. Riesz's contribution (1933 b) is a simplification of an earlier proof.

Fractional potentials

Riesz's best-known theorem will always be the theorem about conjugate functions. His fractional potentials, connected both with Laplace's operator and potential theory and the wave operator and wave propagation, come second. Their origin is without doubt the Riemann-Liouville integral which Riesz had used in his work on summation.

The Riemann-Liouville integral is a convolution

(1) $$I^\alpha f(x) = \int K(\alpha, x-y) f(y) dy$$

of a function $f(x)$ and the kernel

(2) $$K(\alpha, x) = \frac{1}{\Gamma(\alpha)} x_+^{\alpha-1}$$

---

[2] *Communications from the mathematical seminar of Lund University*
[3] *Fysiografiska Sällskapets i Lund förhandlingar*
[4] *Acta Szeged Litterarum ac Scientiarum*

where $\operatorname{Re}\alpha > 0$. Its Fourier-Laplace transform is

$$\int e^{-ix(\xi+it)} K(\alpha, x) dx = (i(\xi + it))^{-\alpha}, \quad t \geq 0. \tag{3}$$

This explains why the operator I^α can be seen as a fractional power $D^{-\alpha}$ of the operator $D = d/dx$ of differentiation with the property that $D^{-\alpha} f(x)$ only depends on the restriction of f to the interval $(-\infty, x)$. Note that the right side of (3) is an analytic function of $\xi + it$ when $t > 0$. The properties (4) below of the operator I^α follow from (3) and direct computation,

$$DK(\alpha, x) = K(\alpha - 1, x), \quad K(\alpha + \beta, x) = \int K(\alpha, x - y) K(\beta, y) dy. \tag{4}$$

It is obvious that $I^\alpha f(x)$ is an analytic function of α for fixed x if $\operatorname{Re}\alpha > -1$ and the integral (1) converges. If f is infinitely differentiable and $= O(|x|^{-N})$ for every $N > 0$ and negative x, then $I^\alpha f(x)$ is an entire analytic function of α. This follows by induction from the formula $I^{\alpha+1} Df(x) = I^\alpha f(x)$.

In his three main papers from his time in Lund, (1937 a,b, 1949), Riesz constructed two variants of the Riemann-Liouville integral which represent fractional powers $\Delta^{-\alpha}$ of Laplace's operator

$$\Delta = \partial_1^2 + \ldots + \partial_m^2$$

and the wave operator

$$\square = \partial_1^2 - \partial_2^2 \ldots - \partial_n^2.$$

In the first case, the kernel $K(\alpha, x)$ is simply the inverse Fourier transform of $|\xi|^{-\alpha}$ with the usual Euclidean norm and $0 < \operatorname{Re}\alpha < m$. By a computation,

$$K(\alpha, x) = |x|^{\alpha - m} / H_m(\alpha), \tag{5}$$

where

$$H_m(\alpha) = \pi^{n/2} 2^\alpha \Gamma(\alpha/2) / \Gamma((m - \alpha)/2).$$

We see here that $\Delta K(\alpha, x) = -K(\alpha - 2, x)$. In the sequel we restrict ourselves to real α. For $\alpha = 2$ we get a real and positive kernel which for $m > 2$ is proportional to Newton's potential $|x|^{2-n}$ (for $m = 2$ an analytical continuation gives the logarithmic potential). What became known as the α-potential of a measure $d\mu(x)$ with compact support,

$$U^\alpha \mu(x) = \frac{1}{H_m(\alpha)} \int |x - y|^{\alpha - m} d\mu(y)$$

equals a constant times the Newtonian potential when $\alpha = 2$.

Frostman (1935) exploited the possibility to do potential theory with the new kernel. To begin with, Riesz's formula

$$K(\alpha + \beta, x - y) = \int K(\alpha, x - z) K(\beta, z - y) dz$$

shows that the energy integral

$$\int U^\alpha \mu(x) d\mu(x) = \int (U^{\alpha/2} \mu(x))^2 dx$$

is non-negative, which makes it possible to minimize the the energy under suitable side conditions. In addition,

$$\Delta |x|^{\alpha-m} = (\alpha - m)(\alpha - 2)|x|^{\alpha-4} < 0,$$

so that the potential kernel is subharmonic when $m \geq \alpha \geq 2$. These features make it probable that the new theory will not be very different from ordinary potential theory. This idea was realized by Frostman (1935) in a well-known thesis. In his paper (1937 a), which will be reviewed in connection with Frostman's thesis, Riesz extended Frostman's results.

The wave operator

For the wave operator $\Box = \partial_1^2 - \partial_2^2 - ... - \partial_n^2$ it is natural to introduce a Lorentz metric with the interior product

$$(x, y) = x_1 y_1 - x_2 y_2 - ... - x_n y_n,$$

in which the wave operator corresponds to the polynomial

$$(\xi, \xi) = \xi_1^2 - \xi_2^2 - ... - \xi_n^2.$$

Let $C: x_1 \geq 0, (x, x) \geq 0$ be the forward light-cone. In analogy with the elliptic potential, Riesz (1937 b) constructed a new integral of the Riemann-Liouville type where the kernel $K(\alpha, x)$ vanishes outside the light-cone $x_1 \geq 0$, $(x, x) \geq 0$ and has the form

(6) $\quad K(\alpha, x) = (x, x)^{(\alpha-m)/2} / H_m(\alpha),$

$$H_m(\alpha) = \pi^{(m-2)/2} 2^{\alpha-1} \Gamma(\alpha/2) \Gamma((\alpha + 2 - m)/2)$$

in the light-cone. When $\alpha - m < 1$ the kernel is no longer locally integrable in the light-cone so that the integral $I^\alpha f(x)$ is no longer defined. It was this difficulty which was overcome by analytical continuation.

When $\operatorname{Re} \alpha > m/2$, the Fourier-Laplace transform of the kernel

$$\int e^{-i(x, \xi + i\theta)} K(\alpha, x) dx, \quad \theta = (1, 0, ..., 0)$$

is simply equal to

$$(\xi + i\theta, \xi + i\theta)^{-\alpha}.$$

Since the basis of this exponential, viz.

$$(\xi, \xi) + 2i(\xi, \theta) - (\theta, \theta),$$

is never zero or positive, there are no problems with the complex exponent α. The new kernel $K(\alpha, x)$ has the same properties as the kernel of the Riemann-Liouville integral. We need only exchange D for the wave operator $\Box$ which in the new situation diminishes the index by two. Hence the corresponding $I^\alpha f(x)$ is an entire analytic function of α if f is indefinitely differentiable and small enough for large negative x_1. It follows that I^0 is the identity and that $\Box I^2 f(x) = f(x)$. We can say that $K(2, x)$, defined by analytical continuation, is a fundamental solution of $\Box$ with support in the light-cone.

This account *a posteriori* gives the Fourier-Laplace integrals too much weight compared to Riesz's original motives where he required that the convolution of his kernel with exponential functions, decreasing in time, should reproduce the exponential modulo a factor. Since the kernel $K(\alpha, x)$ and the convolution $I^\alpha f(x)$ are analytic in α when $\operatorname{Re}\alpha$ is large and since

$$I^\alpha \Box f(x) = I^{\alpha-2} f(x)$$

when f is indefinitely differentiable and vanishes suitably for large negative x_1, it is clear that $I^\alpha f(x)$ is entire analytic for a large class of functions. An essential problem in this situation is now to prove by explicit calculation that I^0 is the identity. For then $I^2 f(x)$ solves the equation

$$\Box I^2 f(x) = f(x),$$

and $K(2, x)$ is in a sense a fundamental solution.[5] The properties of the Γ-function show that $K(\alpha, x)$ vanishes inside the light-cone when

$$\alpha = m - 2k, \quad \alpha = -2k, \quad k = 0, 1, ...,$$

and this in turn means that the fundamental solution has its support on the boundary of the light-cone when m is even and ≥ 4 but not otherwise. Hence the formula for the convolution $I^\alpha f$ exhibits Huyghens's principle.

As we said above, Riesz's main problem was to show that $I^0 = 1$ and to find an explicit formula for $I^2 f(x)$. He also wanted to use analytical continuation to solve Cauchy's problem for the wave equations and space-like surfaces.

A surface $S : s(x) = 0$ is said to be space-like when $\operatorname{grad} s(x)$ or $-\operatorname{grad} s(x)$ belong to the interior of the light-cone C when $s(x) = 0$. The simplest examples are the planes $x_1 = \operatorname{const}$. Let us imagine a space-like surface S whose projection along lines parallel to the x_1-axis covers the entire plane $x_1 = 0$. Such a surface divides space into two parts, a positive part S^+ consisting of points x such that $S(x) = (x - C) \cap S$ is not empty and a negative part S_- for which $(x + C) \cap S$ is not empty. Note that if $x \in S^+$ then $C(x) = (x - C) \cap S^+$ is a cut-off cone with the bottom $S(x)$.

In $C(x)$, Green's formula is as follows :

(7) $\quad I^\alpha u(x) - \int_{C(x)} K(\alpha - 2, x - y) \Box u(y) dy =$

$$\int_{S(x)} (K(\alpha, x-y) u D_\nu u(y) - u(y) D_\nu K(\alpha, x-y)) dS$$

when $\operatorname{Re}\alpha$ is large so that all integrands vanish at the boundary of $x - C$. Here D_ν is differentiation in the direction of $\operatorname{grad} s(y)$ and dS is a surface element on S. An analytical continuation of this formula to $\alpha = 2$ gives the value of the function u at the point x as a linear function of $\Box u$ restricted to $C(x)$ and the derivatives of order < 2 of the function restricted to the part $S(x)$ of S. In addition: in this formula $\Box u(x)$ can be given arbitrarily in $C(x)$ and $u(y)$ and $D_\nu u(y)$ on S. This gives an explicit solution of Cauchy's problem

$$\Box u = f \quad \text{outside of} \quad S, \quad u = g, \quad D_\nu u = h \quad \text{at} \quad S.$$

The first proof that $I^0 = 1$ and an analytic continuation of (7) under explicit regularity conditions on u was given in the thesis (1946) by Nils Erik Fremberg.

---

[5]This was before the theory of distributions so that Riesz did not give a precise meaning to $K(2, x)$.

The reason that Riesz delayed publication of a proof of his statements in his Paris lecture (1937b) was that his esthetics demanded that all proofs should be as invariant as possible under Lorentz transformations. This requirement made the proof that $I^0 = 1$ especially difficult. But the goal is attained in Riesz's final paper (1949). All proofs satisfy strict criteria of invariance, for instance the analytical continuation to $\alpha = 2$ of the formula 7) when $m = 4$, announced in (1937b). Riesz shows that the right side becomes an integral along the intersection S^x between S and the boundary of $x - C$,

$$u(x) = \int_{S^x} (F(x,y)u(y) + u_\nu(y))dS^x.$$

Here dS^x is the Lorentz surface element on S^x, F is a geometrically defined weight and u_ν is differentiation of u in the direction ν of a ray of light from x to y which is reflected in y by a reflection in the surface S. Since $(\nu, \nu) = 0$ Riesz remarks that this is a certain strengthening of Huyghens's principle: the value of the solution $u(x)$ at the point x depends in the first approximation only on the restriction of the solution to S^x. That Riesz was fascinated by the geometry and differential geometry of Lorentz space shines through on almost every page of (1949).

The source of Riesz's interest in the wave equation was a couple of seminars in the early 1930's together with Zeilon. The subject was Hadamard's classical book (1932) about Cauchy's problem for the wave operator with variable coefficients. In this book the author uses his method of *partie finie* to master the divergencies which appear already for constant coefficients. In an appendix of his Paris lecture Riesz gave an elegant alternative to Hadamard's method. He introduced fractional potentials for the wave operator with variable coefficients

$$\Box = \frac{1}{\sqrt{g}} \partial_i \sqrt{g} g^{ik} \partial_k, \quad \partial_i = \partial/\partial x_i,$$

where the metric form $g_{ik} dx^i dx^k$ has Lorentz signature $+, -, ..., -$ and where the summation convention of tensor calculus has been used. The geodesic lines with nonnegative line elements which start form a point Q constitute two cones. Let C_Q be one of them and let r_{PQ} be the geodesic distance from $P \in C_Q$. For large $\operatorname{Re} \alpha$ Riesz constructs a function

$$K(\alpha, P, Q)$$

which vanishes when P is outside of C_Q and has a principal part

$$r_{PQ}^{\theta-m}/H_m(\alpha)$$

such that $\Box_P K(\alpha P, Q) = K(\alpha - 2, P, Q)$ and

$$\int K(\alpha, P, R) K(\beta, R, Q) dv_R = K(\alpha + \beta, P, Q),$$

where dv_R is the volume element of the metric. By analytic continuation, the function $E(P, Q) = K(2, P, Q)$ turns into a fundamental solution, $\Box_P E(P, Q) = \delta(P, Q)$. The details of the construction do not fit into a short review.

Spinors

Very soon after the birth of quantum mechanics Dirac found an equation for the free electron in which the wave operator

$$\Box = \partial_1^2 - \partial_2^2 - \partial_3^2 - \partial_4^2$$

is the square of the linear Dirac operator

$$D = e_1\partial_1 + e_2\partial_2 + e_3\partial_3 + e_4\partial_4$$

where the Clifford numbers $e_1, ..., e_4$ have the property that

$$e_j e_k + e_k e_j = 0, \quad e_1^2 = 1, \quad e_2^2 = e_3^2 = e_4^2 = -1$$

when $j \neq k$. Dirac represented these entities as 4 by 4 matrices. In this form they generate an algebra over the complex numbers which is isomorphic to the full matrix algebra of order four. If we write Dirac's equation as

$$Du = 0,$$

the wave function u transforms linearly, $u \to L(g)u$, when $e_1, ...e_4$ are subject to a Lorentz transformation g. The map $g \to L(g)$ is called the spin representation of the Lorentz group and the wave function u is called a spinor.

Riesz's interest in the Clifford numbers lasted for decades. One of his first contributions was to explain to the Lund physicists what a spinor is: a left ideal in the Clifford algebra. This is quite correct, a left ideal carries the spin representation, but the physicists might have appreciated a more down-to-earth explanation. At the congress of Scandinavian mathematicians in Copenhagen in 1946 Riesz gave a lecture on the Dirac equation. One difficulty with the relativistic wave equations is the construction of a positive energy density. Riesz proved that the difficulty can be mastered using Dirac's equation. Some years later he wrote a paper (1953) about spinors in general relativity where the metric form has variable coefficients. Riesz's work on spinors left the physicists unimpressed but his construction in this paper is a first local construction of what later became known as a spin manifold, a notion which now plays an important part in topology.

Obituary

Riesz's interest in all aspects of mathematics, his ability to find the right word and his natural poise made a lasting impression on his colleagues in the faculty and his students. Some excerpts from Gårding's obituary (1970) give a fuller picture.

> As in Stockholm Riesz very soon came to play an important role for young mathematicians and physicists. He was always working on something but he was also generous with his time. He could talk for hours and liked to have an audience. When he was in good form his grip on the listeners never slipped. Many of them experienced the conversations with Riesz as life-giving injections. The problems appeared in a new light, they became purer and clearer and at the same time more important. A sense of dejection often comes close in mathematical research. The conversations with Riesz helped to keep this feeling at a distance. Riesz was not generous with ideas for research, but if somebody had a promising idea he did not spare his efforts.

Riesz's good international contacts broke Lund's relative isolation. Among others, his brother Fredric, Rolf Nevanlinna, Lars Ahlfors and Harald Bohr lectured in Lund. Lectures by John von Neumann before the war and by Laurent Schwartz shortly afterwards made a lasting impression on the listeners. The brothers Riesz liked each other and Riesz admired his older brother Fredric, who was one of the great names of functional analysis. They were outwardly similar but had different temperaments. Fredric appeared to be measured compared to the lively Marcel. When Fredric lectured in Lund, Marcel took upon himself to explain the salient points to the audience who he knew best. ... Riesz lived alone. He worked at home and liked to work late. His windows were lit long into the night. His long private lessons, sometimes called oral examinations for the licentiate degree, took place in his home or in his favourite café, Håkanssons. They could also take the form of telephone conversations. When he was busy with something that interested him he became enthusiastic and intense and then needed a conversation partner, something which led to many thought provoking telephone contacts. Riesz had many friends in his private life and his two daughters visited him regularly. He was widely read and his interests outside mathematics were literature and politics. Riesz had an exceptionally good memory and he could give penetrating analyses of people and events. This in combination with his ability to create an atmosphere of expectation made him a unique teller of stories. Unfortunately he never wrote his memoirs.

After he retired Riesz spent ten years in the United States as a research professor at the Universities of Chicago and Maryland. Afterwards he returned to Lund.

Zeilon

In 1926, when Zeilon applied for a position as professor at Lund University he could add papers in hydrodynamics to his earlier work (see pp. 141-145). The new papers were inspired by the work of Carl Wilhelm Oseen (1879-1944), who after studies in Lund (p. 180-81) had received the professorship of mathematical physics at Uppsala in 1909. Through his book *Hydrodynamik* (1927) Oseen was a known specialist in the field. To understand Zeilon's papers we must first review Oseen's theory.

Oseen's wake theory

The classical equations of motion in a friction free, incompressible fluid are

(1) $$u_t + (u, \text{grad})u = -\varrho^{-1} \text{grad}\, p, \quad \text{div}\, \varrho u = 0.$$

Here $u = u(t,x)$ with three components is the velocity of the fluid at time t at the point x, $p = p(t,x)$ is the pressure and ϱ the density. The last equation says that no fluid is added or disappears. In the sequel we assume that $\varrho = 1$.

In order to illustrate the difficulties of this system, consider the initial value problem of Burger's equation $u_t + uu_x = 0$, which may be seen as a simplified hydrodynamics in one dimension. If $u(0,x) = f(x)$, the solution $u = f(x_0)$ is constant on the lines $x = x_0 + f(x_0)t$. If $f(x)$ is not monotone, these lines will meet and the solution ceases to exist. The difficulty with the system (1) is the presence of the quadratic terms, but if these are neglected, nothing interesting remains. Another physical difficulty is the hydrodynamical paradox: the system says that a body moving in a fluid does not encounter resistance. Everyday experience, for instance with water, says that this is almost true only if the velocity is very small.

In the Navier-Stokes equations for incompressible and viscous fluids, the right side gets a damping term of order two,

$$(2) \qquad u_t + (u, \text{grad})u = -\text{grad}\, p + \mu \Delta u, \quad \text{div}\, u = 0,$$

where Δ is Laplace's operator and $\mu > 0$ a viscosity constant. These equations are supposed to be a good model for currents of moderate velocity.

In the stationary case, u_t vanishes and the local situation does not change with time. If the quadratic terms of (2) are neglected, the result is a linear system which Stokes used to compute the velocity distribution in a fluid under the influence of a sphere which moves with a constant velocity U. Stokes found that the fluid glides along the sphere but also that it is symmetric before and after the sphere, something which contradicts everyday experience, at least for boats in water.

Around 1910 Oseen started an ambitious project, accounted for in his book *Hydrodynamik* (1927), whose object was to solve the riddles of currents in fluids. He succeeded in constructing solutions of Stokes's problem where the movement has an unsymmetric 'wake' around the body and he could also perform a plausible passage to the limit for vanishing viscosity resulting in new boundary conditions at the moving body.

In view of the result by Stokes it seems natural to consider $u = U$ in Stokes's problem as a first approximation and put $u = U + v$. For simplicity let us assume that $U = (1, 0, 0)$. The equation for v with neglect of the quadratic terms is then

$$(3) \qquad \partial_1 v = -p_x + 2\kappa \Delta v, \quad \text{div}\, v = 0, \quad \kappa = 1/2\mu.$$

Hear ∂_k means differentiation with respect to $x_k, k = 1, 2, 3$; the center of the sphere is at the origin and v is the velocity of a fluid which moves around the fixed sphere and vanishes at infinity.

Oseen's calculations start from the system (3) where, in comparison with (2), the Laplace operator has been replaced by the operator

$$\Delta_1 = \Delta - 2\kappa \partial_1.$$

This operator has a fundamental solution

$$F(x) = e^{-\kappa(|x|-x_1)}/|x|, \quad \Delta_1 F(x) = -4\pi \delta(x),$$

which reduces to the potential $1/|x|$ when $\kappa = 0$ and is very small for large negative x_1 if $\kappa > 0$.

We can now solve the equation (3) with $p = 0$ and $x \neq 0$ if we put $v_k(x) = P_k(\partial) F(x)$ where P_k is a polynomial in the differentiation $\partial = \partial/\partial x$. A simple solution, which also has the property that $\text{div}\, v = 0$, is the following one

$$v_1 = F - \partial_1 F/2\kappa, \quad v_2 = -\partial_2 F/2\kappa, \quad v_3 = -\partial_3 F/2\kappa.$$

Adding a gradient $\text{grad}\, \varphi$ with $\Delta \varphi = 0$, the functions

$$v = A(UF - \partial_1 F/2\kappa) + B\, \text{grad}\, \varphi, \quad p = -B\partial_1 \varphi$$

solve (3) for all A, B when $x \neq 0$. With Oseen's choice

$$\varphi(x) = \frac{1}{|x|} + C\partial_1 \frac{1}{|x|}$$

and a small κ, i.e., a large μ, it was possible to choose $A > 0, B, C$ such that v is close to $-U$ at a small sphere $|x| = a$. When $\kappa = 0$ Oseen's solution reduces to that of Stokes, but in the general case Oseen's solution, in contrast to that of Stokes, is not the same before and after the sphere. Instead the velocity on the first axis is close to $AF(x) \sim A/|x|$ when x_1 is large and positive and to $\operatorname{grad} \varphi(x) = O(1/|x|^2)$ if x_1 is large and negative. More generally, the velocity has the order of magnitude $1/|x|^2$ far away from a 'wake' given by $|x| - x_1 \geq a$ which is the interior of a backwards directed paraboloid which surrounds the sphere and the negative x_1 axis.

With these formulas Oseen had enriched hydrodynamics by the first theoretical example of the common occurence of a wake around moving bodies in a fluid.

Oseen's theory got a great deal of attention (pages 609, 617, 688) in Lamb's encyclopedic book (1932). It also received verification *a posteriori* by later work with existence theorems for nonlinear equations. The first one to solve Stokes's problem was Leray (1933) and he was followed by, among others, R. Finn (1961), who proved that there is a time independent solution of (2) with square integrable derivatives outside of the moving body with a limit u_∞ at infinity. Babenko (1973), who considered Stokes's problem with a boundary value J at a body with a smooth boundary such that the out-flow vanishes, proved that $u - u_\infty = O(1/|x|)$ just as in Oseen's solution.

Oseen could also make the passage to the limit for other bodies than the sphere, in particular for a thin, circular disk which moves with a constant velocity. He found that the movement of the fluid is special in the wake, i.e., the volume covered by the moving disk. In the general case, when the disk is replaced by a body K, naturally convex in the direction U, the situation is as follows. There is a harmonic function outside of K which is small at infinity and such that $u = \operatorname{grad} \varphi(x)$ ouside of Ω. The fluid glides on the surface of the forward part of the body so that u and U have the same component in the normal direction of the body but inside the wake, one has

$$u = U + \operatorname{grad} \varphi(x) - \operatorname{grad} \varphi(x_K)$$

where x_K means the projection of x along U. Together with the condition that

$$\frac{d}{dn}(\operatorname{grad} \varphi, U) = 0,$$

this means that $u = U$ on the backside of the surface of the body so that the fluid is attached to the body. Over the boundary of the wake u has the jump $U - \operatorname{grad} \varphi(x_K)$, inside the wake the fluid velocity u has a rotation $\operatorname{rot} u = -\operatorname{rot} \varphi(x_K)$ which in general does not vanish. By the conditions above, the harmonic function $\varphi(x)$ is in general uniquely determined.

Zeilon's papers

Oseen's program for vanishing viscosity was carried out in detail by Zeilon (1923), both in two and three dimensions. In many cases Zeilon made explicit computations, for instance when K is a half-sphere which moves with the flat or curved side first. Comparisons with experiments showed that the theory can only be qualitatively correct.

In two appendices to Oseen's book (1927) Zeilon modified Oseen's theory in two ways. He treated cases when the unperturbed movement need not be constant, and he eliminated Oseen's discontinuity at the boundary of the wake. Both cases deal with the problem in a plane, i.e., the flow of a fluid around a cylinder. Zeilon's computations are here more convincing than strict.

In two variables the stationary equations (1) can be written as
$$u^* \operatorname{rot} u = -\operatorname{grad}(p + |u|^2/2) + \mu \Delta u, \quad \operatorname{div} u = 0,$$
where $u^* = (u_2, -u_1)$. Then the quantity $\operatorname{rot} u$ satisfies the equation
$$(u, \operatorname{grad}) \operatorname{rot} u = \mu \Delta \operatorname{rot} u.$$
If u is a given potential flow with $\operatorname{div} u = \operatorname{rot} u = 0$, induced by the cylinder, and we replace u by $u + v$ and neglect quadratic terms in v, the result is a linear equation
$$(u, \operatorname{grad})w = \mu \Delta w, \quad w = \operatorname{rot} v$$
for w, a function which, according to Oseen, should be $\neq 0$ in the wake of a body which, in the first approximation, is subject to a flow $-u$.

In order to analyze the wake, Zeilon changes variables. Since $\operatorname{grad} u$ and $\operatorname{rot} u$ both vanish, u is locally the gradient of a harmonic function U with a conjugate function V such that $\partial_1 U = \partial_2 V, \partial_2 U = -\partial_1 V$. Some computation shows that
$$((U, \operatorname{grad}) - \mu \Delta)w = 0$$
where Δ is referred to the coordinates U, V. Exactly as in the three-dimensional case this differential operator has a fundamental solution with pole at U_0, V_0 which tends strongly to zero when $U < U_0$. With vanishing viscosity μ, the wake will consist of regions $U < U_0, V = V_0$ where U_0, V_0 are the values of U, V at the boundary leeward of the flow U, V.

Zeilon illustrates this when the body is an infinite cylinder whose base is the unit disk $x^2 + y^2 < 1$ and lets the flow induced by the body have the form $u = \operatorname{grad} U$ where
$$U + iV = x + iy + iB\log(x + iy), \quad x + iy = re^{i\theta},$$
whence $U = Ax - B\theta, V = y + B\log r$ so that
$$u = (A - B\theta_x, B\theta_y).$$
This means that the cylinder moves with the velocity $(A, 0)$ at the same time that it rotates with angular velocity B.

When $A > 0, B = 0$, the wake is the projection of the body in the direction of the negative x_1-axis. This is Oseen's case. But if $B > 0$ is small it is easy to see that the wake turns in a positive direction and gets narrower far away. When $B \sim 3$ the wake has changed into a finite corona around the unit disk with its cusp in the direction of the negative y-axis, which, when B increases, shrinks to the unit circle. This phenomenon has been observed for air in which case it is called the Magnus effect. The air pressure on the cylinder in the direction of the wake attains almost ten times the pressure against the direction of movement. In the 1920's, when Zeilon wrote his papers, Germany had a so-called rotor ship, a boat with two large, rotating cylinders aimed at using the Magnus effect as a driving force. But the ship was not economical, and the tests were abandoned.

In Lund, Zeilon used his expert knowledge of hydrodynamics when Torsten Gustafson (1904-1987), professor of theoretical physics from 1938, wrote a thesis (1933) on the Magnus effect.

Zeilon combined his scientific activity with painting in a classic-realistic style. During his time in Lund Zeilon's health deteriorated and his artistic and scientific activity waned. His last paper (1946) is an improvement of a paper by Oseen on the Stark effect. Torsten Gustafson is the author of a very appreciative obituary (1958) of Zeilon where he writes about Zeilon's wide culture, his esthetic and humanistic interests and his firm position against the German Nazis during the war.

Mathematicians, theses, and papers 1925-50

Hössjer

During his time as a docent, from 1929 till 1936, Gustav Hössjer (1897-1977) wrote papers in which the general theme is boundary values of analytic functions. In 1936 he became professor of mathematics at Chalmers Technological Institute where he served as rector from 1943.

Hössjer's thesis (1929) starts with a quotation from Riemann in which he sees the boundary value problems for analytic functions in a region D in such a way that the value of unknown function $f(z)$ at a point p at the boundary is required to belong to a curve $T(p)$ in the complex plane. In the simplest case $T(p)$ is a straight line $\operatorname{Re} z = \varphi(p)$ parallel to the imaginary axis. This classical problem can also be formulated otherwise: find a harmonic function with given values at the boundary. If $u(z)$ is the solution, we may put $f(z) = u(z) + iv(z)$ where $dv = u_y dx - u_x dy$ is uniquely determined up to an imaginary constant.

Hössjer, who restricted himself to a simply connected region D, more specifically the unit disk, carried out Riemann's program in an able and imaginative way. His curves $T(p)$ are assumed to bound simply connected open sets $Y(p)$ which contain the origin and are bounded and uniformly continuous functions of p.

At first Hössjer lets all $Y(p)$ be convex and considers the set B of functions which are analytic in D and such that $f(p) \in Y(p)$ for all p. Then B is convex and also the restrictions $B(z)$ to points $z \in D$. The main result is that the extreme points of B give analytic functions which solve the boundary value problem: every such point is an analytic function $f \in D$ such that $f(z)$ is extreme in every $B(z)$, in particular in $B(p) = Y(p)$ so that $f(p) \in T(p)$. In addition, these extreme points are in one-to-one correspondence with the points of the unit circle. Similar results can be obtained when the condition of convexity is relaxed.

At the end of his thesis Hössjer found a connection with the theory of analytic functions of several complex variables. He proved that the region G in C^2 consisting of all z, w such that $w \in B(z)$ for all $z \in D$, belongs to the analytic hull of the subset Δ for which $z = p$ belongs to the boundary C of D and $w \in T(p)$. This means that an analytic function $f(z,w)$ in G which is bounded in Δ has the same bound in all of G.

One of Hössjer's results is an elegant extension of Rouché's theorem: two analytic functions f, g whose boundary values at every point $p \in C$ are inside or outside of $Y(p)$, have the same number of zeros in D. The proof is a simple topological argument.

A similar question has to do with the boundary values of an analytic function $f(z)$ which is bounded in an open connected sector V between two continuous simple curves C_1, C_2 meeting at a single point. Let Z_1, Z_2, W be the sets of limit points of f at a from, respectively, C_1, C_2, V. All three are closed and it is easy to see that they are intersections of closed, connected sets and hence themselves connected.

By a classical result by Lindelöf, the three sets coincide when Z_1, Z_2 are points. Hössjer proved that the two minimal, simply connected sets which contain Z_1 and Z_2 have a common point. He had then not observed a better formulation by Beurling (1933) which says that Z_1, Z_2 cannot be separated by a line through infinity.

In another paper (1935) Hössjer studies a simply connected Riemannian surface R with boundary whose projection is contained in the interior of the unit

Bergström

Viktor Bergström's thesis (1935) is a contribution to the theory of modules of finite dimension and diophantine approximation. It contains few new results but it is a thorough review of an interesting part of mathematics still of interest (see Hlawka (1984)).

Let R^n with elements $x = (x_1, ..., x_n)$ have the interior product $(x, y) = \sum x_k y_k$ and let Z be the set of integers. In this context, a module M is an additive part of R^n which admits multiplication by integers. A module M is said to be discrete when $M \backslash 0$ has a positive distance to the origin. It is easy to see that every module is the direct sum of a discrete module M_d and a non-discrete module M_i, defined as the intersection of the modules generated by those parts of M which have a distance at most $1, 1/2, 1/3, ...$ to the origin. The closed hull $\overline{M_i}$ of this module is a linear subspace L of R^n. The module M_d is the intersection of M and the orthogonal complement of L. If this module is not zero, it can be shown that it has the form

$$M_d = Ze_1 + ... + Ze_k$$

where $e_1, ..., e_k \in M_d$ generate the same linear space as M_d.

The dual M' of a module M is the module of all x such that (x, y) is an integer for all $y \in M$ (Riesz (1937c)). It is clear that M' is closed and the above structure property of M shows that $M'' = \overline{M}$. It is also immediate that $(M + N)' = M' \cap N'$, in particular,

$$(M + Z^n)' = M' \cap Z^n.$$

Hence, if the right side vanishes, the hull of the left side must be equal to R^n. From this follows a famous theorem by Kronecker (1884): if the coordinates of $y = (y_1, ..., y_n)$ are independent irrational numbers, i.e., no nontrivial linear combination of them with integral coefficients vanishes, then R^n is the closed hull of $Zy + Z^n$. Explicitly, the theorem says that to every $x \in R^n$ and $\varepsilon > 0$ there is an integer m and whole numbers $k_1, ..., k_n$ such that

$$|my_1 + k_1 - x_1| < \varepsilon, ..., |my_n + k_n - x_n| < \varepsilon.$$

An infinite sequence $c = (c_1, c_2, ...)$ of real numbers is said to be equally distributed modulo 1 if the remainders mod 1, $C_1 = [c_1], ..., C_N = [c_N]$, are more and more equally distributed in the interval $(0, 1)$ when N gets larger. As remarked by Weyl (1916) the sense of this is that, if $N \to \infty$, the means

(1) $$M_N(c, f) = N^{-1} \sum f(C_k)$$

tend to the integral

(2) $$\int_0^1 f(x) dx$$

for all continuous functions with the period one. Example: the sequence $c_k = ky$ is equally distributed mod 1 when y is an irrational number. In fact, it is immediate that the mean

$$\frac{1}{N} \sum_{k=1}^{N} e^{2\pi i m k y}$$

is 1 for $k = 0$, but tends to zero when $N \to \infty$ and $m \neq 0$ is an integer. This shows that the means (1) converge towards the integral when f is a trigonometric polynomial and hence also when f is a continuous function of period one.

In the second half of his thesis, Bergström writes about equal distribution mod 1 of modules in the sense that the coordinates of the elements have this property. Most of the results have to do with the velocity with which equal distribution is attained. In the final part the author treats results by Hardy and Littlewood which say that the sequence

$$(ky_1, k^2 y_2, ..., k^n y_n), \quad k = 1, 2, ...,$$

is equally distributed mod 1 when $y_1, ..., y_n$ are independent irrational numbers. He also comments on Weyl's extension (1916) of this result to sequences

$$kv_1 + k^2 v_2 + ... k^n v_n, \quad v_1, ..., v_n \in R^n,$$

when the closed hull of the module generated by $v_1, ..., v_n$ is all of R^n.

Frostman

One of Riesz's successful students was Otto Frostman (1907-1977), from 1953 a professor at Stockholm University. There are several reasons why Frostman's work in potential theory in the 1930's received attention. The theory was extended from the Newtonian case to Riesz's new potential kernels and the Lebesgue-Stieltjes integral was used systematically. Frostman's love of the subject and his talent as a writer also contributed.

If $U(x)$ is the Newtonian potential of a mass or electric charge K distributed in space then $-\operatorname{grad} U(x)$ is proportional to the force excerted on a small test body or test charge at the point x. This is the physical motivation of the notion of potential. By Newton's theory of gravitation and electrostatic theory, the potential has the following simple form

$$U(x) = \int_K \varrho(y) dy / |x - y|.$$

Here dy is the euclidean volume element, $\varrho(y)$ is the density of the mass or charge and $|x - y|$ the distance between x and y. By experience, bodies, surfaces, and lines can all carry electrical charge. This made it necessary for nineteenth century potential theory to study potentials of the corresponding densities. In his lectures (1899) on potential theory Poincaré had to treat the three cases separately. One of the first pages has the following statement: 'Consider now distributions of attracting masses; there are three kinds of continuous distributions, on bodies, on surfaces and on lines.' The word continuous here means of course restriction to continuous densities.

In Frostman's thesis (1935) potential theory is modernized in such a way that densities are replaced by distributions, more precisely countably additive functions $\mu(A)$ acting on Borel sets $A \subset R^m$. A measure is a distribution μ for which $\mu(A) \geq 0$ for every A. Every distribution has a support, defined as the complement of the largest open set in which the distribution vanishes. The support can be any closed set.

Modern integration theory entered into Frostman's arsenal in the form of the Lebesgue-Stieltjes integral. Every distribution μ definies a Stieltjes integral

$$\int f(x) d\mu(x)$$

of continuous functions $f(x)$ with compact supports and the corresponding generalized integrals which possess all the essential properties of Lebesgue integrals.

Let $K(x) \geq 0$ be a general potential kernel which is a point-wise bound of increasing continuous functions ≥ 0. Then, if μ is a measure with compact support, the convolution or potential

$$(1) \qquad U^\mu(x) = \int K(x-y) d\mu(y), \quad d\mu(x) \geq 0,$$

is a function of x which is semicontinuous from below and may have the value $+\infty$.

Frostman extended classical potential theory to potentials with Riesz's kernels

$$(2) \qquad K_\alpha(x) = |x|^{\alpha-m}, \quad 0 < \alpha \leq 2,$$

in R^m with $m = 3$. We shall consider the general case $m > 2$. By computation,

$$\Delta K_\alpha(x) = (\alpha - m)(\alpha - 2) K_{\alpha-2}.$$

Hence the bounds on α are chosen so that the kernel is subharmonic outside the origin and locally integrable. As in the classical case, the kernel is positive definite in the sense that the double energy

$$E_\alpha(\mu) = \int K_\alpha(x-y) d\mu(x) d\mu(y)$$

of a distribution μ is ≥ 0 and equal to zero only if $\mu = 0$, at least when the double integral is absolutely convergent. This was first proved by Riesz (1938 p. 5) but is also follows by taking the Fourier transform of the kernel. In fact, it is easy to see that

$$K_\alpha(x) = \text{const} \int e^{i(x,\xi)} |\xi|^{-\alpha} d\xi$$

which means that

$$\int K_\alpha(x-y) d\mu(x) d\mu(y) = \text{const} \int |M(\xi)|^2 |\xi|^{-\alpha} d\xi$$

where $M(\xi)$ is the Fourier transform

$$\int e^{i(x,\xi)} d\mu(x)$$

of the distribution $\mu(x)$.

Without outside influence, an electric charge μ on a conducting body is distributed in such a way that the (double) energy

$$(3) \qquad E(\mu) = \int U^\mu d\mu$$

is minimal. The potential is then constant inside the body and the charge is distributed on the surface of the body. The last statement excepted, Frostman's main theorem is a mathematical version of this physical fact. An earlier mathematical version, based on the existence of a minimizing distribution, was proved by Gauss, but because he only dealt with distributions with continuous densities, his proof was not complete. With these new tools Frostman could improve the variational principle used by Gauss.

To understand the proof, we need the concept of α-capacity: a closed set F is said to have positive α-capacity if there is a measure with support in F with finite energy (3) when the kernel is $K(x) = K_\alpha(x)$. When the value of the parameter α is clear from the context we write only capacity. The capacity of an open set is defined as the supremum of the capacities of its closed subsets. It is easy to see that capacity is a subadditive function on disjunct sets. Sets of capacity zero play the part of exceptional sets in the theory.

Frostman's main result is the following one.

THEOREM. *For every compact set F of positive capacity there is a unique equilibrium measure μ for which the energy $E_\alpha(\mu)$ has its least value M among all measures ν of total mass one, $\nu(F) = 1$. The corresponding equilibrium potential $U^\mu(x)$ is $\leq M$ everywhere and equal to M on F, except on a set of capacity zero.*

Remark. In the classical case $\alpha = 2$, the support of μ is contained in the boundary of F for the potential is harmonic in every open part of F.

Remark. The number $1/M$ measures the capacity of F. The theorem can be formulated otherwise: there is a measure on F whose potential is equal to 1 on F except for a set of capacity zero.

We shall sketch the proof which gives an idea of the precise world of potential theory. To begin with, there is a sequence of measures μ_k, $k = 1, 2, \ldots$ on F with $\mu_k(F) = 1$, such that

$$E(\mu_k) = \int U^{\mu_k} d\mu_k$$

converges to the largest lower bound M of the energy. We may also assume[6] that the sequence has a limit, a measure μ such that

$$\int f(x) d\mu_k(x) \to \int f(x) d\mu(x)$$

for all continuous functions with compact supports. In particular, $\mu(F) = 1$. But Fubini's theorem shows that

$$\liminf E(\mu_k) \geq E(\mu).$$

Hence μ has the energy M. What remains is to show is that μ is unique and that $U^\mu = M$ on F except for a set of capacity zero.

Consider the equation

$$E(\mu + h\nu) = E(\mu) + 2h \int U^\mu d\nu + h^2 E(\nu)$$

where $h > 0$ is small. We shall see that the following situation cannot occur: there are disjunct closed subsets F_1, F_2 av F such that $\mu(F_1) > 0$, F_2 has positive capacity and

$$\min_{F_1} U^\mu(x) > \max_{F_2} U^\mu(x).$$

In fact, under these circumstances there is a measure $\nu_1 \leq \mu$ with support in F_1 and a measure ν_2 with support in F_2 such that $\nu_1(F_1) = \nu_2(F_2)$, which implies that

$$\int U^\mu(x) d\nu(x) < 0, \quad \nu = \nu_2 - \nu_1.$$

Hence $E(\mu + h\nu) < E(\mu)$ for small $h > 0$ and a measure $\mu + h\nu$ of total mass 1, which is impossible.

The impossible situation above is certain to occur when $U^\mu(x) \leq M_1 < M$ in a part F_1 of F with positive capacity. For since $\int U^\mu d\mu = M$, there is certainly a part F_2 of F with $\mu(F_2) > 0$ where $U^\mu(x) > (M + M_1)/2$. Hence the potential U^μ is at least equal to M, except on a set of capacity zero. On the other hand. the potential cannot be larger than M on a subset G of F such that $\mu(G) > 0$.

---

[6] In contrast to measures, the space of continuous functions is not closed under reasonable limits.

It is possible to show that if the inequality $U^\mu > M$ holds at a point, it also occurs in a set of positive measure which is impossible.

Finally, if μ, μ' are two measures which realize the minumum, it is immediate that their difference has vanishing energy and hence itself vanishes. This proves the theorem.

Frostman's thesis also deals with an equilibrium problem under side conditions which can be formulated as follows where it is understood that the kernel of the potential is K_α.

THEOREM A. *Let $U^\mu(x)$ be the potential of a measure and let $F \subset R^n$ be a compact set of positive capacity. If the support of μ is outside of F, the expression*

$$\int_F (U^\nu(x) - 2U^\mu(x))d\nu(x)$$

attains its minimum for measures ν of finite energy and support in F. If ν is a minimizing measure, $U^\nu(x) - U^\mu(x)$ vanishes on the support of ν except on a set of capacity zero.

The proof is entirely analogous to the previous one.

Remark. If μ has finite energy, the problem is to minimize the energy of $\nu - \mu$ when ν is ≥ 0 on F and vanishes outside. The process of writing the restriction of the potential U^μ to F as the potential of a mass on F was called by Riesz and Frostman, after Poincaré, to sweep (*balayer*) the mass μ onto F. For Poincaré the term had had a real sense, but it was not clear in the new surroundings and therefore the two authors sometimes used the word *balayage* within quotation marks.

Remark. Frostman used the above theorem to prove a maximum principle which carries his name: The potential of a measure $\mu \geq 0$ attains its maximum on $F = \text{supp}\,\mu$. This is true for all kernels for which the equilibrium theorem holds. The proof depends on a simple identity which used the fact that the potential of a unit mass outside of F is majorized by the potential of a mass supported in F.

When $\alpha = 2$ in the theorem above, all integrals are harmonic outside of the supports of the corresponding measures so that maximum or minimum has to be attained at the supports. This proves that

(4) $$U^\nu(x) < U^\mu(x) \quad \text{outside } F$$

with equality on F except for a set of capacity zero. At the same time this shows that the the measure ν is unique. But the variational set-up of Theorem A does not permit this reasoning in the general case. It was Riesz (1938) who found a way out in the important case that

$$U^\mu(x) = K_\alpha(z - x)$$

is the potential of a unit mass at the point z. He introduced an inversion in the point z,

$$x' = z + \frac{x - z}{|x - z|^2}.$$

It is easy to see that

$$|x' - y'| = |x - y|/|x - z||y - z|.$$

Hence, if F' is the image of F, we have

$$\int K_\alpha(x'-y')|y-z|^{\alpha-m}d\nu'(y') = |x-z|^{m-\alpha}\int K_\alpha(x-y)d\nu(y)$$

where ν, ν' are corresponding measures on F and F' respectively. Now we need only choose $\nu = \nu_z$ such that $|y-z|^{\alpha-m}d\nu'(y')$ is a measure which gives the value 1 to the left side above when $x' \in F'$ except on a set of capacity zero. Such a measure exists by Frostman's first theorem. In this case $U^{\nu_z}(x) = \int K_\alpha(x-y)d_z(y)$ equals $|x-z|^{\alpha-m}$ on F except for a set of capacity zero and outside of F the sign of equality changes to a $<$. Hence Theorem A, enlarged by (4), holds with a unique ν_z for α-potentials when μ is a point-mass outside of F.

In a lengthy paper (1938) Riesz reaped the fruits of the enlarged theorem A. In analogy with the classical theory he named

$$G(z,x) = K_\alpha(z-x) - U^\nu(x), \quad U^\nu(x) = \int_F K_\alpha(x-y)d\nu_z(y)$$

Green's function with a pole at z of the complement of F. In fact, it vanishes on F except for a set of capacity zero, and $K_\alpha(z-x)$ is the dominating part at the pole. The measures $\nu_z(y)$ were named Green masses. Among the applications there are an extension of Harnack's inequality and the formula

(5) $$U^\mu(z) = \int U^\mu(y)d\nu_z(y),$$

which prolongs a potential U^μ of a measure μ of finite energy, supported on F, to points outside, a formula entirely independent of the measure μ. The proof is immediate by integration with respect to $d\mu(x)$ in

$$\int_F K_\alpha(x-y)d\nu_z(y) = K_\alpha(z-x).$$

The formula (5) is analogous to the solution of Dirichlet's problem in the classical case.

Riesz's paper was followed by a series of carefully worked out papers by Frostman about the details of the theory. Among other things he proved in (1939a) a complete version of Theorem A, i.e., $U^\nu \leq U^\mu$ with equality on F, except on a set of capacity zero, and a definition (1938) of the Green masses when the pole belongs to the set which carries these masses.

Let $G(z,y)$ be Green's function with Green masses ν_z on a closed and bounded set F of positive capacity. A point $y \in F$ is said to be regular if $G(z,y)$ tends to zero when z tends to y outside of F. This means that the Green masses ν_z then tend to a unit mass at the point y. A known criterion by Kellogg says that a point y is irregular if the equilibrium potential U^μ of the intersection F_r of F and a ball with center y and radius r has the property that $U^\mu(y) \to 0$ if $r \to 0$. Frostman's paper (1939b) has an alternative formulation:

(6) $$\int_0^1 c(r)r^{\alpha-m-1}dr < \infty$$

where $c(r)$ is the capacity of F_r. For $\alpha = 2$ this is a classical criterion by Wiener for y to be an irregular point for Dirichlet's problem with continuous boundary data on F. Frostman's simple proof rests on the identity

$$\int U^\nu(x)d\mu(x) = \int U^\mu(x)d\nu(x) = c(s)$$

where $U^\nu(x) = \int K_\alpha(x-y)d\nu(y)$ is the potential of the equilibrium distribution ν on F_s, $s < r$. The last equality sign follows since $U^\mu(x) = 1$ on F_r except on a set of capacity zero. Hence we have

(7) $$c(s) \leq \mu(F_{ks}) + \int_{F_r \setminus F_{ks}} U^\nu(x) d\mu(x)$$

for all $k \geq 1$ with equality when $k = 1$. On the other hand,

(8) $$U^\mu(y) = \int_0^r s^{\alpha-m} d\mu(F_s) = \mu(F_r) r^{\alpha-m} + (m-\alpha) \int_0^r \mu(F_s) s^{\alpha-m-1} ds,$$

By (7) with k=1, $\mu(F_s) \leq c(s)$ so that (8) shows that y is irregular if (6) converges. Conversely, if y is irregular, $U^\mu(r)r^{\alpha-m}$ tends to zero with r. If we choose r sufficiently small and, for instance, $k = 2$ in (7), it is easy to see that the second term on the right is at most $c(s)/2$ so that $c(s) \leq 2\mu(F_{2s})$ for all small s. Hence (6) converges.

A review of Frostman's works would be incomplete without the potential theoretic contributions to the theory of meromorphic functions which constitute the final part of his thesis. The most important ones all have to do with Rolf Nevanlinna's theory of meromorphic functions $w(z)$. In the version by Shimizu and Ahlfors the spherical metric is used,

$$|z, z'| = |z - z'|/\sqrt{1+|z|^2}\sqrt{1+|z'|^2}.$$

The main ingredients of the theory are the mean

$$m(r, a) = \frac{1}{2\pi} \int_0^{2\pi} \log \frac{1}{|w(re^{i\varphi}), a|} d\varphi$$

and the area

$$s(r) = \frac{1}{\pi} \int_{|z|<r} |w'(z)|^2 dx dy / (1+|w(z)|^2)$$

of the spherical image of the disk $|z| \leq r$ under $z \to w(z)$. In all this, a is a fixed complex number. Gauss-Jensen's formula with $|z-a|$ replaced by $|z, a|$ gives the following equality

$$n(r, a) + r\partial_r m(r, a) = s(r)$$

where $n(r, a)$ means the number of times w equals a in the disk and and the second term on the left side is nonnegative. If

$$T(r) = \int_0^r s(t) dt/t, \quad N(r, a) = \int_0^r n(t, a) dt/t + \text{const},$$

division by r and an integration gives the fundamental formula

(9) $$N(r, a) + m(r, a) = T(r)$$

where the right side is Nevanlinna's characteristic. That the right side is independent of a is the content of Nevanlinna's first main theorem. The number

$$\delta(a) = \liminf \frac{m(r, a)}{T(r)} = 1 - \limsup \frac{N(r, a)}{T(r)}$$

is called the defect of a for the function w. The number $1 - \delta(a)$ measures the part of the total growth $T(r)$ which depends on the number of times that $w(z)$

equals a. A vanishing defect means that the value a is assumed at a normal rate. When the defect is 1, the function w assumes the value a only exceptionally.

Frostman's potential theoretic contributions to Nevanlinna's theory all come out when the fundamental formula (9) is integrated with a measure $\mu(a)$ of total mass 1,
$$T(r) = \int N(r,a) d\mu(a) + \frac{1}{2\pi} \int U^\mu(w(re^{i\varphi})) d\varphi$$
where U^μ is the potential of μ with the kernel $\log 1/|w,a|$. Here the potential can be made to equal the equilibrium potential of different closed sets. This method permitted Frostman to prove that the boundary values of the function w in the disk $|z| < r$ constitute a set of positive capacity when $T(r) < \infty$. He could also show that the set of values a with positive defect have the capacity zero. Frostman's results form an essential complement to Nevanlinna's great theory.

Berg Gårding Fremberg Malmheden Lannér

Olof Thorin's contemporary Erik Berg did not write a thesis, but he wrote two number theoretic papers (1935, 1936). The first one received the honour of being quoted in v. d. Waerden's classic *Moderne Algebra I*. In his thesis (1947) which already has been quoted, Nils-Erik Fremberg carried out the analytic continuation of Riesz's hyperbolic potentials for the wave equation with constant coefficients.

Dirac's successes with his equation for the electron aroused the interests of physicists for two decades. The general requirements for similar systems of equations in vacuum were that they had to be linear with constant coefficients,
$$L(\partial)u(x) = 0$$
where $x \in R^n$, $\partial = \partial/\partial x$. The system should also be invariant under a group G of linear transformations g of ∂, corresponding to the Lorentz group. This means that there are two representations $g \to T(g), g \to S(g)$ such that $u(gx) = T(g)u(x)$ and
$$L(g\partial) = S(g)L(\partial)T(g)^{-1}.$$
Finally, corresponding to Dirac's factorization of the wave operator $\Box$, it was required that all components of the wave function should satisfy the wave equation $\Box u = 0$. This can be interpreted as the existence of a left inverse $M(\partial)$ with $M(\partial)L(\partial) = E\Box$, where E is the unit matrix. As formulated, this program gave no physical information and the interest waned with time.

In his thesis (1944), Lars Gårding took up the problem on a larger scale when G is one of the classical groups and their representation theory can be used. In his paper (1947) he constructed Riesz kernels and analytic continuation for the differential operators
$$\partial_x = \left(\frac{(1+\delta_{jk})}{2}\partial/\partial x_{jk}\right), \quad \partial_z = (\partial/\partial z_{jk}),$$
associated to symmetric and hermitian matrices of order $n \times n$ when, with a natural restriction, their elements are considered as independent variables. The operators $\det \partial_x$ and $\det \partial_z$ turned out to have fundamental solutions with supports in the set of matrices ≥ 0 of rank two and one, respectively, This gives instances of Huygens's principle where the dimensions of the supports are much less than the dimension of the surrounding space.

Harry Malmheden made his debut in (1934) with an elementary solution of Dirichlet's problem for balls. In his thesis (1947) he treated linear matrix-valued differential operators $L(\partial)$ which, by a left inverse, can be diagonalized to a multiple of the wave operator. Using Riesz's theory for the wave operator he could solve Cauchy's problem for the operator $L(\partial)$ plus lower terms with bounadry data on a time-like surface.

Folke Lannér's thesis (1950) is purely topological and was therefore outside Riesz's sphere of influence. His goal was to generalize Plato's regular bodies. This was carried out in the form of combinatorial complexes with a transitive group of symmetries. These groups were found among the Coxeter groups. Finally Lannér could realize his combinatorial complexes as geometrical complexes in euclidean space or on the sphere.

At the end of the 1940's Marcel Riesz had yet another student, Lars Hörmander (1931-), who very soon published papers in analysis and later became an outstanding mathematician.

BIBLIOGRAPHY

Riesz.

BOHR H.
1932. *Fastperiodische Funktionen*, Ergebnisse der Math. bd 1, 1932, no 5.
FREMBERG N.-E.
1946. *A study of generalized hyperbolic potentials*, Ak. Avh. Lund 1946.
FROSTMAN O.
1935. *Potentiel d'équilibre et capacité des ensembles avec quelques applications à la théorie des fonctions*, Ak. Avh. Lund 1935, Medd. LMS no 3..
GÅRDING L.
1970. *Marcel Riesz*, Kungl Fys. Sällsk. i Lund Årsbok 1970, 65-73.
HADAMARD J.
1923. *Lectures on Cauchy's Problem*, Yale 1923.
RIESZ M.
1924. *Les fonctions conjugées et les séries de Fourier*, C.R. Paris 178, 1924, 1464-1467.
1927. a. *Sur les maxima des formes bilinéaires et sur les fonctionelles linéaires*, Acta Math. 49, 1927, 465-497.
1927. b. *Sur les fonctions conjuguées*, Math. Zeitschr. 27, 1927, 218-244.
1931. *Sur certaines inégalitées dans la théorie des fonctions avec quelques remarques sur les géométries non-euclidiennes*, Fysiografiska Sällsk. i Lund Förhandl. I, 1931.
1933. a. *Sur les ensembles compacts de fonctions sommables*, Acta Szeged Sect. Math. 6, 1933, 136-147.
1933. b. *Zum Eindeutigkeitssatz der fastperiodischen Funktionen*, Fys. Sällsk. i Lund Förh. 3, 1933, no 10.
1936. a. *Volumes mixtes et facteurs invariants dans la théorie des modules*, Congr. int. math. Oslo 1936.
1036. b. *Modules réciproques*, Congr. int. math. Oso 1936.
1937. a. *Intégrales de Riemann-Liouville et potentiels*, Acta Szeg. Sect. Math. 9, 1937-38.
1937. b. *L'intégrale de Riemann-Liouville et le problème de Cauchy pour l'équation des ondes*, Conf. à la Réunion des Math. Paris Juillet 1937.
1937. c. *Modules réciproques*, C.R. Congres Int. des Math. Oslo 1937, II, 36-37.
1946. *Sur certaines notions fondamentales en théorie quantique rélativiste*, 10 Skand. Mat.-kongr. Köpenhamn 1946.
1949. *L'intégrale de Riemann-Liouville et le problème de Cauchy*, Acta Math. 81, 1949, 1-223.
1953. *L'équation de Dirac en relativité relativiste*, 12 Skand. Mat.-kongr Lund 1953.
THORIN O.
1938. *An extension of a convexity theorem due to M. Riesz*, Fys. Sällsk. Förh. 8, 1938, no 14.
1948. *Convexity theorems generalizing those of M. Riesz and Hadamard*, Ak. Avh. Lund 1948.

Zeilon.

BABENKO K.I.
1973. *On stationary solutions of the problem of flow past a body in a viscous fluid*, Math. USSR Sbornik vol. 20, 1973, no 1.
FINN R.

1961. *On the steady-state solutions of the Navier-Stokes equations III*, Acta Math. 105, 1961, 197-244.

GUSTAFSON T.

1933. *Über den Magnuseffekt nach der asymptotischen hydrodynamischen Theorie*, Ak. Avh. Lund 1933.

1958. *Nils Zeilon*, Kungl Fys. Sällsk. i Lund Förh. 28, 1958, 45-49.

LAMB H.

1932. *Hydrodynamics, sixth edition*, Dover 1945.

LERAY J.

1933. *Etudes de diverses équations intégrales non linéaires et de quelques problèmes que pose l'hydrodynamique*, J. Math. Pures et Appl. (9)12, 1933, 1-82.

OSEEN C.W.

1927. *Neuere Methoden und Ergebnisse in der Hydrodynamik*, Leipzig 1927.

ZEILON N.

1923. *On potential problems in fluid resistance*, KVA Handl. 3, 1923, bd 1, no 1.

Mathematicians 1930-1950

HÖSSJER G.

1929. *Über ein Riemannsches Problem in der Funktionentheorie*, Ak. avh. 1929, Lunds U. Årsskrift N.F. Avd 2. Bd 24 nr 9.

1934. *Über funktionstheoretische Nullmengen und das Maximumprinzip bei mehrdeutigen analytischen Funktionen*, Skand. Mat. Kongr. Stockholm 1934.

1935. *Über die Funktionszweige einer allgemeiner Klasse mehrdeutiger Funktionen*, Fys. Sällsk. i Lund Förh. 5 nr 4, 1935.

1936. *Bemerkung über einen Satz von Lindelöf*, Fys. Sällsk. i Lund Förh. bd 6 no 3, 1936.

BERGSTRÖM V.

Beiträge zur Theorie der endlichdimensionalen Moduln und der diophantischen Approximation, Ak. Avh. 1935, MLS bd 2.

BERG E.

1935. *Über die Existenz einer Euklidischen Algoritmus in quadratischen Zahlkörpern*, Fys. Sällsk. Förh. 5, 1935, no 5 MLS 4 no 2..

1936. *On a Theorem of Hardy and Littlewood concerning Diophantine Approximation*, Fys. Sällsk. Förh. 6, 1936, no 4, MLS 4 no 7.

FROSTMAN O.

1935. *Potentiel d'équilibre et capacité des ensembles avec quelques applications à la théorie des fonctions*, Ak. Avh. Lund 1937, LMS no 3.

1937. *La méthode de variation de Gauss et les fonctions sousharmoniques*, Acta Litt. Sci. Szeged 8, 1937, LMS bd 4..

1938. *Sur le balayage des masses*, Acta Szeged 9, 1938, LMS bd 4.

1939. a. *Les fonctions surharmoniques d'ordre fractionnaire*, Arkiv 26A, 1939, nr 16.

1939. b. *Les points irréguliers et le critère de Wiener*, Fys. Sällsk. Förh. 9 nr 2, 1939, LMS bd 4.

GÅRDING L.

1944. *On a class of linear transformations connected with group representations*, Ak. avh. Lund 1944, LMS no 6.

1947. *The solution of Cauchy's problem for two totally hyperbolic linear differential equations by means of Riesz integrals*, Ann. Math. 46.4, 1947.

LANNÉR F.

1950. *On complexes with transitive groups of automorphisms*, Ak. avh. Lund 1950, LMS no 11.

MALMHEDEN H.

1934. *Eine neue Lösung des Dirichletschen Problems für sphärische Bereiche*, Fysiogr. Sällsk. förh. 4, no 24, 1934, LMS 2 no 5.

1947. *A class of hyperbolic systems of linear differential equations*, Ak. Avh. Lund 1947, MLS no 8.

References.

NEVANLINNA R.

1953. *Eindeutige Analytische Funktionen, zweite Auflage*, Springer 1953.

BEURLING A.

1933. *Études sur un problème de majoration*, Ak. avh. Uppsala 1933.

HARDY G. LITTLEWOOD J.E.

1914. *Some problems of diophantine approximation*, Acta Math. 37, 1914, 155-239.

HLAWKA E.

1984. *The Theory of Uniform Distribution*, AB Ac. Publ. Reading 1984.
KRONECKER L.
1884. *Näherungsweise ganzzahlige Auflösung linearer Gleichungen*, Werke III:1, 47-110.
POINCARÉ H.
1899. *Théorie du potentiel Newtonien*, Paris 1899.
WEYL H.
1916. *Über die Gleichverteilung von Zahlen mod Eins*, Math. Ann. 77, 1916, 313-352.

CHAPTER 16

Uppsala 1930-1950

Nagell   Beurling

Wiman's chair of mathematics at Uppsala University was specified to algebra and number theory. When he retired in 1932 he was succeeded by the Norwegian number theorist Trygve Nagel (Nagell in Swedish) (1895-1988). Holmgren's successor, Arne Beurling (1905-1986), continued the Swedish analytical tradition.

Nagell

Analytic number theory had been represented at Stockholm University but otherwise Sweden had no tradition in number theory. Norway, on the other hand, had a number of mathematicians who, since the turn of the century, were successful in classical number theory, for instance Axel Thue (1863-1922), and Thoralf Skolem (1887-1963), Viggo Brun (1885-1970) and Atle Selberg (1917-).

Thue is known for the first sharpening of Liouville's result (p. 135) that an algebraic number, a root of an equation of degree n with integral coefficients, can not be approximated by infinitely many rational numbers $\frac{x}{y}$ such that

$$|\xi - \frac{x}{y}| \leq \text{const}/y^\mu$$

when $\mu > n$. In (1908) Thue improved this result to $\mu > n/2+1$, Siegel (1922) to $\mu > 2\sqrt{n}$. In the end Roth (1955) found the ultimate result $\mu > 2$. Skolem wrote both in logic and number theory, Brun improved the sieve of Erathostenes and proved that the sum of the inverses of the twin primes converges. Selberg, who became a permanent member of the Institute for Advanced Study, Princeton in the United States, worked in analytic number theory. He was first with an elementary proof of the prime number theorem and he did fundamental work in harmonic analysis with what is known as Selberg's trace formula.

Nagell, who started writing papers in the early 1920's, devoted to the theory of rational solutions of polynomial equations with rational coefficients.

Generalities about rational solutions

In the theory of rational solutions of polynomial equations with rational coefficients there is scarcity of general results and an surplus of problems. This situation dominates the entire theory and makes it difficult to review papers in the field in detail. When the problem is the existence of rational zeros of polynomials $f(x,y)$ in two variables, two polynomials are naturally equivalent when they are mapped to each other by a birational transformation $x, y \to x', y'$ such that x, y are rational functions of x', y' with rational coefficients and conversely.

A classical case with a classical solution is the set of Pythagorean numbers, integral solutions of $x^2 + y^2 = z^2$. These solutions are all known and lists of them have been found in tablets from the second Babylonian empire, around 600 B.C. In another formulation, the Pythagorean numbers are rational numbers x, y such that $x^2 + y^2 = 1$. From one solution, for instance $x = 1, y = 0$, the others may be deduced as follows. Every straight line

$$x = 1 + at, \ y = bt$$

with rational coefficients a, b meets the circle $x^2 + y^2 = 1$ in yet another rational point, determined by a linear equation in t. One gets $t = -2a/(a^2 + b^2)$, a formula which produces all Pythagorean numbers. The same argument shows that if $f(x, y)$ is a polynomial of degree two with rational coefficients and the curve has one rational point, then it has infinitely many which are everywhere dense on the curve.

A similar situation prevails for cubic curves, defined by an equation $f(x, y) = 0$ where f is an irreducible polynomial of degree three with rational coefficients. If the curve has a rational point P, one gets another one at the intersection of the curve and the tangent at P, and, if the curve has two rational points P, Q, a line through these two meets the curve in a third rational point $P \star Q$. In fact, this kind of a product makes the rational points on a cubic curve form a commutative group. If one rational point E is called the neutral element, the formula $P + Q = E \star (P \star Q)$ defines addition in the group. The rank of the group, i.e., the minimal number of generators, is a birational invariant. Mordell (1922) proved the important result that the rank is finite. In the proof, the curve is mapped birationally to a surface of the type $f_4(x, y) = f_2(z)$ where the indices denote homogeneity. Mordell proved that every sufficiently large entire solution is a rational function of a smaller one, which proves the theorem. The same kind of proof, called finite descent, was used by Fermat to prove that the equation $x^4 - y^4 = z^2$ has no integral solutions. The theory of rational points on cubic curves is one of the most studied fields of arithmetic (see for instance Lang (1978)).

For polynomials of degree > 3 the situation is more involved with very few general results. To begin with it is not the degree of the curve but rather its genus that is important. For a birational transformation may change the degree but does not change the genus. Generically, the genus of a curve of degree m is $(m-1)(m-2)/2$.

Poincaré proved that a curve of genus 1 with a rational point is birationally and rationally equivalent to a cubic. For irrational curves $f(x, y) = 0$ of higher genus it was conjectured by Mordell (1922) that there are only finitely many rational points. For equations of the type

$$f(x, y) = g(x, y)$$

where $f(x, y)$ is irreducible and homogeneous of degree n and g has a lower degree m, this follows from the results by Thue and Siegel in the following way. Large solutions x, y with coprime x, y can be chosen so that x/y approximates a zero ξ of $f(x, y) = 0$ and such that $y \to \infty$, $|y| \geq \text{const}\, |x|$. If η is the corresponding root of the equations above, it is easy to see that

$$\left| \eta - \frac{x}{y} \right| \leq \text{const}\, \frac{1}{|y|^{n-m}}.$$

According to Thue this is impossible if $m < n/2 - 1$, in particular if $m = 0$ and $n > 2$ and, according to Roth, if $n - m > 2$. If $n > 2$ and $f(x, y)$ is reducible but not a power, Runge (1887) and Skolem (1922) proved that the equation has finitely many rational solutions. Mordell's conjecture was proved by Faltings (1989).

Algebraic number theory

In its later stages, the theory of rational solutions of polynomial equations employed the theory of algebraic numbers whose basic terminology will be explained below.

An algebraic number α of degree n over a field K is a zero of an irreducible polynomial P of degree n with coefficients in K. Its conjugates are all the zeros $\alpha_1, ..., \alpha_n$ of P. The norm $N(\alpha)$ is the product of the conjugates and hence equal to $\pm$the constant term of P when the highest coefficient equals one. All linear combinations

$$h(\alpha) = a_0 + a_1\alpha + ... + a_{n-1}\alpha^{n-1}, \quad a_0, ..., a_{n-1} \in K,$$

constitute a field $K(\alpha)$, the field generated by α. In general it is assumed below that K is the field Q of rational numbers.

When $h(x)$ is a polynomial with coefficients in K, the number $h(\alpha)$ has the conjugates $h(\alpha_1), ..., h(\alpha_n)$, some of which may be equal. The discriminant of n numbers $\beta_1 = h_1(\alpha), ..., \beta_n = h_n(\alpha)$ in $K(\alpha)$ is the rational number

$$D(\beta_1, ..., \beta_n) = (\det(h_j(\alpha_k)))^2.$$

With α as above, $D(1, \alpha, ..., \alpha^{n-1})$ is the classical discriminant of the irreducible polynomial with the zero α.

An algebraic number α of order n is said to be an algebraic integer when it is a zero of a polynomial with integral coefficients and highest coefficient equal to 1. If its norm is one, i.e., the constant coefficient is ± 1, the number is said to be a unit. Under multiplication, all units of an algebraic number field constitute a commutative group which may be very large. All algebraic integers of an algebraic number field of degree n constitute a ring which has the form

$$Z\omega_1 + ... + Z\omega_n$$

where Z denotes the integers and $\omega_1, ..., \omega_n$ are algebraic integers which form a basis of the ring. The discriminant $D(\omega_1, ..., \omega_n)$, which is independent of the choice of basis, is called the discriminant of the ring.

Nagell's work

Nagell's predecessor was the Russian mathematician Boris Delauney. In papers since 1915, written in Russian, he had studied integral solutions of the equation $x^3 + dy^3 = 1$ where d is a noncubic integer. A paper in German (1928) has the complete result. His theorem says that the equation has at most one solution except the trivial one $x = 1, y = 0$. If x, y is a solution, then $\theta = x + y\sqrt[3]{d}$ is a unit of the field $Q(\alpha)$, $\alpha = \sqrt[3]{d}$, since its norm is $x^3 + dy^3$. The proof is a series of detailed verifications that only one unit has the same form as θ, i.e., has no term containing α^2.

Nagell's main result is an inventory of integral solutions of equations $f(x, y) = 0$ where $f(x, y)$ is a homogeneous cubic form with integral coefficients and negative discriminant. According to Thue's theorem, any such equation has only finitely many solutions. If these are known, a theorem by Lagrange makes it possible, at least in principle, to majorize the number of integral solutions of $f(x, y) = N$ where N is an integer.

In his paper (1928) Nagell found that the equation above has in general three solutions and in exceptional cases at most five. This happens when $f(x, y)$ is equivalent to the form $x^3 - xy^2 + y^3$. Since $f(x, y)$ may be factored into three linear factors, the proofs, like those of Delauney, rest on the existence of units in cubic number fields. Nagell's more general situation leads to several cases and more complicated proofs. In the review (1929) Nagell collected the most important results about rational solutions of equations of higher degree, including his own.

Nagell's papers (1930, 1931 a,b) are an attempt to treat elementary number theory in a constructive way by the introduction of algorithms for the construction of the units and the ring of integers of an algebraic number field. Nagell shows for instance that every such field of order n has finitely many rings with a given discriminant. When $n = 3$ the result is explicit: at most $10|D|$ rings with the discriminant D.

In the beginning of the 1930's Nagell had established himself in algebraic number theory, in particular the arithmetic theory of plane cubics of genus one, which was the subject of many of his papers. One of the papers (1934) gives algorithms to decide whether of not a given cubic has rational points.

When a cubic of genus one has a rational point, it can be written in Weierstrass's normal form in which rational points correspond to certain points in the period parallelogram. In the paper (1935) this situation is used for a proof that every rational point of finite order must have integral coordinates. This is an important result, known as the Nagell-Lutz theorem (see Lang (1978)), and extends to cubics over algebraic number fields.

A rational point P on a cubic gives rise to a sequence $F(P) = P, Q, \ldots$ of rational points where each point is the intersection of the curve and the tangent through the preceding one. A point P is said to be exceptional if $F(P)$ has finitely many elements. It can be shown that all exceptional points constitute a subgroup of the group of rational points. In (1935) and (1943) Nagell studied exceptional points on cubics over the rationals and algebraic number fields K. A method is sketched to determine groups of exceptional points when their orders are bounded. In the paper (1946) this work is carried out in detail for certain cubics over quadratic algebraic number fields.

Summary

By his later papers Nagell became a known specialist in the arithmetic of cubic curves of genus one over quadratic number fields, a much studied and still current field of research in algebraic number theory. In addition to his scientific work Nagell wrote a book on elementary number theory and a textbook of algebra.

Nagell's students

Before 1950, Nagell had three thesis students, Bergström (1937), Billing (1938), and Lind (1940). Billing and Lind wrote about rational points on cubic curves of genus 1. Bergström chose another arithmetic subject in class field theory. This is a complicated theory, founded by Hilbert, which deals with sets of ideals in algebraic number fields and resists every attempt at simple explanation. Afterwards, Bergström switched to mathematical statistics and got a chair in this subject at Chalmers Technological Institute in Gothenburg.

Beurling

Arne Beurling (1905 - 1986) studied in Uppsala and held a chair there from 1937. His thesis (1933) was followed by a number of noted papers in classical and harmonic analysis. In 1954 these made him Herman Weyl's successor at the Institute of Advanced Study, Princeton N.J. USA.

Beurling's thesis

In Uppsala Beurling had two teachers, Wiman and Holmgren, but it is only Wiman's papers on the maximal and minimal moduli, in particular Wiman's conjecture (p. 160), which seem to have made an impression on Beurling. Other similar problems which were not solved at the time were Denjoy's conjecture about asymptotic paths (p. 197) and Milloux's problem (1925): if $f(z)$ is analytic in the unit disk, if $|f(z)| \leq M$ everywhere and if $|f(z)| < m$ in at least one point on every circle $|z| = r < 1$, how large can $|f(z_0)|$ be at a given point z_0? All these problems could be reduced to constructions of harmonic majorants and harmonic measures (see p. 196). Beurling's thesis (1933) is a collection of inventive procedures to majorize harmonic measures by geometric methods. It is obvious that Denjoy's conjecture and Milloux's problem had been important driving forces.

Extremal distance and extremal length

The first main point of Beurling's book is a definition of *extremal distance*, a metric in simply connected domains which is invariant under conformal mappings. The construction starts from the the interior euclidean distance $\varrho(z, z_0, D)$ between two points z, z_0 in a region D. To make it invariant under rotation, dilation, and translation, this distance is divided by R_D where πR_D^2 is the euclidean area of D,
$$l(z, z_0, D) = \varrho(z, z_0, D)/R(D).$$

For a disk this is now a normalized euclidean distance. Finally, the extremal distance is
$$\lambda(z, z_0, D) = \sup l(z^*, z_0^*, D^*)$$
under all conformal bijections $z \to z^*$, $D \to D^*$.

By construction, the extremal distance is invariant under conformal maps. A classical object with the same property is Green's function $G(z, z_0, D)$. Beurling's first theorem says that the two are connected by the formula
$$e^{-2G} + e^{-\lambda^2} = 1.$$

The proof, which takes place in the unit disk, consists of some simple estimates. The second theorem is about the extremal distance from a point z to a part γ of the boundary S of D,
$$\lambda(z, \gamma, D) = \inf_{s \in \gamma} \lambda(z, s, D).$$

The following two important inequalities by Beurling show that a harmonic measure is majorized by effective minorants of the extremal distance,
$$\omega(z, \gamma, D) \leq e^{1-\lambda^2},$$
and
$$\omega(z, \gamma, D) > \frac{2}{\pi} e^{1-\lambda^2}$$
when γ is an arc on S. Here the extremal distance appears as a supple and effective tool. In his thesis Beurling used it to give a new proof of Denjoy's conjecture ((1933) pp. 104-106).

Ten years after the thesis Beurling improved his tool to a still better one, a similarly defined extremal length. The conformal invariance is now replaced by a variable surface density $\varrho(z) \geq 0$ in a given domain D with corresponding length and surface elements

$$\varrho|dz|, \varrho\,dx\,dy.$$

A conformal map $z \to w$ of D can then be transported back to D by a change of density to $\varrho|dz/dw|$. This metric gives a surface element

$$A(D,\varrho) = \int_D \varrho\,dx\,dy$$

and a length

$$L(\gamma,\varrho) = \int_\gamma \varrho|dz|$$

of a rectifiable arc γ. The extremal length of a family Γ of rectifiable curves γ is now defined by

$$\lambda(\Gamma) = \sup_\varrho \sup_{\gamma \in \Gamma} L(\gamma,\varrho)^2/A(D,\varrho).$$

The construction of extremal length depends only on the function ϱ and this makes it especially easy to find minorants. Beurling presented his construction to the Scandinavian Congress in 1946, but his talk was not printed. The properties of the extremal length and its use are described, for instance, in Ahlfors (1974), Chapter 4.

In their joint paper (1946) Beurling and Ahlfors used the extremal length to find conformal invariants of simply connected regions whose boundaries have been divided into n pieces. When $n = 4$ the extremal distance between two opposite sides is the sole conformal invariant. As normal form one can take a rectangle with a given ratio between the sides. The authors give a complete set of invariants in the case $n = 5$.

Beurling's lemma

The second main point of Beurling's thesis is a geometric majorant of certain harmonic measures, presented in a key lemma which dominates a large part of the thesis. It is clear that the lemma is geared to Milloux's problem. At the same time it minorizes the size of intervals where the minimal modulus of a power series is at least a fraction of the maximal modulus.

Let E be a collection of finitely many, closed and connected subsets of the unit disk $S : |z| < 1$ and let $\Omega(z,E)$ be a harmonic function in $S \backslash E$ which equals 1 on the unit circle C and vanishes on E. The problem envisaged is to majorize $\Omega(z,E)$ on circles $|z| = r$. Beurling's lemma has an explicit majorant of this kind. In a first step it is shown that

(1) $$\Omega(z,E) \leq \Omega(-|z|,E^*), \quad 0 \leq |z| \leq 1$$

where E^* is a circular projection $z \to |z|$ of E onto the interval $0 < z < 1$. Since the inequality (1) is invariant under rotations, it suffices to show that

(1') $$\Omega(-r,E) \leq \Omega(-r,E^*), \quad 0 \leq r \leq 1.$$

The set E^* consists of finitely many intervals

$$I_k : r_k \leq x \leq r'_k, \quad k = 1,...,n,$$

ordered so that $I_1 < I_2 < ....$

We shall sketch Beurling's proof of (1'). It uses the function $U(z,E) = 1 - \Omega(z,E)$ which vanishes on the unit circle C and equals 1 on E. It can be seen as the equilibrium potential of a distribution $\mu(w)$ on E whose kernel is Green's function $G(z,w)$ in the unit disk,

$$U(z,E) = \int_E G(z,w)d\mu(w), \quad G(z,w) = \log|1-\bar{z}w|/|z-w|.$$

In fact, this potential is harmonic in $S\setminus E$, equals 1 on E and 0 on the unit circle C.

Now we note the elementary inequalities

(2) $\quad G(-r,w) \leq G(-r,z) \leq G(-r,-w) = G(r,w), \quad r, w > 0, |z| = w.$

If $w \to h(w)$ is a measurable map of $F = E^*$ with $|h(w)| = w$ and μ_F is the equilibrium measure on F, an integration shows that

$$U(-r,F) \leq \int G(-r,h(w))d\mu_F(w) \leq \int G(-r,-w)d\mu_F(w) = U(r,F) \leq 1.$$

Here $V(z) = \int G(z,h(w))d\mu_F(w)$ is harmonic outside $h(F)$, zero on C and ≤ 1 for $G(z,h(w)) \leq G(|z|,w)$. Hence $V(z) \leq U(z,E)$ when $h(F) \subset E$ and it follows that $U(-r,E^*) \leq U(-r,E)$ which is the same as (1').

The set E^* is a union of intervals $I_1, ..., I_n$. By a dilation I_1 to aI_1 with $a > 1$ the interval aI_1 can be made to border on I_2. Repeated operations of this kind transform E^* to a single interval $J = (s,1)$ where

$$s = \prod_1^n \frac{r'_k}{r_k},$$

which means that the logarithmic measures

$$m_l(E^*) = \sum(\log r'_k - \log r_k), \quad m_l(J) = \log 1/s$$

are equal. By a simple argument Beurling shows that $U(0,E^*) \leq U(0,J)$. Hence

$$\Omega(0,E) \leq \Omega(0,J).$$

Here the right side is the value at $z = 0$ of the harmonic measure $\omega(z,C,D)$ where C is the unit circle and D is the unit disk cut along J. This measure is computed in an appendix at the end of this chapter and gives Beurling's estimate

$$\Omega(0,E) \leq \frac{4}{\pi}\arctan e^{-m_l(E^*)/2}$$

for $s = e^{-m_l(E^*)}$. A conformal map of S which maps the point 0 to $-r$ and some additional computation shows that

(3) $\quad\quad\quad\quad\quad\quad \Omega(-r,E) < 2e^{-m_l(E^*)/2}.$

It is obvious that this estimate when transported to arbitrary circles can minorize the logarithmic measure of intervals where the minimal module has a given growth. For entire functions of order less than 1/2 this sheds new light on Wiman's theorem (see Kjellberg (1948)).

Milloux's problem

Beurling's lemma (1) immediately solves Milloux's problem: if $f(z)$ is analytic in the unit disk, if $|f(z)| \leq M$ on the unit circle and. finally, if $|f(z)| \leq m$ on a curve E from the origin to the boundary, then

$$\log \frac{|f(z)|}{m} \leq (\log \frac{M}{m})\frac{4}{\pi} \arctan \sqrt{\frac{1-|z|}{(1+|z|}}.$$

In fact, the left side is subharmonic, ≤ 0 on E and $\leq \log M/m$ when $|z| = 1$ and, by the appendix, the remainder of the right side is the harmonic measure at the point $-|z|$ of the unit circle in a domain consisting of the unit disk cut along the interval $0 \leq z \leq 1$. In another form this inequality says that

$$|f(z)| \leq m(M/m)^{\frac{4}{\pi} \arctan \sqrt{(1-|z|)/(1+|z|)}}.$$

Apart from the two main points Beurling's thesis contains many applications, variants and inventive remarks. Ahlfors (1930) and Carleman (1933) came before Beurling with proofs of Denjoy's conjecture and Nevanlinna (1933) with a solution of Milloux's problem. In this way Beurling could not really reap the fruits of his work. It is clear that this he was troubled by this. In both cases he remarks that he had the proofs already in 1929.

Complex analysis

Exceptional sets, outer capacity

In (1940) Beurling studied exceptional sets for convergence of a class S of Fourier series

$$f(\theta) \sim \sum_{-\infty}^{\infty} a_n e^{2\pi i n \theta}$$

such that $a_0 = 0$ and

$$\sum |n||a_n|^2 < \infty.$$

This class is a complex Hilbert space with a natural interior product (f, g). Every function f in the class is the boundary value at $r = 1$ of the harmonic function

$$f(r, \theta) = \sum r^{|n|} a_n e^{2\pi i n \theta}$$

in the unit disk. The class is related to the logarithmic potential

$$u(z) = \int \log \frac{1}{|z - e^{2\pi i \theta}|} d\mu(\theta)$$

of a measure $d\mu$ on the unit circle. The connection is given by the identity

(1) $$(f, u) = \sum a_n \overline{c_n} = \int f(\theta) d\mu(\theta)$$

where

$$c_n = \int_0^{2\pi} e^{-2\pi i n \theta} d\mu(\theta)$$

are the Fourier coefficients of $\mu \geq 0$. It is easy to see that c_n/n, $n \neq 0$ are the Fourier coefficients of the logarithmic potential restricted to $|z| = 1$.

The norm square of the logarithmic potential,

$$(u,u) = \int u(\theta)d\mu(\theta),$$

defines an energy of the measure μ and hence a capacity $c(E)$ of closed subsets E of the unit circle given by $1/M$ where M is the infimum of the energy (u,u) for measures μ with supports in E and of total mass 1. The capacity of an open set is defined as the upper bound of the capacities of the closed sets which it contains.

If E has positive capacity then a function $f \in S$ cannot be positive and very large on E for if it is $> m$ on E, (1) shows that

$$m^2 \leq (f,f)(u,u).$$

As yet this says nothing about convergence, but Beurling proved that the function f in this inequality can be replaced by the function

$$F(r,\theta) = \int_0^r |f_t(t,\theta)|dt$$

which turns out to be subharmonic with $(F,F) \leq (f,f)$. Since F increases with r, the function $F(1,\theta)$ is continuous from below so that $F(1,\theta) > m$ defines an open set. Hence $F(1,\theta) < \infty$ means that the limit

$$\lim_{r \to 1} f(r,\theta)$$

exists and is finite and this happens in particular when the Fourier series of f converges at θ. This gives the main result of the paper: When $f \in S$, the limit above exists except for a set D of logarithmic capacity zero. More precisely, D is contained in a decreasing sequence of open sets whose capacities tend to zero. Beurling says that D has outer capacity zero which is a sharper measure than just capacity zero.

In a note at the end of this paper Beurling hints that corresponding results hold when the factors $|n|$ of the interior product are exchanged for $|n|^\alpha, 0\alpha < 1$. The proof was carried out in detail by A. Broman (1947).

Outer and inner functions

In one of Beurling's best-known papers (1949 b) he created the notions of outer and inner functions. The background is a known factorization of functions in the Hardy class $H^p, p > 0$, consisting of functions which are analytic in the unit disk and have the property that

$$\limsup_{r \to 1} \int |f(re^{i\theta})|^p d\theta < \infty.$$

Every function $f(z) \neq 0$ in this class is a product $f(z) = B(z)C(z)D(z)$ of three factors. The factor $B(z)$ is a Blaschke product

$$B(z) = z^m \prod \frac{\overline{a_k}}{|a_k|} \frac{a_k - z}{1 - \overline{a_k}z}$$

with $0 \neq |a_k| < 1$ and $\sum(1 - |a_k|) < \infty$ so that $\prod |a_k| < \infty$, the function $C(z)$ is analytic without zeros in the unit disk,

$$C(z) = \exp\left(-\int \frac{e^{i\theta} + z}{e^{i\theta} - z} d\mu(\theta)\right),$$

where $\mu \geq 0$ is a singular measure and $D(z)$ is a function of the same form with an absolutely continuous measure $d\nu(\theta)$. The product is unique modulo constant factors, the factors $B(z)$ and $C(z)$ are of absolute value 1 almost everywhere on the unit circle. Beurling called $D(z)$ the outer factor of $f(z)$ and $g(z) = B(z)C(z)$ the inner factor.

In the introduction of (1949 b) the reader meets a bounded operator T on a Hilbert space H and its adjoint $S = T^*$. The problem is to characterize the closed hulls of $C_f = (f, Tf, T^2f, ...)$ and $C_g^* = (g, Sg, S^2g, ...)$ with given $f, g \in H$. A couple of extra assumptions reduce the situation to the following one: H has an orthonormal basis $e_0, e_1, ...$ such that T and T^* are shift operators,

$$Te_k = e_{k-1}, Te_0 = 0, \quad T^*e_k = e_{k+1}.$$

Then H can be identified with the space of analytic functions in the unit disk with square integrable boundary values and an interior product

$$(f, g) = \int_0^{2\pi} f(e^{i\theta})\overline{g(e^{i\theta})}d\theta$$

so that

$$Tf(z) = (f(z) - f(0))/z, \quad T^*f(z) = zf(z).$$

The eigenfunctions of T have the form $1/(1-\lambda z)$, $|\lambda| < 1$ where λ is the eigenvalue. Beurling's main result says that the hull of $C_f^* \neq 0$ has the form Hg where g is the inner factor of f, a result which can be extended to the Hardy class H^p with $p > 1$.

Halmos (1959) has remarked that Beurling's theorem for H^2 can be somewhat extended and that the proof can be conducted entirely in H^2. We shall reproduce both the theorem and its proof.

THEOREM. *Any closed subspace $M \neq 0$ of H which is invariant under multiplication by bounded functions has the form*

$$M = Hg$$

where g is an inner function.

PROOF: If $zM = M$ then $z^n M = M$ for all $n = 1, 2, ...$. This means that every $f \in M$ has a zero of infinite order at 0, which is impossible since $M \neq 0$. Hence zM has an orthogonal complement $N \neq 0$ in M so that $(N, z^n M) = 0$ for all $n > 0$. In particular, $(g, z^n g) = 0$ when $0 \neq g \in N$ which means that

$$\int e^{-in\theta}|g(e^{i\theta})|^2 d\theta = 0, \quad n > 0.$$

A conjugation gives the same equality for $n \neq 0$ so that $|g(z)|$ is a constant almost everywhere on the unit circle. We may then assume that $|g(z)| = 1$ almost everywhere, i.e., g is an inner function. It follows that

$$g, zg, z^2g, \ldots$$

is an orthonormal system. We shall see that it is complete in M. In fact, if $f \in M$ is orthogonal to this system, then $(f, z^n g) = 0$ for all $n \geq 0$. On the other hand, all $z^m f$, $m \geq 0$ also have this property so that $(z^{m-n}f, g) = 0$ for all $m > 0, n \geq 0$. Hence $f\bar{g} = 0$ on the unit circle and f must vanish. Since

$$(P(z)g, P(z)g) = (|P(z)|^2 g, g) = (P(z), P(z))$$

for every polynomial $P(z)$, Hg is closed and equals M.

It follows from the theorem that if f is an outer function and g is the corresponding inner function, then the hull of Hf equals Hg. With practically the same proof, the theorem also holds when the elements of H are analytic functions whose values are operators of the Hilbert-Schmidt type on a Hilbert space h. In this case the function $g(z)$ is to be replaced by a function $G(z)$ such that $G^*(z)G(z)$ is a fixed orthogonal projection on h when $|z| = 1$.

Beurling's primes

Beurling's perhaps most inventive paper is an analysis (1937) of the prime number theorem

$$\pi(x) \sim x/\log x, \quad x \to \infty,$$

where $\pi(x)$ denotes the number of primes at most equal to x. The idea is to show that the classical analytical proof of this theorem depends very little on the arithmetic properties of the primes. Below, after a sketch of the classical proof, follows a sketch of Beurling's theorem.

Riemann's ζ-function is defined by the formula

$$\zeta(s) = \sum_1^\infty n^{-s} = 1/\prod(1 - p_n^{-s}), \quad \operatorname{Re} s > 1,$$

where the product runs over the primes 2,3,5,... . From this follows that

(1) $$\log \zeta(s) = \sum_{n,m} p_n^{-sm}/m = \int_1^\infty x^{-s} d\Pi(x)$$

where

$$\Pi(x) = \pi(x) + \pi(x^{1/2})/2 + \pi(x^{1/3})/3 + ...$$

so that $\Pi(x) - \pi(x) = O(x^{(1+\varepsilon)/2})$ for every $\varepsilon > 0$ since $\pi(x) = O(x)$. Hence, if the desired order of magnitude of $\pi(x)$ is $x/\log x$, it suffices to work with $\Pi(x)$. Differentiation of (1), a change of variables $x = e^t$ and an integration by parts gives

$$f(s) = -\zeta'(s)/s\zeta(s) = \int_0^\infty e^{-ts}\Pi(e^t)dt.$$

In this situation, where $\Pi(x)$ is an increasing function and the integral converges when $\operatorname{Re} s > 1$, a theorem by Ikehara says that $e^{-t}\Pi(e^t) \sim A$ for large t provided that

$$f(s) - \frac{A}{s-1}$$

has a locally uniform limit when $\operatorname{Re} s$ decreases to 1. Hence, in order to prove the prime number theorem, it suffices to verify this last property which in turn follows if $\zeta(s)$ is regular analytic on the line $\operatorname{Re} s = 1$, apart from a pole at $s = 1$. The behavior at the pole determines $\Pi(x)$ for large x.

In his paper (1937) Beurling gives sharp variants of Ikehara's theorem under suitable assumptions about $\zeta(s)$, which no longer represents Riemann's ζ-function but a function $\zeta(s)$ which is analytic for $\operatorname{Re} s > 1$ and has the form

$$\log \zeta(1+s) = \int_0^\infty e^{-sy}e^{-y}d\Pi(e^y), \quad \operatorname{Re} s > 0,$$

where $\Pi(x)$ is non-decreasing. Beurling's method is to pass to harmonic functions in the upper half-plane, more precisely to

$$v(\sigma, t) = \arg \zeta(1 + \sigma + it)$$

and its derivative $u = v_t$. The result is

(2) $$u(\sigma, t) = \int_0^\infty e^{-\sigma y} \cos ty \, d\mu(y),$$

where $d\mu(y) = ye^{-y} d\Pi(e^y)$. Note that $u(+0, t) = \pi \delta(t)$ when $d\Pi(x) = dx/\log x$.

In this new situation, the problem is to find conditions on u at $t = 0$ which bound the growth at infinity of $\mu(x)$. The basic requirement that $v(+0, t)$ has a bounded variation in an interval around 0 entails that $\mu(x) = O(x)$. A sharpening of the behavior of $v(+0, t)$ at $t = 0$ gives more precise results which cannot be given in a short review. The proofs use the following identity which had been used in similar situations by Wiener

$$2 \int_{-\infty}^\infty e^{ixt} h(t/\lambda) u(\sigma, t) dt = \int_{-\infty}^\infty \lambda H(\lambda(x-y)) e^{-\sigma|y|} d\mu(y).$$

Here $\mu(-y) = -\mu(y)$, $h(t)$ is a non-negative function with support in $|t| \leq 1$, H is its Fourier transform and $\lambda > 0$ is a parameter.

Beurling's general ζ-function can have the form

$$\zeta(s) = \sum x_k^{-s} = 1/\prod(1 - y_k^{-s}), \quad \text{Re}\, s > 1$$

where $y_1, y_2, \ldots$ are artificial primes whose products constitute the similarly artificial integers $x_1, x_2, \ldots$. If $N(x)$ is the number of x_k which are $\leq x$ and $\Pi(x)$ the number of y_k which are less than $\leq x$, Beurling shows for instance that $\Pi(x) \sim x/\log x$ if

$$N(x) = Ax + O(x/\log^\gamma x)$$

and $\gamma > 3/2$ but not if $\gamma = 3/2$. The announcement of this and similar results constitute an impressive introduction to the paper.

Spectral analysis

Beurling wrote his best-known papers in harmonic analysis, for instance the paper on outer and inner functions. His first paper in the field, a lecture (1938) at the Scandinavian congress of mathematicians in Helsingfors, was inspired by Wiener's famous paper (1932) on Tauberian theorems.

Wiener's simplest theorem says that if

$$f(t) = \sum_{-\infty}^\infty e^{int} a_n$$

is an absolutely convergent power series, $\sum |a_n| < \infty$, and $f(t)$ never vanishes, then also $1/f(t)$ has an absolutely convergent power series. Analogously, if

$$f(t) = \int_{-\infty}^\infty e^{ixt} F(x) dx$$

is the Fourier integral of an integrable function F and $f(t)$ never vanishes, then all linear combinations of the translates of F are dense in the space of integrable functions. One of the applications is the following so-called Tauberian theorem: if $g(x)$ is a bounded function and there is a number A such that

$$\int F(x-y)g(y)dy \to A \int F(x)dx, \quad x \to \infty,$$

for a function F as above, then the same is true for all integrable functions F. A suitable choice of such functions then shows that $g(x) \to A$ when $x \to \infty$ except on a null set.

Starting from Wiener's theorems, Beurling studied corresponding results when the space L of integrable functions is replaced by the space V of functions F of bounded variation on the real line or, rather, measures dF on the line of finite total mass,

$$\| dF \| = \int |dF(x)| < \infty.$$

According to Lebesgue every F is a unique sum $F_a + F_d + F_s$ of three measures which are, respectively, absolutely continuous, discrete and singular. The corresponding subsets of V are denoted by V_a, V_d, V_s of which V_a and V_s are rings without a unit.

With the norm above, the space V is a Banach space with a commutative and associative convolution,

$$dF * G(x) = \int dF(x-y)dG(y)$$

which means that

$$\int h(x)dF * G(x) = \int h(x)dF(x-y)dG(y)$$

when h is continuous and has compact support. The function $E(x)$ which is 1 when $x > 0$ and zero otherwise is the unit of the convolution $F * G(x) = \int F(x-t)G(t)dt$ with the property that

$$\| F * G \| = \| F \| \| G \|.$$

In the sequel, the product $F * ... * F$ with n factors is written as F^n. Every $F \in V$ has a uniformly continuous Fourier transform

$$f(t) = \int e^{ixt} dF(x).$$

We define the spectrum $S(F)$ of F as the range of its Fourier transform. Let T be the space of Fourier transforms of V

The first main item of Beurling's congress lecture is the theorem that

(1) $$\sup |f(t)| = \lim_{n \to \infty} \sqrt[n]{\| F^n \|}$$

when $F \in V_a + V_d$. The proof is a direct computation. As in Wiener's paper this can then be used to show that F is invertible in V when $|f(t)|$ has a positive lower bound. Beurling also states without a proof that if $g(z)$ is analytic in a neighborhood of $S(F)$, then $g(F)$ belongs to V.

One of Beurling's notions, which we shall only define, is the minimal extrapolation. A function $g \in T$ of minimal norm $\| dG \|$ which equals another function $f \in T$ on a closed set E is said to be a minimal extrapolation f_E of f outside of E. The minimum is realized and the norm of the minimal extrapolation is a norm on the space of restrictions of T to E.

Wiener's and Beurling's papers can be seen in the light of Gelfand's theory (1941) of normed rings, i.e., complex Banach spaces $R = (a, b, \ldots)$ with norm $\| a \|$ and a commutative and associative multiplication $a, b \to ab$ such that $\| ab \| \leq \| a \| \| b \|$. In addition it is required that the ring has a multiplicative unit e of norm 1. In such a ring there are maximal ideals $M \neq R$ which may be identified with homomorphisms $a \to a(M)$ from R to the complex numbers such that $|a(M)| \leq \| a \|$. The set $S(a)$ of all $a(M)$ with varying M is called the spectrum of a. An element x which does not belong to any maximal ideal M, i.e., such that $x(M) \neq 0$ for all M, turns out to be invertible.

More generally, if z is outside of $S(a)$ the formula

$$\frac{1}{2\pi i} \int_{\partial D} f(z) dz/(a - z),$$

where D is an open set containing $S(a)$ and $f(z)$ is analytic in $\overline{D}$, defines an element $f(a) \in R$ such that $fg(a) = f(a)g(a)$ and $f(a) = 1$ when $f(z) = z$. In particular, if $a = 0$ and $f(z) = 1/z$ then $f(a) = a^{-1}f$.

The abstract counterpart of Beurling's theorem (1) is the well-known formula for the spectral radius $\varrho(a) = \max |a(M)|$, i.e.,

$$\varrho(a) = \lim \sqrt[n]{\| a^n \|},$$

now obtained as a consequence of the fact that $1/(a-z)$ is analytic when $|z| > \varrho(a)$.

The general results about normed rings imply Wiener's theorem for Fourier series[1] and Beurling's theorem above. In fact, a maximal ideal of the ring $R = V_a \cup V_d$ consists of all F for which the Fourier transform f vanishes at some point t. To see this, consider the elements $E_a(x) = E(x - a)$. They are invertible and have the property that $E_a * E_b = E_{a-b}$ which means that every homomorphism φ of E_a to the complex numbers has the property that

$$\varphi(E_a) = e^{ita}$$

for some real t so that

$$\varphi(\sum k(a) E_a) = \sum k(a) e^{ita}$$

for finite sums with complex coefficients $k(a)$. With such sums it is possible to approximate the Fourier transform $f(t)$ of $F \in V$ when F does not have a singular component. Hence

$$\varphi(F) = f(t).$$

Wiener and Pitt (1938) and Beurling (1938) proved simultaneously a theorem which includes the singular component F_s: F is invertible when $|f(t)| \geq c > 0$ and $\| F_s \| < c$.

In the second part of his congress lecture Beurling introduced weight functions $p(x) \geq p(0) = 1$ such that

$$p(x + y) \geq p(x)p(y), \quad p(rx) \geq p(x), \ r > 1.$$

---

[1] The theorem for Fourier integrals does not follow immediately for lack of a unit.

It follows from these properties that the numbers

$$\alpha = \lim \log p(x)/x,\ x > 0, \quad \beta = \lim \log p(x)/|x|,\ x < 0$$

exist, possibly as $\pm\infty$, but both these cases cannot occur simultaneously. With the norm $\| F \|_p = \int p(x)|dF(x)|$ for measures and

$$\| F \|_p = \int p(x)|F(x)|dx, \quad \| F \|_{1/p} = \int |F(x)|dx/p(-x)$$

for locally integrable functions, Beurling obtained new Banach spaces and convolution rings denoted by, respectively, $V_p, L_p, L_{1/p}$ where the last two do not have units. The Fourier transform

$$f(t) = \int e^{ixt} dF(t), \quad F \in V_p$$

and a corresponding one for L_p are continuous in the region $D_p : \alpha \leq \operatorname{Im} t \leq \beta$ and analytic in the interior. When D_p is the line and

$$\int \log p(x) dx/(1+x^2) < \infty,$$

the class L_p is said to be nonquasianalytic.

In this new situation Beurling proved an extension of (1),

(1p) $$\sup_{D_p} |f(t)| = \lim \sqrt[n]{\| F^n \|}$$

when F does not have a singular part. After this he added a number of extensions and variants of Wiener's Tauberian theorem: if $f(t) \neq 0$ for all t and the class L_p is not quasi-analytic, then $F * M$ is dense in M when $M = L_p$ or $L_{1/p}$ or if $M = C_p$, a space defined by the property that

$$\sup |G(x)|/p(-x) < \infty$$

and that the translation $G(x) \to G(x+h)$ is continuous in this norm.

Beurling's paper (1938) is a massive effort that shows the author's ability to see the essential content of a theory and to find variations. The theorems (1) and (1p) are entirely Beurling's own. They appeared later in Gelfand's striking theory (1941 a) of normed rings where the proofs are simpler and the general impact bigger. Beurling may have been the first one to introduce weight functions for convolution algebras. They occur also in Gelfand (1941 b).

The mystery of the singular parts of the elements of the convolution algebra on the real line was solved in papers by Sjreider (1950, 1951). The corresponding homomorphic images can have absolute value less than one.

In his paper (1945) Beurling gave a new aspect of Wiener's Tauberian theorem. The problem is shifted to nonvanishing functions $f(x)$ which are bounded and uniformly continuous on the real line. The theorem says that there exists at least one exponential e^{itx}, t real, which can be approximated locally uniformly by linear combinations of the translates of f,

$$g(x) = \sum c_k f(x - a_k),$$

and such that, at the same time, $\| g \| = \sup |g(x)|$ tends to $\| f \|$. Beurling referred to the corresponding topology as narrow. From this theorem follows that of Wiener for if $K \in L$, $0 \neq g \in L^\infty$ and g is orthogonal to the translates of K, i.e.,

$$K * g(x) = \int K(x)g(x+t)dx = 0$$

for all t, then also $K * f = 0$ where, with suitable h, $f = g * h \neq 0$ is bounded and uniformly continuous. Hence Beurling's theorem shows that

$$\int K(x)e^{itx}dx = 0,$$

for some t, i.e., the Fourier transform of K has a zero. This presentation does not fully use Beurling's theorem. We may exchange K for a linear functional $L(f)$ which is continuous with respect to the narrow topology and such that $L(f_t) = 0$ for all translates f_t of a non-vanishing and uniformly continuous function in L. Then L vanishes on some exponential. An example is

$$L(f) = \int (K(x+z) - K(x))f(x)dx$$

where z is fixed and the function within parentheses is integrable although K need not have this property.

Beurling defined the spectrum of a bounded function f by properties of the harmonic transform

$$\int_{-\infty}^{\infty} e^{ixt-\sigma|x|}f(x)dx.$$

The open parts of the t-axis where the limit for $\sigma \to 0$ vanishes constitute the complement of the spectrum. In the theory of distributions there is a simpler, equivalent definition. The Fourier transform of f is a distribution and we may identify the spectrum of f with the support of its Fourier transform. This definition provides the following simple proof (orally from Lars Hörmander) of Beurling's theorem.

The proof uses the fact that an entire function $h(x)$ whose absolute value on the real axis is at most 1 and satisfies the inequality

$$\log^+ |h(z)| = \varepsilon|z| + O(1), \quad z = x + iy,$$

has the property that

$$|h'(x)| \leq \varepsilon e.$$

For by the Phragmén-Lindelöfs principle, $\log^+ |f(z)| \leq \varepsilon|y|$, $z = x + iy$ and Cauchy's inequality gives

$$|f'(x)| \leq e^{\varepsilon r}/r$$

for every $r > 0$, in particular when $r\varepsilon = 1$.

This theorem applies to a function $g(x)$ whose Fourier transform $G(t)$ is smooth with support in $|t| \leq \varepsilon$. Then $g(x+iy)$ is entire analytic, $\log^+ |g(x+iy)| = \varepsilon|y| + O(1)$ and the same inequality holds for $\int |g(x+iy-t)|dt$.

In the proof of Beurling's theorem we may assume by a translation that the support of f contains the point $t = 0$. Then for every $\varepsilon > 0$ there is smooth function $G(t) = G_\varepsilon(t)$ supported in $|t| \leq \varepsilon$ with Fourier transform $g = g_\varepsilon$ such that $h = g * f$ does not vanish. Exchanging f for a translate of f and multiplying g by a constant we may also assume that

$$1 - \varepsilon < g * f(0) < 1 = \| g * f \|_\infty.$$

Now, as we have seen above, $h = g * f$ is an entire function such that $\log^+ |h(x + iy)| = \varepsilon|y| + O(1)$. Hence $|h'(x)| \leq \varepsilon e$ for all x. Letting ε tend to zero we see that h converges to 1 in the narrow topology. At the same time we may approximate g in the narrow topology by Beurling's sums. This proves the theorem.

Spectral synthesis

The space L_1 of integrable functions on the real line is a convolution algebra in which the maximal ideals consist of functions f whose Fourier transforms

$$F(t) = \int e^{ixt} f(x) dx$$

vanish at a single point. A closed ideal $I \subset L_1$ is said to have spectral synthesis if I is the intersection of all maximal ideals which contain I. This intersection consists of all functions whose Fourier transforms vanish at the set N of points which are common zeros of the Fourier transforms of functions in I. A condition for spectral synthesis turns out to be that $\int f(x)g(x)dx = 0$ for all $f \in L_1$ and bounded functions g whose spectra $S(g)$ are contained in N. Malliavin (1959) proved that spectral synthesis is not possible for some complicated sets N.

Spectral synthesis in L_1 corresponds to spectral synthesis in the space of bounded functions which can be formulated as a vague question: is a bounded function g contained in the linear hull of the exponentials of its spectrum $S(g)$? According to Beurling (1949 a) the answer is positive if the linear hull is taken in a norm

$$\| g \|_w = \int_{-\infty}^{\infty} |g(x)| w(x) dx$$

where $w(x) = w(|x|)$ is a positive integrable function such that $w(|x|)$ does not increase.

Important features of this paper, and later ones, are contractions $T(z)$ of the complex plane with the property that

$$|T(z_1) - T(z_2)| \leq |z_1 - z_2|, \quad T(0) = 0.$$

The circular projection on the positive real axis which was important in his thesis and may have been at the origin of Beurling's interest in contractions. The one used in the paper (1949 a) is a radial contraction $T = T_s$ onto a disk of radius $s > 0$, so that $Tz = z$ when $|z| \leq s$ and $T(z) = sz/|z|$ otherwise.

When f is an integrable function with the Fourier transform F, let f_s be the function whose Fourier transform is $T_s F(t)$. Beurling's point of departure is a class of integrable functions f which are uniformly contractible in the sense that

(A) $$\int |f_s(x)| dx = o(1), \quad s \to 0.$$

Such a function has the property that

$$F(t) = 0, t \in S(g) \implies \int f(x)g(x)dx = 0.$$

For $F - F_s$ vanishes in a neighborhood of $S(g)$ so that $\int (f - f_s)(x)g(x)dx = 0$, whence the property above.

The dual of the space L_w of measurable functions g with $\| g \|_w < \infty$ consists of functions which are pointwise majorized by a constant times w. It is clear that Beurling's above result follows if all these functions have the property (A).

The greater part of the paper is a study of the condition (A) with the result that the desired approximation in the norm $\| g \|_w$ is possible when $w(x) = w(|x|)$ and $w(|x|)$ does not increase with $|x|$.

Later papers and summary

The papers which have been reviewed above are only part of Beurling's many inventive contributions to the subject of spectral synthesis, which culminated in the paper (1964) with constructions of convolution algebras with problem free spectral synthesis. The idea of using contractions was employed in the papers (1959) and (1960) (with Jacques Deny) for the creation of an elegant general potential theory without an explicit kernel.

After 1950 Beurling wrote many other important papers, but a detailed analysis is outside the scope of this book, which covers the time up to 1950. What has been said about Beurling does not give a complete picture but it covers major papers where Beurling's power and inventiveness is made clear. He wrote many things that he did not publish and liked to indicate their existence. His strong and self-assertive personality is reflected in his papers where the reader may get the impression of being led by a strong hand through an exciting landscape. Beurling's collaborator Lars Ahlfors once said something essential about his friend: there is a touch of genius in everything he does.

During his time in Sweden, Beurling was Sweden's leading mathematician. His own work, which he often talked about in seminars, made his teaching in Uppsala lively and exciting. Towards the end of the 1940's abstract harmonic analysis was a popular subject and Beurling's fame attracted many American mathematicians to Uppsala, John Wermer, Henry Helson, and Edward Hewitt. During a period of the war Beurling was drafted to work with the Swedish Secret Service and managed to break the German code, something which was very useful for the Swedish state. Beurling got a decoration for his feat. Privately he used to say that the state owed him several million because it could read German secret messages and hence knew in advance Germany's best bid in the yearly trade negotiations.

Beurling's students

Before 1950 seven mathematicians wrote theses in Uppsala of which six were more or less inspired by Beurling. All of them found university positions.

Esseen

The first one, Carl-Gustaf Esseen, later professor of mathematical statistics in Uppsala, is known together with C. Berry for an optimal error term in the law of large numbers. The essential point is to estimate

$$(1) \qquad F(x) - G(x) = \frac{1}{2\pi} \int e^{ixt} i(f(t) - g(t)) dt/t$$

the difference between two probability distributions with Fourier transforms

$$f(t) = \int e^{-ixt} dF(t), \quad g(t) = \int e^{-ixt} dG(t),$$

normalized so that $f(t) \sim 1 - t^2/2$ for small t and the same for $g(t)$.

In probability theory, the problem is to estimate the difference above when $G = \exp(-t^2/2)$ corresponding to the normal distribution and $F = F_n$ is the normalized distribution function of a sum $X_1 + ... X_n$ of stochastic variables. In the simplest situation it is assumed that their third moments have a uniform bound $O(1)$. The Taylor series of $f_n(t)$ at the origin then has the bound

$$f_n(t) = e^{-t^2/2}(1 + O(t^3/\sqrt{n}))$$

so that
$$\int_{-T}^{T} |(f(t)-g(t))dt/t| = O(1/\sqrt{n}), \quad T = O(\sqrt{n}).$$

An earlier estimate by Liapounoff was
$$|F_n(x) - G(x)| = O(\log n/\sqrt{n}).$$

That the same estimate holds without the logarithm follows from Esseen's main result in his thesis (1944). It says the following: to every $k > 1$ there is a $c(k) > 0$ such that

(2) $$|F(x) - G(x)| \leq \frac{k}{2\pi}\int_{-T}^{T} |(f(t)-g(t))dt/t| + c(k)/T$$

for large T. We shall sketch Esseen's proof when F and G are continuous.

As is well known the function $h(t) = (1-|t|)_+$ is the Fourier transform of a positive, even function with the total integral 1. The function $h_T(t) = h(t/T)$ is then the Fourier transform of $H_T(x) = TH(Tx)$, a function which approximates $\delta(x)$ when $T \to \infty$.

It follows from (1) that

(3) $$\int H_T(x-y)(F(y)-G(y))dy = \frac{1}{2\pi}\int e^{ixt}i(f(t)-g(t))h_T(t)dt/t$$

where the right side has the majorant

$$\frac{1}{2\pi}\int_{-T}^{T} |(f(t)-g(t))dt/t|$$

and the left side has the limit $F(x) - G(x)$ when $T \to \infty$. This explains what now follows.

Put $\Delta = \sup |F(y) - G(y)|$. By a change of variables and a change of sign in the left side we may assume that $\Delta = F(0) - G(0)$, which, for small x, will give a good minorant of the left side.

Since $|G'(x)| < 1$ we have $F(y) - G(y) > \Delta - y$ when $0 < y < x$, and this gives a first minorant

$$\int_0^{\Delta} H_T(x-y)(2\Delta - y)dy - \Delta \int H_T(x-y)dy.$$

If we write $2\Delta - y = (2\Delta - x) + (x - y)$ and $b = \int H(x)|x|dx$, we get a new minorant

$$(2\Delta - x)\int_0^{\Delta} H_T(x-y)dx - \Delta - b/T.$$

If we now choose $x = m\Delta, 0 < m < 1$, this turns into

$$(2-m)\Delta \int_{-m\Delta}^{(1-m)\Delta} H_T(y)dy - \Delta - b/T.$$

Here the integral tends to 1 when $T \to \infty$. With small $m > 0$ this proves Esseen's inequality.

Borg

Göran Borg's thesis (1946) is about Sturm-Liouville operators

$$L = -D^2 + p(x), \quad D = d/dx, \; p(x) \in L^2(I)$$

in the interval $I : 0 < x < \pi$ with a real potential $p(x)$. It is a classical fact that the equation $Lu = \lambda u$ with real boundary conditions

(4) $\quad R(\alpha, \beta) : u(0) \cos \alpha + u'(0) \sin \alpha = 0, \quad u(\pi) \cos \beta + u'(\pi) \sin \beta = 0$

has nonvanishing solutions only for a discrete set

$$\lambda = \lambda_1 < \lambda_2, \cdots \to \infty$$

of real eigenvalues. Borg's problem is to determine the potential $p(x)$ from these eigenvalues under one or several boundary conditions. His thesis is the first strict treatment of this problem. Similar ones occur in practice when a physical phenomenon can only be observed from its spectra.

In a preliminary study Borg treats the case $\alpha = \beta = 0$. Here only one spectrum is not sufficient since it is obvious that $p(x)$ and $p(\pi - x)$ have the same spectra. Therefore he restricts himself in the beginning to potentials which are invariant under the reflection $J : x \to \pi - x$. If p and q then have the same spectrum $\lambda_1, \lambda_2, \ldots$, the formula

$$-u_n'' + p(x)u_n = \lambda u_n, \quad -v_n'' + q(x)v_n = \lambda v_n$$

and an integration show that

$$\int_I (p(x) - q(x))u_n(x)v_n(x)dx = 0.$$

Uniqueness now follows if $z_n = u_n v_n$ för $n = 1, 2, \ldots$ is a complete system in the part of $L^2(I)$ which is symmetric under J. A large part of the thesis is devoted to a proof of this statement. For large n, the eigenfunctions $u_n(x), v_n(x)$ are aproximately equal to $\sin nx$ when suitably normalized. Hence

$$u_n(x)v_n(x) \sim (\sin nx)^2 = (1 - \cos 2nx)/2$$

for large n. Borg shows that an expansion of $u_n(x)v_n(x)$ in a series according to the right sides above leads to a system of equations with a matrix of the type $E + A$ where A is completely continuous. That the system is invertible then follows from a fortuitous construction of a system which is bi-orthogonal to the system $u_n(x)v_n(x)$. This proves that $p = q$.

Borg's general result, which requires some computation, is roughly that that the spectra of two independent boundary conditions (4) determine the potential uniquely and that the dependence has a certain continuity.

After Borg's pioneering work inverse spectral problems have been studied under various assumptions. One possibility is to deduce the potential from the spectral measure (see Levitan (1987)), another one is to study the isospectral problem. i.e., the set of potentials for which the operator L has a given spectrum (see Pöschel and Trubowitz (1987)). Both references sum up a long development since the 1950's.

Broman, Kjellberg

Arne Broman's thesis (1947) about capacities of order α has already been reviewed in conection with Beurling's paper *Exceptional sets*. Bo Kjellberg's study of minimal modulus (1948) starts with a review of the results and problems, a sharp variant of the Phragmén-Lindelöf principle due to Beurling and an instructive figure about the circular projections that Beurling used in his thesis. In the main chapter Kjellberg studies harmonic functions outside intervals a^n, b^n on the positive axis where $0 < a < b$ and n runs through the integers. Their existence and properties are studied in detail. The last chapter treats the behavior of harmonic functions at complicated boundaries. Kjellberg's thesis is written with a careful attention to both the text and the reader.

Nyman, Hall, Carleson

In 1950 three of Beurling's students defended their theses, Bertil Nyman, Tord Hall, and Lennart Carleson.

Nyman made his debut with a paper (1949) about the connection between the counting functions of Beurling's integers and primes (1936). The main result of his thesis (1950), which uses Beurling's spectral theory, is that all left translations $t \to f(t-s), s > 0$ of a function $f \in L^1(0, \infty)$ are dense in this space provided f does not vanish almost everywhere in any interval $(0, a)$ and its Fourier-Laplace transform

$$\int e^{izt} f(t) dt$$

does not vanish when $\operatorname{Im} z \geq 0$. The rather refined proof is based on the fact that a function $g \in L^\infty(0, \infty)$ which does not vanish in an interval adjoining the origin and has the property that

$$\int f(t)g(s+t) dt$$

vanishes when $s \geq 0$, must be identically zero.

Tord Hall's first paper (1937) is a variant for the half-plane of Beurling's main lemma in his thesis (1933). Hall's thesis (1950) contains a variety of results about complex polynomials with given bounds at given points.

Lennart Carleson, later professor in Uppsala and Stockholm and an outstanding mathematician, defended his thesis at the age of twenty-two. The subject is Rolf Nevanlinna's and Ahlfors's theory of meromorphic functions f. In this theory the spherical area $A(r)$ of the image of the region $|z| < r < 1$ plays a decisive part. If

$$\int^1 A(r) dr/r < \infty,$$

then f is the quotient of two bounded functions and has also other good properties. Carleson studied how the theory is modified when the condition above is replaced by the more demanding one

$$\int^1 A(r) dr/r^a < \infty, \quad 0 \leq a < 1.$$

The results defy a simple review.

The theses of his students mirror Beurling's inventiveness and wide interests in harmonic analysis and function theory. We can also trace a generosity with advice and ideas which was met with admiration and affection.

Appendix

If D is a plane region and γ an arc on the boundary, let $\omega(z,\gamma,D)$ be the value at the point z of the harmonic measure of γ in D. We shall compute an explicit harmonic measure which is important in Beurling's thesis.

If D is the right half of the unit disk and γ is its right boundary, then $\omega(z,\gamma,D) = 2(1-\alpha/\pi)$ where α is the visual angle from z. On the real axis,

$$\omega(t,\gamma,D) = \frac{4}{\pi}\arctan t, \quad 0 < t < 1.$$

The conformal map

$$z \to w = g(z,s) = \sqrt{\frac{s-z}{1-zs}}, \quad 0 \leq s < 1.$$

maps a region D, consisting of the unit disk cut along the interval $\gamma : s \leq z \leq 1$, onto the previous region. Hence

$$\omega(-r,\gamma,D) = \frac{4}{\pi}\sqrt{g(-r,s)}, \quad 0 < r < 1,$$

where γ and D have their new meanings. In particular,

$$\omega(0,\gamma,D) = \frac{4}{\pi}\sqrt{s}.$$

Bibliography

BERGSTRÖM B.E.H.
1937. *Zur Theorie biquadratischer Zahlkörper. Die Arithmetik auf klassenkörpertheoretischer Grundlage*, Ak. Avh. Uppsala 1937.

BILLING C.G.G.
1938. *Beiträge zur arithmetischen Theorie der ebenen kubischen Kurven vom Geschlecht Eins*, Ak. Avh. Uppsala 1938.

DELAUNAY B.
1928. *Vollständige Lösung der unbestimmten Gleichung $X^3q+Y^3=1$*, Math. Zeischr. 28, 1928, 1-9.

LANG S.
1978. *Elliptic Curves Diophantine Analysis*, Springer 1978.

LIND C-E.
1940. *Untersuchungen über die rationalen Punkte der ebenen kubischen Kurven vom Geschelcht Eins*, Ak. Avh. Uppsala 1940.

MORDELL.
1922. *On the rational solutions of the indeterminate equations of the third and fourth degrees*, Proc. Cambridge Phil. Soc. vol 21 (1922).

NAGELL T.
1928. *Darstellung ganzer Zahlen durch eine binäre kubische Form mit negativer Diskriminante*, Math. Zeitschr. 28 (1928).

1929. *L'analyse indéterminée de degré supériur*, Mémorial des Sciences Mathématiques Fasc. XXXIX (1929) 1-62.

1930. *Zur Theorie der kubischen Irrationalitäten*, Acta Math. 55 (1930) 33-65.

1931. a. *Sätze über algebraische Ringe*, Math. Zeitschr. 34 (1931) 179-182.

1931. b. *Zur algebraischen Zahltheorie*, Math. Zeitschr. 34 (1931)183-193.

1934. *Zur Theorie der rationalen Punkte auf ebenen Kurven dritten Grades*, Forh. Norske Vidensk. Selsk. i Oslo 7, 1934, 140-42.

1935. *Solution de quelques problèmes dans la théorie arithmétique des cubiques planes du premier genre*, Norske Vidensk. Selsk. i Oslo, Skrifter 1935 no 1.

1946. *Les points exceptionels sur les cubiques planes du premier genre*, I,II Nova Acta Soc. Sci. Upsal. 14, 1946 no 1, 14, 1947 no 3.

ROTH K.F.
1935. *Approximation by rationals to algebraic numbers*, Mathematika 2, 1955,1-20.
RUNGE C.
1887. *Ueber ganzzahlige Lösungen einer Klasse von Gleichungen mit zwei Unbestimmten*, Jour. de Math. bd 100, 1887.
SIEGEL C.
1922. *Ueber den Thueschen Satz*, Skrifter, Norske Vidensk. Selsk. i Kristiania, Skrifter 1922.
SKOLEM TH.
1922. *Approximation algebraischer Zahlen*, Math. Zeitschr. 10, 1921.
THUE A.
1908. *Bemerkungen über gewisse Näherungsbrüche algebraischer Zahlen*, Norske Vidensk. Selsk. i Kristiania, Skrifter 1908.

Beurling.

AHLFORS L.
1929. *Untersuchungen zur Theorie der konformen Abbildung und der ganzen Funktionen*, Acta Soc. Sci. Fenn. NS A1 9, 1930, 1-40.
1974. *Conformal Invariants. Topics in Geometric Function Theory*, Mc Graw Hill 1974.
BEURLING A.
1989. *Collected Works*, I,II. ed. Carleson L., Malliavin P., Neuberger J.,Wermer J. Birkhäuser 1989.
1933. *Études sur un problème de majoration*, Ak. Avh, Uppsala 1933.
1934. *Sur les fonctions limites quasi analytiques des fractions rationelles*, Åttonde Skand. Mat.-kongressen, Stockholm 1934.
1937. *Analyse de la loi asymptotique de la distribution des nombres premiers généralisés I.* Acta Math. 68, 1937, 255-291.
1938. *Sur les intégrales de Fourier absolument convergentes et leur applications à une transformation fonctionelle*, Nionde Skand. Mat.-kongressen Helsingfors 1938.
1940. *Ensembles exceptionnels*, Acta Math. 72, 1940, 1-13.
1945. *Un théorème sur les fonctions bornées et uniformément continues sur l'axe réel*, Acta Math. 73, 1945, 127-136.
1946. *Sur quelques formes positives avec une application à la théorie ergodique*, Acta Math. 78, 1946, 319-334.
1949. a. *On the spectral synthesis of bounded functions*, Acta Math 81, 1949, 225-238.
1949. b. *On two problems concerning linear transformations in Hilbert space*, Acta Math. 81, 1949, 225-238.
1949c. *Sur les spectres des fonctions*, Analyse harmonique CNRS Int. Coll. 15, Paris 1949, 9-29.
1959. *Espaces de Dirichlet* 1. Le cas élémentaire, Acta Math. 99, 1958, 203-224 (with J. Deny).
1960. *Dirichlet spaces*, Proc. Nat. Aca. Sc. USA 45, 1959, 208-215 (with J. Deny).
1964. *Construction and analysis of some convolution algebras*, Ann. Inst. Fourier, 14, 1964, 1-32.
1950. *Conformal invariants and function-theoretic null sets*, Acta Math.83, 1950, 101-129 (med Ahlfors).
BORG G.
1946. *Eine Umkehrung der Sturm-Liouvilleschen Eigenwertaufgabe. Bestimmung der Differentialgleichung durch die Eigenwerte*, Acta Math. 78, 1946, 1-96.
BROMAN A.
1947. *On two classes of trigonometrical series*, Ak. avh. Uppsala 1947.
CARLEMAN T.
1921. *Sur les fonctions inverses des fonctions entières*, Arkiv 15, 1921, no 10.
1933. *Sur une inégalité différentielle dans la théorie des fonctions analytiques*, CR Paris 196, 1933, 995-997.
CARLESON L.
1950. *On a class of meromorphic functions and its associated exceptional sets*, Ak. avh. Uppsala 1950.
ESSEEN C.G.
1944. *Fourier analysis of distribution functions. A mathematical study of the LaplaceGaussian law*, Acta Math. 77, 1944, 1-125.
GELFAND I.
1941. a. *Normierte Ringe*, Mat. Sb. NS 9, 1941, 1-24.
1941. b. *Über absolut konvergente trigonometrische Reihen und Integrale*, Mat. Sb. NS 9, 1941, 51-66.

HALL T.
1937. *Sur la mesure harmonique de certains ensembles*, Arkiv Mat. Fys. 25A, 1937.
1950. *On polynomials bounded at an infinity of points*, Ak. avh. Uppsala 1950.

KJELLBERG B.
1948. *On certain integral and harmonic functions. A study in minimal modulus*, Ak. avh. Uppsala 1948.

LEVITAN B.M.
1987. *Inverse Sturm-Liouville Problems*, VSP Publishers, Zeist, Nederl. 1987.

MALLIAVIN P.
1959. *Impossibilité de la synthèse spectrale sur les groupes abéliens non compacts*, Publ. Math. IHES 1, 1959.

MILLOUX H.
1925. *Sur le théorème de Picard*, Bull. Soc. Math. France B 53, 1925, 181-207.

NEVANLINNA R.
1933. *Über eine Minimumaufgabe in der Theorie der konformen Abbildung*, Göttinger Nachr. I37, 1933, 103-115.

NYMAN B.
194. *A general prime number theorem*, Acta Math. 81, 1949, 299-307.
1950. *On the one-dimensional translation group and semi-group in certain function spaces*, Ak. avh. Uppsala 1950.

PÖSCHEL J., TRUBOWITZ E.
1987. *Inverse Spectral Theory*, Pure App. Math, 130, Academic Press, Boston 1987.

SJREIDER IO. A.
1950. *Construction of maximal ideals in convolution rings of measures (Russian)*, Mat. Sb. 27. 1950, 237-318.
1951. *One example of a generalized character (Russian)*, Mt. Sb. 29, 1951, 419-426.

WIENER N.
1932. *Tauberian Theorems*, Ann. Math. 33, 1932, 1-100.
1938. *On absolutely convergent Fourier-Stieltjes transforms*, Duke J. 4, 1938, 420-436 (with Pitt E.R.).

Mathematicians in Sweden 1700–1950

Ivar Bendixson

Alexander Berger

Arne Beurling

Carl Fredrik Björling

Emanuel Gabriel Björling

Karl Bohlin

Erland Samuel Bring

Torsten Brodén

Viktor Bäcklund

Torsten Carleman

Fritz Carlson

Harald Cramér

Herman Th. Daug

Göran Dillner

Matths Falk

Ivar Fredholm

Otto Frostman

Allvar Gullstrand

Hugo Gyldén

Carl Johan Hill

Erik Holmgren

Hjalmar Holmgren

Samuel Klingenstierna

Gustaf Kobb

Helge von Koch

Sonja Kovalevski

Anders Lindstedt

Carl Johan Malmsten

Johannes Malmquist

Gösta Mittag-Leffler

Trygve Nagell

Niels Erik Nörlund

Wilhelm Oseen

Henrik Petrini

Edvard Phragmén

Åke Pleijel

Marcel Riesz

Erik Stridsberg

Anders Wiman

Nils Zeilon

Postscript

Mathematical knowledge is forever reworked by the efforts of mathematicians. Old results are perpetually revised and enter into expanded fields of knowledge. Old areas ramify into new fields and perspectives change. Every new generation of mathematicians gets new visions also of the classical parts of the subject. Once-sensational results fade and may enter into a body of general knowledge which is taught to millions as anonymous learning. Infinitesimal calculus is the prime example.

It is therefore not to be expected that many of the results that have been reviewed in this book can be found in contemporary mathematics. Some examples: Malmsten is entirely forgotten. Mittag-Leffler is visible only through his theorem about meromorphic functions, Bendixson is unknown, Fredholm is a name without known papers. The work of Phragmén is unknown except for the Phragmén-Lindelöf principle, Wiman is unknown and Holmgren's name is only mentioned in connection with the uniqueness theorem. The work of Carleman is not known although several of his papers have grown into very important key areas. Riesz is known only by his theorem about conjugate functions and the Riesz-Thorin theorem. Beurling may still have a certain reputation, but his papers are not known.

The author can only hope that the disappearing of the traces of the work of many Swedish and other mathematicians will not deter the reader from making himself familiar with the mathematics of the past. In any case it puts present mathematics into perspective and it is part of our national and international cultural heritage.

Index

Abel 10,11,13,18,27-29,110,158
Acta Mathematica 77-79
Ahlfors L. 197,241,252,254,264
algebraic geometry 41-44
   in Lund 43-49
algebraic surfaces 44
analytic number theory 138-142, 249

Bäcklund A.V. 51-62
   part. diff. eqs 51-54
   B. on part. diff. eqs 54-55
   Bäcklund transf. 55-60
   after B. 60-61
   obituary 61
Bendixson 109-112
Bendixson's index formula 112
Beurling A. 250-264
   Beurling's lemma 252-254
   Milloux's problem 254
   Exceptional sets, outer capacity 254
   outer and inner functions 255-257
   Beurling's primes 257-258
   Spectral analysis 258-264
   spectral synthesis 263
   later papers, conclusion 264
   students 264
Berg E. 242
Berger 68-70
Bergstedt J. 41,59
Bergström H. 250
Bergström V. 235-236
Bernoulli J. 3
Bernstein S. 144
Billing C.G.G. 250
Birkhoff G.D. 201
Björling E.G. 14-17
Björling C.F.E. 38-39, 45-47
   textbooks 46-47
Block H. 177-178
Bohlin 105
Bohr H. 164
Bonsdorff E. 76
Boltzmann's equation 204
Bolinder E.G. 164
Borel 88, 91
Borel summation 88
Borg G. 264
Bouquet 74
Brill 41
Bring E.S. 5-7

Briot 74
Brodén T. 171-174
Broman A. 267
Brundin C.G. 164
Bucht G. 164

capacity 237-242
Carleman T. 164,169,185-206, 254
   integral equations 185-188
   C. and the abstract theory 189-191
   Jensen-Carleman's formula 191-193
   quasianalytic classes 193-196
   approximation 193
   short papers 196-206
   harmonic majorants 196-197
   asymptotic paths 197-199
   Lindelöf's function 200
   ergodic theorem 201
   asymptotics of eigenvalues 202-203
   kinetic theory of gases 204
   conclusion 206
   students 206-207
Carlson A. 163
Carlson F. 163,181,208-213
   power series 208-210
   Dirichlet series 211-212
   geometry 212
   as censor 213
Cartan E 60
Cassel 134
Castelnuovo 43
Cayley 41,48
celestial mechanics 97
Celsius 2
Chasles 41,44,74
Chebyshev 68,69,138
Cremona 41
Clebsch 43
Cramér 141-142

Darboux 60
Daug H.Th. 65-66
Delauney B. 220
Denjoy A. 193,197
Denjoy-Carleman's theorem 216-216
Deny J. 264
differential equations of Fuchsian type 172
differential geometry 65
difference equations 175

Dillner 66-67
DiPerna R.J. 205
Dirichlet's problem 122
Dirichlet series 143
discrete subgroups 173-74
distribution theory 126
double refraction 93-95

Edge W.L. 162
Enskog 125
equilibrium distribution 235
Essén M. 164
Esseen C.G. 264-265
Euler 180
extremal distance 251
extremal length 251

Fabry 120
Falk M. 22,67-68
Fatou 144
Fejér 143
Frostman O. 236-242
Fourier analysis 142-151
Fuchs 74
fundamental solutions
  generalities 126
  homogeneous equations 128
  Gelfand-Shilov formula 128
  variants of 129
  hyperbolic equations 129
Franco-Prussian war 74
Fredholm 119-124
  after Fredholm 124
  fundamental solutions 120-21
  integral equations 121-24
  obituary 128
  alternative 124

Galois E. 29
Galois group 27
Gårding L. 27,94,242
Gellerstedt S. 168
generating function 100
Goursat 54,60
Green's function 240
Grönwall 135
  lemma 136
Grünwald 150
Gullstrand 106
Gustafson T. 233
Gyldén 81,97,105

Hadamard 123
Hall T. 267
Halmos P. 256

Hamilton 97,98
Hardy G.H. 144
Hardy class 255
Helson H. 264
Herglotz 150
Hermite 74
Hewitt E. 264
Hilbert 186,213-214,250
Hill C.J.D. 10-12
Hille 138
Hlawka E. 235
Holmgren Hj. 26
Holmgren E. 164-168,186
  uniqueness theorem 165
  other papers 166
  obituary 168
  students 168
Hörmander 167,200,243
Hössjer 234
Huygens 151

infinite determinants 115
inversion 239

Jacobi 23,41
Jonzon B.A.S. 164
Jordan C. 156,164

Kjellberg B. 267
Klein F.156
Klingenstierna 1-5
Kobb 133-134
von Koch 115-119,140
von Koch's star 119
Kovalevski S. 79-80,93-95
Kronecker 76,81
Kummer 76

lacunas 151
Lagrange 98-99
Lamé 149
Landau 164
Lannér F. 243
Laplace 180
Lebesgue 143
Lie S. 48,53
Lind C.-E. 250
Lindelöf L. 74,76
Lindman C.F. 17
Lindstedt 102-103
Lindwall B.A.G. 164
Lions P.L. 205
Liouville 45
logarithmic potential 254
Lundberg E.178

Lundell E. 164

Magnus effect 233
Malmheden 242
Malmrot R. 164
Malmsten 17-26
   functional equations 19
   continued fractions 21
   differential equations 23
   time after the professorship 25
Malmquist 136-138
Mattson R. 163
Maxwell 149
Mittag-Leffler G. 70,73-84,158
   family 73-74
   studies 74-76
   time in Helsingfors 76-77
   eventful decade 77-81
   royal prize 81
   after age of fifty 82
   will and foundation 83
   students 83
   meromorphic functions 86
   M.L.'s star 87-92
Mittag-Leffler S. 77,81
Möbius group 173
modules 235
moment problem 145-146
Monge 53
Monge-Ampère 54
Muncaster R.G. 205

Nagell T. 247-250
   rational solutions 247
   algebraic number theory 249
   Nagell's work 249
Nagell-Lutz theorem 250
von Neumann 189,212-215
Nevanlinna F. 197,241
Nevanlinna R. 197,252
Newton 97-98
Nörlund N.E. 174-177
Nyman B. 267

Oseen C.W. 1,5,61,180
   Oseen's wake theory 231-232

Persson G.P. 163
Petterson E.L. 164
Petrén L. 180
Petrini 134,185
Petrovski 150,151
Pleijel 163, 207
Phragmén E. 81,112-115,139
Phragmén-Lindelöf principle 93,113-115,262
Plücker 41,44
Poincaré 45,81,100-101,104, 115,236
Poincaré-Bendixson theory 112
potential theory 236-242
professorship fight 34-37
projective geometry 41

quaternions 66

Riccati equation 137
Riemann B. 21,41,74,234
Riemann surface 41
Riemann's problem 172
Riesz F. 125
Riesz M. 141-146, 235,239
   Riesz summation 143
   Riesz-Thorin theorem 219-221
   conjugate functions 222-223
   short papers 223-224
   fractional potentials 214-226
   wave operator 226
   spinors 229
   obituary 229
rotor ship 233
ruled surfaces 47-49

Salmon 44
Salmon-Fiedler 44,49
Schering 70
Schubert 44
af Schultén 76
singularities of
   differential equations 116
Sjöstedt C.E. 164
Sjöstrand O. 168
Skau C. 27
solvable equations 27-29
space-like surface 227
spectral theorem 213-215
Stadler G. 178
Stadler S. 179
Stenström O. 164
Stone M. 189
Stridsberg 140
Strömer 2
summation of series 87-88
Svanberg 13-14
Svensson B. 180

Täcklind S. 168
Tidskrift f. mat. o. fysik 33-34
Tricomi 167
Truesdell T. 205

Uhler A. 181
university life 155-156

Wahlund A. 163
Weierstrass 74
Wermer J. 264
Weyl H. 235
Wennberg S. 164
Wiener N. 258,259,261
Wigert 140-141
Wiman 41,49,156-164
   group theory 156-157
   solvable equations 158
   entire functions 158-161
   differential equations 162
   algebraic geometry 162
   obituary 163
   students 163
Volterra 150

Zeilon 146-150,230-233
   fundamental solutions 147
   doubly refracting crystals 148
   hydrodynamical papers 232-233
Zeuthen H. 41,70

QA 27 .S85 G37 1998
Gårding, Lars, 1919-
Mathematics and
   mathematicians